Diana Preston is an Oxford-trained historian, writer and broadcaster who lives in London. She is the author of *The Road to Culloden Moor: Bonnie Prince Charlie and the '45 Rebellion*; *A First Rate Tragedy: Captain Scott's Antarctic Expeditions*; *Wilful Murder: The Sinking of the Lusitania* and *A Pirate of Exquisite Mind: The Life of William Damper* (written with her husband, Michael Preston). *Before the Fall-Out* was longlisted for the 2006 BBC Four Samuel Johnson Prize for Non-Fiction and won the 2005 *Los Angeles Times* Science and Technology Award.

'In a swashbuckling spirit, armchair adventureres will savor *a Pirate of Exquisite Mind*. But armchair historians will too. The Prestons, having personally retraced Dampier's routes, have a fine appreciation of his stamina . . . Happily for his curious contemporaries, Damper made his far-flung voyages. Luckily for us, the Prestons have sailed in his wake'
Wall Street Journal

'This eloquently enthusiastic biography, besides charting Dampier's astonishing achievements, offers fascinating information about his times'   *The Age*, Melbourne

Acclaim for *Wilful Murder*:

'A complex story of heroism and courage . . . compulsively readable'
Independent on Sunday

'The most comprehensive and accessible account of the sinking there has been or perhaps will be'   *Sunday Telegraph*

'A fitting monument to a multitudinous loss'
John Updike, *The New Yorker*

'It is not easy, nowadays, to write an original book on the First World War . . . but Preston has succeeded'   Norman Stone, *Sunday Times*

'Very good . . . Preston has done an extraordinary amount of work, particularly in tracing the memories of surviviors'   *Sunday Times*

'Sets a standard which other books have not achieved'
Irish Independent

'Clear and effective . . . benefits from exhaustive research'   *TLS*

Also by Diana Preston

THE ROAD TO CULLODEN MOOR:
Bonnie Prince Charlie and the '45 Rebellion
A FIRST RATE TRAGEDY: Captain Scott's Antarctic Expeditions
BESIEGED IN PEKING: The Story of the 1900 Boxer Rising
WILFUL MURDER: The Sinking of the Lusitania

with Michael Preston
A PIRATE OF EXQUISITE MIND: The Life of William Dampier

BEFORE THE FALL-OUT

The Human Chain Reaction from
Marie Curie to Hiroshima

Diana Preston

CORGI BOOKS

BEFORE THE FALL-OUT
A CORGI BOOK : 0552770868
9780552770866

Originally published in Great Britain by Doubleday,
a division of Transworld Publishers

PRINTING HISTORY
Doubleday edition published 2005
Corgi edition published 2006

1 3 5 7 9 10 8 6 4 2

Set in 11/14pt Sabon by
Falcon Oast Graphic Art Ltd.

Corgi Books are published by Transworld Publishers,
61–63 Uxbridge Road, London W5 5SA,
a division of The Random House Group Ltd,
in Australia by Random House Australia (Pty) Ltd,
20 Alfred Street, Milsons Point, Sydney, NSW 2061, Australia,
in New Zealand by Random House New Zealand Ltd,
18 Poland Road, Glenfield, Auckland 10, New Zealand
and in South Africa by Random House (Pty) Ltd,
Isle of Houghton, Corner Boundary Road & Carse O'Gowrie,
Houghton 2198, South Africa.

Printed and bound in Great Britain by
Cox & Wyman Ltd, Reading, Berkshire

Papers used by Transworld Publishers are natural, recyclable
products made from wood grown in sustainable forests. The
manufacturing processes conform to the environmental
regulations of the country of origin

To Michael,
my husband and partner in writing

CONTENTS

ACKNOWLEDGEMENTS

FIRST I SHOULD ACKNOWLEDGE THE COLLABORATION OF MY husband Michael in both the research and the writing of this book. Without him I could not have undertaken it.

I am also indebted to many others: Hans Bethe, Robert Christy, Bertrand Goldschmidt, Philip Morrison and Sir Joseph Rotblat for talking to me about their personal experiences of working on the Allied bomb programme; Lorna Arnold, formerly the UK Atomic Energy Authority's official historian, for her generous help and advice; and Arnold Kramish and Carl-Friedrich von Weizsäcker for corresponding with me and answering my questions.

In the UK, the staff and archivists of many libraries and organizations gave me their help: the BBC Written Archives Centre; the Bodleian Library, Oxford; the British Library; the Cambridge University Library; the archives of Churchill College, Cambridge; Liverpool University Physics Department (and Peter Rowlands in particular); the London Library; the UK National Radiological Protection Board; the Royal Society; and the UK National Archive.

In the US, I must thank the American Institute of

Physics, in particular Julie Gass, for their generosity in sending me transcripts of oral interviews; the US National Archives and Records Administration; the Library of Congress; and the Bancroft Library of the University of California.

Elsewhere, I am grateful to the Niels Bohr Archive in Copenhagen, and especially Finn Aaserud, for making the recently released post-war letters from Niels Bohr to Werner Heisenberg so accessible; to Aubrey Pomerance of the Jüdisches Museum, Berlin, for information about Fritz Strassmann's concealment in his apartment of the Jewish pianist Andrea Wolffenstein; to Yad Vashem in Jerusalem for a copy of the citation acknowledging Strassmann's courage; and to the Deutsches Museum in Munich for access to formerly secret documents about the German atom bomb project from 1938 to 1945.

In Japan, I was touched by the kindness and hospitality of the many people we met there: Miho Nakano, for translation and research, and for welcoming us to her city; Kazuhiko Takano, deputy director of the Hiroshima Peace Memorial Museum, for insights into the pre-war life and history of the city; Yoko Kono, for guiding us around the Hiroshima Peace Memorial Museum; Emiko Ono, for sharing with us her family history; Masanori Ishimoto of the Hiroshima City Museum of History and Traditional Crafts, for telling us about the city's artisans; Jun Fujita and Toshie Kawase, for their childhood reminiscences; and Margaret Irwin of the Radiation Effects Research Foundation's Archive Office, Hiroshima, for information about the early history of radiology in Japan.

I must also thank family and friends: Ulrich Aldag, Rhys Bidder, St John Brown, Clinton Leeks, Kim Lewison, Graeme Low, Neil Munro and Oliver Strimpel for their insights on the text; Eric Hollis for the loan of books; my aunt Lily Bardi-Ullmann for newspaper research in the US; and my mother and parents-in-law for their support.

Lastly, the help of our agents Bill Hamilton and Michael Carlisle was invaluable, and it has been a pleasure working with Michèle Hutchison and the team at Doubleday in London and with George Gibson and his team at Walker Books in New York.

PROLOGUE

ON 6 AUGUST 1945, THE CHRISTIAN FEAST OF THE Transfiguration, the Festival of Light, a young mother, Futaba Kitayama, looked up to see 'an airplane as pretty as a silver treasure flying from East to West in the cloudless pure blue sky'. Someone standing by her said, 'A parachute is falling.' Then the parachute exploded into 'an indescribable light'.

The American B-29 bomber *Enola Gay* had just dropped 'Little Boy', a four-ton bomb which detonated with the explosive power of 15,000 tons of TNT over the Japanese city of Hiroshima. Pilot Paul Tibbets, who had the day before named his plane after his own mother, struggled to hold the aircraft steady as the first shock waves hit. Bathed in a bright light, he looked back and saw 'a giant purple mushroom boiling upward like something terribly alive'. He switched on the intercom and announced to his shaken crew, 'Fellows, you have just dropped the first atomic bomb in history.'

On the ground, Futaba Kitayama felt her face become strangely damp; 'When I wiped my face the skin peeled off.' Her eyes began to mist over and close as her face swelled. 'Suddenly driven by a terror that would not permit inaction' she staggered past writhing, flayed bodies as she tried to escape. To one doctor in the doomed city, the pervasive stench of burnt flesh was like 'dried squid when it is grilled – the squid we like so much to eat'. By December 1945, about 140,000 inhabitants of Hiroshima would be dead, either as a result of the blast and the fires that followed or of the insidious, silent effects of nuclear radiation.

When news of the bombing was announced, young Allied soldiers preparing for the invasion of Japan 'cried with relief and joy. We were going to live. We were going to grow up to adulthood after all.' President Truman told a group of sailors aboard the cruiser on which he was returning from the Potsdam Conference, 'This is the greatest thing in history.' Winston Churchill struck a more reflective note: 'This revelation of the secrets of nature, long mercifully withheld from man, should arouse the most solemn reflections in the mind and conscience of every human being capable of comprehension.' Only three days after Hiroshima, and within days of giving birth to her second son, a New York mother wrote, 'torturing regrets that I have brought children into the world to face such a dreadful thing as this have shivered through me. It seems that it will be for them all their lives like living on a keg of dynamite which may go off at any moment.'

Soon, worries were widespread that the invention of the bomb had unleashed a Frankenstein's monster capable of striking back at its creators in a wholesale and in-discriminate fashion. Although over the past sixty years such concerns have wavered in intensity and the source of the perceived threat has varied, the fear that a single plane or a single person with a suitcase can obliterate a city haunts us today.*

The destructive flash that seared Hiroshima into history was the culmination of fifty years of scientific creativity and more than fifty years of political and military turmoil. Generations of scientists had contributed to that moment in physics. Yet, when they first began to tease out the secrets of matter not even future Nobel Prize winners could have predicted how their pioneering insights would combine with exterior events to produce such a defining moment in history. Like all in this story, they were only human.

For the scientists of many nations, the journey of discovery had begun in the 1890s when dedicated researchers such as Marie Curie, working alone or in small teams with rudimentary equipment, intent on achieving a fuller understanding of nature, started to identify the minute building blocks forming the world around them. Blinding discoveries were matched by blind alleys. People rushed to publish their results, not for profit nor for national prestige and power, often not even for personal glory, but rather for the pure joy of knowledge.

For a long time no-one realized their work could unlock immense energy to furnish a devastating new weapon, or, indeed, if properly harnessed, to provide a city with electricity. At the beginning of the twentieth century,

---

* In 1998 a Russian general revealed that the Soviet Union had previously developed a portable atomic bomb and that, by then, fewer than half of the more than a hundred manufactured could be accounted for. Despite subsequent official Russian denials that any were missing and assurances that all would be destroyed by 2000, experts remain concerned.

radioactivity was seen as only producing benefits to health through the use of X-rays for diagnosis and the use of radioactive materials to treat many diseases including cancer. Physics was a new subject. The 1910 *Encyclopaedia Britannica* devoted fifty pages to chemistry, but physics did not feature. Around that time there were, perhaps, a thousand physicists worldwide, of whom maybe 10 per cent were engaged in the study of radio-activity. Consequently, all those involved knew one another. At a time of intense national rivalry and of com-petition for empire, trade and natural resources, results were pooled internationally, as further pieces in a communal jigsaw puzzle for which no-one had the master picture or pictures. Scientists studied at one another's institutes. North Americans and Japanese visited Germany; Germans came to Britain; Britons went to North America; Russians studied in France. Colleagues skied, hiked and made music together. Allegiances and rivalries stemmed from where and with whom people had studied, rather than from nationality or race.

All met at conferences, where results were shared, contacts maintained and gossip exchanged. Albert Einstein called them 'witches' sabbaths'. Few conferences were as marked by gossip as that in Brussels in 1911, when Marie Curie was forced to withdraw as a result of an alleged affair with Paul Langevin, a close colleague and a married man. However, personalities were strong, and debate often heated. This was particularly the case when entirely novel concepts such as relativity or quantum theory were discussed, which undermined the Newtonian con-cept of a predictable, mechanical world whose ordered processes could be measured and whose future behaviour could be as accurately forecast as its past could be determined. Those involved were, as they recalled, undertaking 'wholly new processes of thought beyond all the previous notions in physics', and 'filled with such tension that it almost took [their] breath away ...' 'It was an heroic time ... not the doing of any one man' but 'the collaboration of scores of scientists from many different countries ... a period of patient work in the laboratory, of crucial experiments and daring action, of many false starts and many untenable conjectures ... It was a time of creation ...'

Yet when, in 1933, despite the great advances already made, one of the world's leading physicists, Ernest Rutherford, dismissed the idea of harnessing energy from atoms as 'moonshine', the physicists' world was changing. Hitler was in power. Scientists who had once travelled simply to where the best science was were now compelled to flee his and other totalitarian regimes because of their race or political views. Ernest Rutherford himself became one of those who did most to welcome them and find them work. Their knowledge and brain power were to prove vital to their hosts in the impending conflict.

In Berlin in 1939, on the eve of the long-feared war, German scientists, with considerable secret help from one of their exiled Jewish former colleagues, Lise Meitner, discovered nuclear fission – a way to unleash the power of the atom. Scientists across the world recognized that an atomic weapon might be a possibility. The personal experience of the émigrés gave added urgency to their efforts to stimulate the democracies to action so that Germany could not blackmail the world into submission by her possession of a unique and uniquely destructive weapon. The success of their advocacy meant that what had for more than forty years been an open quest for knowledge became, almost overnight, a race between belligerent nations, working in secret with large teams, for high and sinister stakes, using all available means of sabotage, espionage and disinformation to thwart their opponents.

The scientists' fears of their German colleagues' potential led one British physicist, during the 1940–1 Blitz, surreptitiously to take a Geiger counter from his laboratory to monitor bomb craters in case the enemy had mixed radioactive materials with conventional explosives to contaminate whole areas and poison their inhabitants. Allied scientists remained so concerned about what are now called 'dirty bombs' that they warned General Dwight D. Eisenhower that the Germans might well use them against the Allied troops under his command during the D-Day landings in Normandy in June 1944.

Well before D-Day, nuclear physics had become big science and big engineering. No other country was able to replicate the resources put into the American Manhattan Project. It cost $2 billion and was as big as the US car industry. The Project employed 130,000 people, from American and British scientists to security guards and process workers, not counting the military and government staff and politicians.

A fortnight after Hiroshima, an editorial in *Life* magazine commented, 'Our sole safeguard against the very real danger of a reversion to barbarism is the kind of morality which compels the individual conscience, be the group right or wrong. The individual conscience against the atomic bomb? Yes, there is no other way. No limits are set to our Promethean ingenuity provided we remember that we are not Jove.' The very success of the bomb project in its own terms retrospectively sharpened the moral searchings among those involved. To some it came to symbolize science's loss of innocence. Sound sense and acute sensibility coexisted uneasily in the character of Robert Oppenheimer, the scientific leader of the Manhattan Project. For as long as it took to complete his task, he subdued his humanist principles to achieve the most inhumane of weapons, but he would later state that 'physicists had known sin' and that he, personally, was 'not completely free of a sense of guilt'. Another leading scientist said that the bomb had 'killed a beautiful subject'.

However, even before the bomb was dropped, a sense of individual responsibility had compelled other key staff to speak out. Joseph Rotblat, a future winner of the Nobel Prize for Peace, actually left the Manhattan Project when he realized that the weapon would become a permanent part of military arsenals which politicians were prepared to contemplate using against their then ally Russia, as well as against Germany. The Dane Niels Bohr and the Hungarian refugee Leo Szilard both argued for international co-operation and control of the discovery, and for a demonstration of the bomb's explosive power before all nations, rather than its immediate use in combat.*

For most of the war, the moral dilemmas posed to scientists in Axis countries and in those under German

---

* Szilard personifies the complex character of many of the scientists. One of the brightest minds and sharpest and most liberal analysts of the moral dilemma, he had such an opinion of himself and aversion to physical labour that he employed others to do his experimental work and was thrown out of his residential apartment at Chicago University for habitually refusing to empty his bath water or flush the lavatory on the grounds that this was 'maid's work'.

occupation, such as Denmark and France, were starker and entailed immediate personal vulnerability. The ambiguities and uncertainties of the Copenhagen meeting in 1941 between the leading German nuclear physicist Werner Heisenberg and Niels Bohr have been widely explored, but others also strove to reconcile personal conscience and patriotic sentiment. Fritz Strassmann, one of the discoverers of fission, hid a Jewish pianist in his Berlin apartment while working on nuclear calculations for the Nazi government. Before later joining the resistance and helping to liberate Paris, Marie Curie's son-in-law, Frédéric Joliot-Curie, had to decide how far he could acquiesce in German use of his nuclear institute in Paris at a time when the prospects of Allied victory seemed remote.

The majority of Allied scientists involved would maintain that Oppenheimer's apologia was unwarranted. Knowledge was neutral; the use to which politicians put it was the dilemma. In any case, the Allies could not have neglected the weapon's potential when they knew that the Germans had embarked on a weapons research programme. That an Allied team had won the race on behalf of the democracies was preferable to any other outcome.

Whichever view the scientists took, the final decision to use the bomb was a political one, and one which the American and British public supported overwhelmingly on the grounds that it saved Allied lives and brought the war to a speedier end than would otherwise have been the case. With hindsight, and with distance from the feelings of individuals in war-weary nations who were apprehensive of the cost in terms of the lives of their loved ones of an invasion of Japan, historians have questioned the political judgements. They have suggested that there were alternatives to the use of the atomic bomb to end the war which would have saved Japanese lives without sacrificing Allied ones.

The moral issues that faced both the physicists in advising on the use of the bomb and the politicians in deciding upon it were, in fact, at least half a century old. Alfred Nobel, the inventor of nitroglycerine and the founder of the Nobel Prizes, not least for peace, had justified his invention as putting an end to war. In 1899, at the time of Marie Curie's pioneering work on radium, the nations of the world met at the Hague to discuss how to avoid conflict by the creation of systems for arbitration. They also laid down in the Hague Convention rules for the conduct of war if it could not be avoided. Among them, four years before the first powered flight, was a prohibition against bombarding 'by whatever means ... undefended' civilian towns or buildings, and another prohibition against the dropping of bombs from balloons 'or other kinds of aerial vessels'.

A second conference was held at the Hague in 1907 at the instigation of President Theodore Roosevelt to review the provisions of the first. Only twenty-seven countries, including Britain and the US, supported renewal of the ban on aerial warfare. Seventeen, including Germany and Japan, did not, so the provision fell. All could agree, however, on a definition of targets permitted to be bombarded by whatever means. Civilian targets were still excluded, but aerial bombardment had gained legitimacy.

The First World War brought science and warfare together in a way no other conflict had. On the evening of 22 April 1915, Germany launched the world's first poison gas attack. The German scientist in charge of the programme defended the use of gas as a means of shortening the war and thus saving lives. After initially condemning the attacks as further breaches of the rule of civilized law by the barbarous 'Hun', Britain, France and later the United States, after her entry into the war, did not long delay in following suit. By the Armistice, Allied production of chemical weapons far exceeded Germany's. The 'Great War' would also come to be known as the 'Chemists' War'. By the end of the conflict, about 5,500 scientists on all sides had worked on chemical weapons alone, and there had been a million casualties from gas attacks. Among them was Lance Corporal Adolf Hitler, who, temporarily blinded by a British gas grenade on 13 October 1918, was still in hospital the day Germany surrendered nearly a month later. Yet this 'war to end wars' would not do so, and the next world conflict, precipitated by that lance corporal, would be the physicists' war.

The First World War had seen the death of some ten million men, the fall of three empires, the establishment of a major communist state and the emergence of the aeroplane as a weapon. Yet, at post-war conferences, countries were lukewarm about defining further rules for the conduct of air warfare. No agreement was ever ratified. Over the years, the definition of what in the previously agreed documents was 'civilian' and thus free from attack became blurred. At the beginning of the Second World War, President Franklin Roosevelt pleaded with the belligerents to refrain from 'bombardment from the air of civilian populations or unfortified cities'. The 1940 memorandum from two émigrés to the British government arguing that an atomic bomb was feasible and urging an immediate start to a research programme suggested that the very likely high number of civilian casualties 'may make it unsuitable as a weapon for use by this country'.

Yet, over the next five years of increasingly total war the Allied air forces followed the precedents set by their enemies and attacked whole cities such as Hamburg, Dresden and Tokyo, in the latter attack using the newly developed 'sticky fire' – napalm. Even before 6 August 1945 any distinction between civilians and combatants had been eliminated in practice, if not in presentation.

Today, we still experience the scientific, political and moral fall-out from 6 August 1945. Against the tumultuous background of the history of the first half of the twentieth century, *Before the Fall-Out* explains how joy in pure scientific discovery created a beautiful science which was suddenly transmuted into a wartime sprint for the ultimate weapon. Through the stories and voices of those involved it tells how individuals responded to the questions of personal responsibility posed by the results of their compulsive curiosity, and why the bomb fell on Hiroshima and its people and changed our world for ever.

CHAPTER ONE

'BRILLIANT IN THE DARKNESS'

TOWARDS MIDNIGHT IN A PARIS GARDEN ON A WARM JUNE night in 1903, attentive guests watched Pierre Curie take a phial from his pocket and hold it aloft. The radium inside shone 'brilliant in the darkness'. Curie's gesture was a tribute to his wife, Marie, the discoverer of radium. Earlier that day this slight woman with her high-domed forehead and intense, grey-eyed gaze had become the first female in France to receive a doctorate. The occasion was an impromptu celebratory dinner party at the villa of the Curies' friend, scientist Paul Langevin.

Marie Curie, born in 1867, was the youngest child of a progressive-minded Polish teacher of physics and mathematics, Wladislaw Sklodowski. She had left her native Warsaw, where women were barred from the university, for Paris, driven by a determination to study science and to do so in a free society. As a sovereign entity, Poland no longer existed: the three rival empires of Germany, Austro-Hungary and Russia had partitioned Marie's homeland between them. The Sklodowskis, a close-knit, intellectual family, lived in Russian Poland where Polish culture was crudely suppressed and 'Russianized'. In adolescence, Marie had risked prison or deportation to Siberia by studying and then teaching at the clandestine 'Floating University' in Warsaw – a radical Polish night-school for young women. The university's aim was to develop a cadre of committed women capable, in turn, of educating Poland's poor and thereby equipping them to resist Russian oppression. To avoid suspicion, the students gathered in small groups in impromptu class-rooms in the cellars and attics of those bold enough to host them.

Science, particularly mathematics and chemistry, had fascinated Marie from an early age. The Floating University provided her with her first taste of working in a laboratory, albeit an illicit one, concealed from the prying eyes of the authorities in a Warsaw museum. Casting around for a suitable foreign university in which to complete her scientific education, Marie was attracted to the Sorbonne, part of the University of Paris. Not only did it have a high reputation for science, but many of Poland's intellectual elite had settled in Paris.

However, the Sklodowskis were perennially short of money. Marie's chances of achieving her ambition seemed remote until she identified a way of helping both her elder sister, Bronya, and herself. She would work as a governess and send all her wages to fund Bronya's medical studies in Paris; then, as soon as she had qualified as a doctor, Bronya would send for her younger sister and, in turn, support her through her own studies. Refusing to listen to Bronya's objections, the eighteen-year-old Marie secured a post with the Zorawski family fifty miles north of Warsaw and set out in the depths of winter for their manor house. As she later wrote, that cold, lonely journey remained 'one of the most vivid memories of my youth'. The final leg was a chilling five-hour sleigh ride across snow-covered beet fields, and she made it with a heavy heart.

Initially, though, Marie found life as a governess bearable, even pleasant. During the day she instructed her employers' daughters and, applying the philosophy of the Floating University, also taught the local peasant children. In the evenings she pursued her own studies by candlelight. As she later recalled, 'during these years of isolated work ... I finally turned towards mathematics and physics, and resolutely undertook a serious preparation for future work'. She also learned 'the habit of independent work'. However, Marie's tranquillity was broken when she and the Zorawskis' eldest son, Kazimierz, fell in love when he came home on vacation from Warsaw University, where he was studying mathematics. Although his parents liked Marie, they refused to contemplate their son's talk of marriage to a woman they considered socially inferior. Eventually Marie left the Zorawskis, where, as she confessed to her brother, the 'icy atmosphere of criticism' had become intolerable. She still hoped that Kazimierz would show the strength of character to defy his parents and marry her, but finally, four fruitless years after their first meeting, she accepted that he would not.

Bronya, by then qualified and married to another Polish doctor, had meanwhile been urging Marie to come to Paris. At last, in November 1891, the twenty-three-year-old Marie bought the cheapest possible train tickets for the forty-hour, thousand-mile journey to Paris, where she enrolled in the Sorbonne's Faculty of Sciences. At first she lived with Bronya, but then found lodgings in an attic room on the Left Bank, sacrificing all comforts to the one essential – solitude to study in peace. She later wrote, her

room was 'very cold in winter, for it was insufficiently
heated by a small stove which often lacked coal'.
Sometimes the temperature fell so low that the water froze
in her hand basin, and 'to be able to sleep I was obliged to
pile all my clothes on the bedcovers'. When that failed
to warm her, she pulled towels and anything else she
possessed, including a chair, on top of her. She survived on
a meagre diet of tea and bread and butter supplemented
by the occasional egg. One day she fainted on the street.
Bronya carried her home, made her eat a large steak and
lectured her on taking better care of herself, but Marie
persisted in her spartan, single-minded existence.

Physical deprivation was unimportant. She had found a
stimulating intellectual challenge: 'It was like a new world
opened to me, the world of science, which I was at last
permitted to know in all liberty.' She passed her *licence ès
sciences physiques* (comparable to a bachelor of science
degree) in 1893, not only top of the class but also the first
woman to receive such a degree. She took her *licence ès
sciences mathématiques* in 1894, coming second in her
class. While she was still preparing for her mathematics
exams, the Society for the Encouragement of National
Industry invited her to perform a study of the magnetic
properties of steels. She was eager to do so but lacked
sufficient room for the necessary equipment in her
laboratory at the Sorbonne. Polish friends in Paris came to
her aid. They invited her to tea to meet French physicist
Pierre Curie, laboratory chief of the Paris School of
Physics and Chemistry. He too was working on
magnetism, and they hoped that he might be able to help
her.

Pierre's background, like Marie's, was radical and pro-
gressive. His father, a determinedly republican doctor,

Eugène Curie, had tended wounded activists during the rising in 1871 of the Paris Commune – the revolutionary council formed by the workers of Paris after France's defeat by Prussia. The Communards had gone to the barricades in defiance of the French government, which had concluded an armistice they considered shameful. The Commune lasted ten weeks before being bloodily suppressed by French government forces, leaving some twenty thousand dead. Eugène Curie sent Pierre, only twelve at the time, and his slightly older brother Jacques out into the streets to search for wounded people in need of medical care and protection from the troops.

Later, as life returned to normal, Dr Curie had encouraged his sons to explore the natural world. Both became scientific assistants at the Sorbonne where, working together in the laboratory of mineralogy, they began studying the structure of crystals. This led them to a remarkable discovery – the phenomenon of piezo-electricity* whereby crystals subjected to pressure produce a current – which became the basis for the gramophone. The two young men had developed a piezo-electric quartz instrument capable of measuring the tiny voltages emitted by the crystals.

When he met Marie, Pierre Curie was thirty-five years old, introspective and unworldly. Many years before he had loved a girl whom he described in a private note as 'the tender companion of all my hours', but she had died. Since then he had devoted himself to his work while striving to avoid emotional though not physical entanglements. He believed that 'a kiss given to one's mistress is less dangerous than a kiss given to one's mother, because

* 'Piezo' comes from the Greek *piezein*, meaning to 'press tight'.

the former can answer a purely physical need'. Perhaps as a defence against intellectual engagement he claimed to believe that 'women of genius are rare' and that 'when, pushed by some mystic love, we wish to enter into a life opposed to nature, when we give all our thoughts to some work which removes us from those immediately about us, it is with women that we have to struggle . . .'

After her experience with Kazimierz Zorawski, Marie was wary of relationships. Young students at the Sorbonne frequently propositioned the gamine ash blonde, excited by her combination of cool intellect and sexual charisma, but none impressed her. Pierre Curie, however, did. As she later wrote, 'his simplicity, and his smile, at once grave and youthful, inspired confidence'. Tall, with cropped auburn hair and a pointed beard, he had an unconscious, loose-limbed grace. He was unable to offer Marie accommodation for her experiments, but their meeting sparked an intense relationship. They quickly discovered what Marie called 'a surprising kinship' in their ideas. Both believed science to be the world's salvation. Both believed that they should devote their lives to make it so.

Pierre was soon broaching marriage. Marie hesitated, knowing that it would put paid to her cherished scheme of one day returning to her homeland to teach. During a visit to Poland in the summer of 1894, despite her feelings for Pierre, she actively explored the prospect of an appointment at the University of Cracow. However, Pierre knew exactly how to woo her, writing to her that, 'It would, nevertheless, be a beautiful thing in which I hardly dare believe, to pass through life together hypnotized in our dreams; your dream for your country, our dream for humanity; our dream for science. Of all these dreams, I believe the last, alone, is legitimate.' Such pleas touched Marie, as did his offer to move to Poland, a sacrifice which she told her sister Bronya she had no right to accept. On 26 July 1895 Pierre and Marie were married at a brief civil ceremony with no white dress, wedding ring or elaborate wedding breakfast. They spent their honeymoon roaming Brittany on bicycles purchased with money given as a wedding present.

By early September, the Curies were back in Paris, living in a tiny three-room apartment which Marie, impatient of domestic distractions, furnished with the bare minimum – two chairs, a table, bookshelves and a bed. Just before their wedding Pierre Curie had been appointed to a new chair of physics, created especially for him, at the Paris School of Physics and Chemistry. Marie was allowed to transfer her work on steels there from the Sorbonne. As a woman working in a laboratory she was an object of curiosity and some animosity, but this did not deter her. Neither did the birth in September 1897 of the Curies' first daughter, Irène, whom Marie delightedly called her 'little queen' in letters home to Poland. She completed her report on steels within three months of the birth and at once began seeking a suitable subject for her doctoral thesis. She chose a newly discovered phenomenon, Becquerel rays.

Becquerel rays owed their discovery to a phenomenon that had caught the public imagination. Two years earlier, in late 1895, Wilhelm Röntgen, a reclusive German physicist at the University of Würzburg, had been following up work by Heidelberg physicist Philipp Lenard on how electrical currents pass through gases at low pressures. Röntgen's prime piece of equipment was a three-foot-long glass tube from which most of the air had been pumped out. Inside the tube were two metal terminals – one positive, called the 'anode', and the other negative, called the 'cathode'. Fine wires passing through the glass connected the terminals to an electrical source.

Lenard had observed that, when the power was on, the negative plate produced a stream of rays which caused the tube walls to glow with a soft green light. Röntgen was prepared for this. What startled him was that, despite the black card with which he had mantled his tube to exclude exterior influences on his observations, a nearby paper screen painted with fluorescent substances (barium platinocyanide) was also glowing brightly. In fact, each time electricity pulsed through the blacked-out tube, the paper screen luminesced. Röntgen moved the screen two metres away from the tube, but still it glowed.

Lenard's experiments had demonstrated that cathode rays were stopped by quite thin barriers, so Röntgen

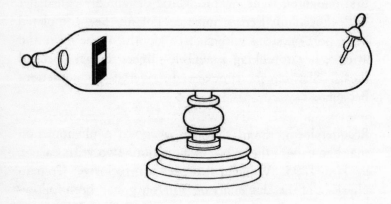

A cathode ray tube

realized that some sort of penetrating rays – hitherto unknown, and which he therefore named 'X-rays' – were escaping through the glass walls of his tube. He further deduced that these 'X-rays' were caused by the impact of the cathode rays on the tube's glass walls. He discovered that although his X-rays could penetrate thick books or decks of cards, they could not pass through denser materials such as metal so easily. When he placed his hand between the tube and the fluorescent screen, Röntgen was staggered to see the shadows of his own bones. The rays had penetrated the soft tissue but the denser bones were sharply delineated on the screen.

Röntgen tested the rays' effects using photographic plates, capturing in the world's first X-ray pictures images of everything from a compass needle in a metal case to his bones. Röntgen realized the implications: his rays could be used to identify fractures in bones and find bullets embedded in tissue. In January 1896 he announced his discovery publicly in Berlin, and before the month was out radiographs were being produced around the world. In 1901 he would become the first recipient of the Nobel Prize for Physics, introduced that year after Alfred Nobel left the bulk of his estate in trust for the annual award of five prizes for services to physics, chemistry, medicine, literature and peace. In the years ahead, the physics and chemistry awards would be dominated by those exploring the new atomic science.

As news of the miraculous rays spread and they were successfully put to work in medical diagnosis, Röntgen became a reluctant celebrity, forced to dodge newspaper reporters. Some people, though, were disturbed by his discovery. Women seriously contemplated buying 'X-ray proof underwear' to repel lascivious

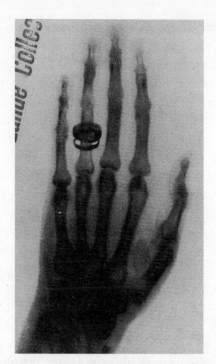

THE NEW ROENTGEN PHOTOGRAPHY.
" LOOK PLEASANT, PLEASE."

1896 X-ray of a hand with a ring by Röntgen; contemporary cartoon

peeping Toms. One rhyme warned:

> I hear they'll gaze
> Through cloak and gown – and even stays
> Those naughty, naughty Röntgen rays.

Punch magazine quipped:

> We do not want, like Dr. Swift,
> To take our flesh off and to pose in
> Our bones, or show each little rift
> And joint for you to poke your nose in.
> We only crave to contemplate
> Each other's usual full-dress photo;
> Your worse than 'altogether' state
> Of portraiture we bar *in toto*!

Meanwhile, puzzled scientists struggled to explain the source of the mysterious X-rays. In Paris, physicist Professor Henri Becquerel decided to investigate whether phosphorescent and fluorescent substances produced these invisible rays.* Becquerel carefully placed successive glowing materials onto photographic plates which he had previously wrapped in thick black paper to see whether rays would penetrate the paper and darken the plates. Nothing happened until he selected the powdery white salts of the rare metal uranium, luminous in sunlight. At last, there was a result. When the plates were developed

---

* Fluorescent substances absorb light of one colour or wavelength and in its place radiate light of another colour. When the source of the light is turned off, that radiation ceases. With phosphorescent materials, the radiation continues after the light source has been removed.

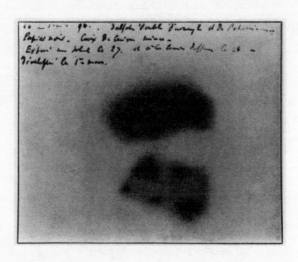

Becquerel's plate showing the image of the copper cross

Becquerel noted faint smudges – evidence of penetrating radiation. He conducted further tests, sometimes adding a coin or metal sheet and observing the faint traces of their outline.

One day he placed uranium salts together with a copper cross onto a photographic plate, but the Paris weather became overcast. Sharing the common belief that substances needed natural sunlight to luminesce, he thrust the plate into a drawer to await a brighter day. Some days later, on 1 March 1896, sheer chance or what another scientist, William Crookes – who was present and saw what happened – admiringly called 'the unconscious pre-vision of genius' caused Becquerel to develop the plate. He found that despite being in darkness the uranium salts had emitted radiation. The image of the copper cross was 'shining out white against the black background'.

Becquerel wrote up his results with both puzzlement and excitement. He had, in fact, discovered 'radioactivity'

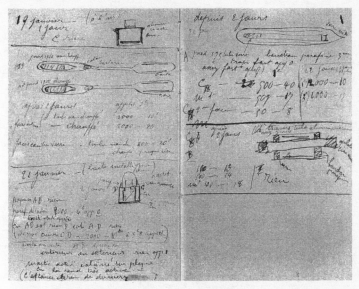

Pages from the notebook Marie Curie kept while working to extract
radium from pitchblende

– the first new property of matter since Newton identified
gravity. Although he did not appreciate the full
significance of his findings, he realized that they were
important and unexpected, and was therefore piqued
when they attracted little comment. Röntgen's X-rays still
commanded all the attention.

Marie Curie read Becquerel's work and was, as she later
wrote, 'much excited by this new phenomenon, and I
resolved to undertake the special study of it'. Since the
subject was 'entirely new' – no-one except Becquerel had
yet written about it – all she needed to do before getting
started on her doctorate was to read his papers. Marie
was offered a small, damp, glass-panelled storage room
on the ground floor of the School of Physics as her

laboratory, and on 16 December 1897 she began work.

Becquerel had noted that his rays released a light electrical charge into the air. Marie therefore decided to measure the electric current emanating from uranium salts. The Curie brothers' piezoquartz electrometer, sensitive to the faintest trace of electrical current, was tailor-made for her purpose. She found the rays' activity to be directly proportionate to the quantity of uranium in the specimens and that it was unaffected by light, temperature or the chemical form the uranium was in.

Wondering whether other chemical elements besides uranium might share these qualities, she plundered her colleagues' shelves for specimens. Her careful examination of these elements revealed that, in addition to uranium, only thorium, the heaviest of the known elements after uranium, was active. Her measurements also showed that pitchblende, a heavy black ore rich in compounds of uranium, appeared nearly four times as active as pure uranium. This was not what she had expected. She repeated her meticulous tests twenty times but her results remained the same. Since she had already tested all known elements for activity, logically this could only mean one thing: the pitchblende contained a new element. She told her sister Bronya, 'The element is there and I've got to find it.'

Marie immersed herself completely in her work, helped by Pierre. As their younger daughter Eve later wrote, he had followed his wife's progress 'with passionate interest. Without directly taking part in Marie's work, he had frequently helped her by his remarks and advice. In view of the stupefying character of her results, he did not hesitate to abandon his study of crystals for the time being in order to join his efforts to hers in the search for the new substance.' They began breaking down the pitch-blende to extract the tiny fragment containing the activity, hoping thereby to solve the puzzle. They did this by extracting from the pitchblende sulphur of bismuth, a substance which, according to their measurements, was far more active than uranium. Since pure sulphur of bismuth was itself inactive, this meant that the new active ingredient had to be present in the bismuth.

It was laborious, painstaking but exciting work. As soon as they had extracted a tiny amount of active material, Marie bore it off to Eugène Demarçay, a specialist in spectrography – the science of identifying elements by the rainbow-coloured 'spectra' they display when energized by an electric current. Despite having lost an eye in a laboratory explosion, his abilities were still acute. He analysed Marie Curie's specimen and declared it was something he had never seen before.

The Curies announced their discovery of what they believed to be a new element in July 1898 in the Academy of Sciences' *Comptes Rendus*, the most influential scientific publication in France. They declared that, if proved correct, they would name it 'Polonium' in tribute to the land of Marie's birth. The title of their paper, 'On a New Radioactive Substance Contained in Pitchblende', coined a new word. The terms 'radioactive' and 'radio-activity', from the Latin word 'radius' meaning ray, were quickly taken up. So was the term 'radioelement' to define any element with this property.

After a cycling trip to the Auvergne with baby daughter Irène, whose first words 'Gogli, gogli, go' Marie recorded with as much delight as her experimental findings, they returned to Paris to resume their investigation. As they laboured, they were astonished to discover a further new radioactive element in the pitchblende. On 26 December 1898, just six months after finding polonium, they announced the likely existence of this second new element, naming it 'radium' and telling the world that its radioactivity 'must be enormous'. Their paper also stated that 'one of us' (probably Marie) had shown that 'radioactivity seems to be an atomic property' – in other words, it derived from some characteristic within the atom, the tiny brick from which all matter is built.

The Curies had made these startling discoveries with tremendous speed – within a year of Marie beginning her doctoral thesis. They next had to convince the many sceptics that radium and polonium were not chimera, but real. So far they had succeeded in isolating only tiny specimens of each. To prove their existence beyond dispute they needed larger samples.

It was already clear that radium was the more active of the two and therefore easier to isolate. Accordingly, Marie Curie focused on extracting pure radium – a formidable task, since radium constitutes less than a millionth part of pitchblende. She needed fifty tons of water and some six tons of chemicals to process just one ton of pitchblende from which the maximum yield would be no more than four hundred milligrams of radium – about one hundredth of an ounce. The task required facilities on an industrial scale. Instead, the School of Physics offered the Curies what Marie called a 'miserable old shed' abutting the narrow Rue Lhomond. This old wooden hangar with a leaking skylight and a rusting cast-iron stove had been used as a dissecting room. A visiting German chemist likened it to a cross between a stable and a potato cellar.

As Marie Curie recalled, she felt 'extremely handicapped by inadequate conditions, by the lack of a proper place to work in, by the lack of money and of personnel'. Nevertheless, the Curies moved in and awaited the delivery of ten tons of pitchblende residue from the St Joachimsthal uranium mines in Bohemia, the principal source of uranium ore in Europe. The valuable uranium salts extracted from pitchblende were used to dye skins for the then fashionable yellow gloves and to stain glass in rich hues of orange and yellow, but the residue was considered worthless. The Curies hoped it would still contain enough radium for their purposes. When horse-drawn carts finally delivered the sacks of ore, Marie impatiently ripped one open, spilling the contents, still mixed with Bohemian pine needles, out on the courtyard. She tested a chunk with an electrometer and to her relief found it highly radioactive.

Marie effectively took charge. Pierre later admitted that, left to his own devices, he would never have embarked on such an enterprise. Day after day the small figure dressed in a baggy, stained linen smock could be seen obsessively filling cauldrons in the courtyard. She processed the pitchblende in batches, pulverizing, crystallizing, precipitating and leaching to purify and extract the precious radium which glowed blue in its glass containers. As she later recalled, 'Sometimes I had to spend a whole day mixing a boiling mass with a heavy iron rod nearly as large as myself. I would be broken with fatigue at the day's end. Other days, on the contrary, the work would be a most minute and delicate fractional crystallization, in the effort to concentrate the radium.'

The hangar lacked any proper ventilation so, unless it was raining, Marie performed her chemical treatments in the courtyard to avoid breathing in the noxious fumes. By the time the work was complete she had shed nearly fourteen pounds in weight. However, there were compensations. As Marie later recalled, 'Our precious products . . . were arranged on tables and boards; from all sides we could see their slightly luminous silhouettes, and these gleamings, which seemed suspended in the darkness, stirred us with ever new emotion and enchantment.'

As the work progressed, with Pierre helping to interpret and present their results, the Central Society of Chemical Products offered Marie facilities to carry out the early stages of purification on a more industrial scale. She accepted gratefully, and the work was overseen by one of Pierre's students, the young chemist André Debierne from the Sorbonne who, in 1899, had isolated a third radio-active element in pitchblende – actinium.

On 28 March 1902, over three years after announcing

Contemporary print showing work on the extraction of radium from pitchblende

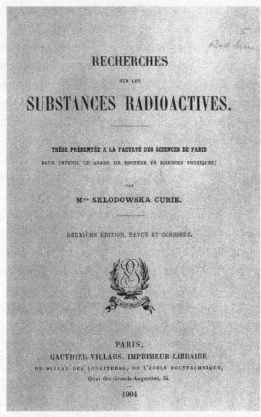

Title page of Marie Curie's published thesis

her belief in its existence, Marie Curie finally had
sufficient radium – one tenth of a gram – for a definitive
test. Once again she hurried to expert spectroscopist
Eugène Demarçay. He confirmed definitively what she
had known intuitively, that radium was indeed a new
element. She weighed it carefully and recorded the result
– 225 times the weight of hydrogen, the lightest element
(and very close to the current agreed weight of 226). By
May 1903, Marie Curie's thesis, 'Researches on

Radioactive Substances', was ready for the printer. In June she appeared before three luminaries of the Sorbonne to be questioned on her work, a pale figure austerely clad in black. But it was a formality. She knew far more about her findings than her inquisitors. With little ado they conferred her degree with the accolade 'très honorable'. Seven months later, in December 1903, the Academy of Science of Stockholm announced the awarding to the Curies of the Nobel Prize for Physics, shared with Henri Becquerel, for the extraordinary services they had rendered by their study of Becquerel rays.

Like Röntgen before them, the Curies became unwilling celebrities. People hailed radium as a 'miracle substance'. It seemed to offer limitless possibilities and quickly became the most costly substance in the world, valued at 750,000 gold francs a gram. An American chemist speculated, 'Are our bicycles to be lighted with disks of radium in tiny lanterns? Are these substances to become the cheapest form of light for certain purposes? Are we about to realize the chimerical dream of the alchemists – lamps giving light perpetually without consumption of oil?' American exotic dancer Loie Fuller, who had arrived in Paris with Buffalo Bill's 'Wild West Show' to become the toast of the Folies Bergère, begged the Curies for shimmering 'butterfly wings of radium'. They had to disappoint her, but Loie nevertheless insisted on performing one of her outré routines in their small house.

The Curies' success had been rapid and dazzling, but there was a price. When Pierre Curie raised his glowing tube of radium aloft at the party to fête his wife's doctorate, a guest noticed that his long, slender hands were in a very inflamed and painful state. This was the result of exposure to radium rays. Sometimes he found it impossible to button his clothes. He also suffered disabling stabbing pains in the legs for which he dosed himself with strychnine – then a recognized treatment for rheumatism – but which in retrospect were probably the result of radiation. Marie's fingertips, too, were hardened and burned. A few weeks later she would suffer a miscarriage. Neither understood the risks they had been taking.

Indeed, alerted by reports from two German scientists that radium appeared to have physiological effects on the body, Pierre Curie had actually begun experimenting on his own body, tying a bandage containing radium salts to his arm for a few hours. The resulting wound, as he observed with interest, took months to heal. In his detailed report on it he added that 'Madame Curie, in carrying a few centigrams of very active material in a little sealed tube, received analogous burns . . .' These effects sparked the thought in Pierre Curie's mind that radium could, perhaps, be used to destroy cancerous cells, and he began to work with physicians. Radium was first used in radiotherapy – known as 'Curietherapy' in France – as early as 1903 to treat cancers but also such conditions as the skin disease lupus, strawberry marks and granulations of the eyelids. A number of treatments evolved, ranging from washing in a solution of radium to injections of radium and drinking radium 'tonics'. The treatment for cancer was to place tiny glass or platinum tubes containing radium directly next to the malignant cells. The Curies, though, derived no personal financial benefit from the 'miracle' substance. They decided not to patent their process for extracting radium, believing it to be against the spirit of science to seek commercial advantage. Knowledge should be available to all.

*   *   *

Marie Curie's discovery of radium was an emphatic push on a door just starting to open on a new sub-atomic world whose implications challenged long-established beliefs. To some they were unthinkable. Unravelling the mysteries would require intuitive skills, a daring but disciplined imagination, physical energy and a first-rate scientific mind. These were exactly the qualities of the guest who had been observing Pierre Curie's damaged hands with such sympathetic interest, the young New Zealand physicist Ernest Rutherford.

CHAPTER TWO

'A RABBIT FROM THE ANTIPODES'

RUGGED, RUDDY AND ROBUST, ERNEST RUTHERFORD LOOKED more like a rugby player than a scientist. His appearance reflected his roots in the still-young British colony of New Zealand, where he was born in 1871, a few miles south of the pioneering town of Nelson on South Island. His grandfather George Rutherford, a craggy-faced wheelwright with mutton-chop sideboards, had arrived in New Zealand from Dundee in Scotland with his family in 1843. The party included his five-year-old son James, who, in 1866, married schoolteacher Martha Thompson. Ernest was their fourth child and second son.

James Rutherford earned his living for a while, like his father, as a wheelwright, but life was hard and the large family struggled. In 1883, after other ventures had failed, James loaded wife, children and possessions onto a paddle steamer bound for Havelock where he worked as a flax-miller, processing flax harvested in the adjoining swamps. The young Ernest enjoyed roaming the countryside, shooting pheasants and wild pigeons for the pot. Newton-like, he also made models of waterwheels and enjoyed taking clocks to pieces and reassembling them.

Rutherford's obvious intelligence coupled with relent-less curiosity and remarkable powers of concentration won him a scholarship to the small but prestigious Canterbury College in Christchurch, part of the University of New Zealand. Here Rutherford excelled in mathematics and physical sciences. In his fifth year, after gaining his BA, MA and B.Sc., he turned to research. The recent discovery in 1888 by German scientist Heinrich Hertz of electromagnetic waves, or radiowaves as they are called today, caught his imagination. He developed a magnetic detector, a prototype radio receiver, to pick up radiowaves.

However, without funds to support himself, an academic career seemed beyond his grasp. His father's flax business had not prospered and he was in no position to help. Rutherford pinned his hopes on winning an 1851 Exhibition Scholarship. The Great Exhibition, an inter-national celebration of industry, science and commerce instigated by Prince Albert and held in London in 1851, had attracted over six million visitors and made a fat profit, some of which had been channelled into scholar-ships to pluck gifted science graduates from across the Empire and bring them to Britain. Rutherford was digging in the family garden when the postman brought the letter announcing he had been awarded a scholarship for his work on magnetism and electricity. He reputedly flung down his spade with the triumphant cry, 'That's the last potato I'll dig.'

In 1895, the year that Röntgen discovered X-rays, Rutherford borrowed money for his passage to England, packed up his magnetic detector and set out. Almost immediately on reaching London he skidded on a banana skin and wrenched his knee. It was several days before he could catch a train to Cambridge and limp into the famous Cavendish Laboratory. His scholarship did not specify which university he should go to. It was up to Rutherford to find somewhere he wanted to work and which was willing to accept him. The Cavendish, with its impressive pedigree, seemed a promising possibility.

The laboratory had been founded in the 1870s by William Cavendish, the gifted seventh Duke of Devonshire who, according to an admiring article in *Vanity Fair*, 'would have been a rare professor of mathematics' had he not been born a nobleman. The first holder of the Cavendish Chair of Physics had been James Clerk Maxwell, a Scottish laird who in 1864 had published his theory of electromagnetic fields showing that electricity and magnetism constituted a single fundamental unity. Taking up his appointment in 1871, he had prophetically warned against the prevailing opinion that 'in a few years all the great physical constants will have been approximately estimated, and that the only occupation . . . left to men of science will be to carry on these measurements to another place of decimals . . . we have no right to think thus of the unsearchable riches of creation, or of the untried fertility of those fresh minds into which these riches will continue to be poured'.

The Cavendish and its amiable director, Professor Joseph John Thomson, impressed Rutherford immediately. Known to his students as 'J. J.', Thomson was a Manchester-born mathematician, the son of an impecunious bookseller. In 1884 he had been appointed head of the Cavendish Laboratory aged just twenty-eight. His reluctance to pay for elaborate or expensive equipment, perhaps the result of his impoverished childhood, had established the legendary 'sealing wax-and-string' tradition of the Cavendish where everyday materials were ingeniously used to make and patch up experimental equipment, sealing wax proving particularly useful for vacuum seals. Thomson was, Rutherford noted, badly shaven, with long hair, a small straggling moustache and a thin, furrowed, clever-looking face. He also had 'a most radiating smile' and, at just forty, was 'not fossilised at all'.

Rutherford decided that he would indeed like to work at the Cavendish. He was fortunate that Cambridge University had just opened its doors for the first time to research students who had graduated elsewhere and was prepared to accept him. With characteristic optimism he hoped he would quickly make enough money from developing his magnetic detector to enable him to marry his fiancée, Mary Newton, the eldest daughter of his erstwhile landlady in Christchurch. Soon he was bustling vigorously around Cambridge, setting up experiments and receiving radio signals from more than half a mile away. As a 'colonial' he was perceived as something of an oddity and was sometimes the object of clumsy jokes, but his robust good humour, undoubted ability and passion to find things out impressed his colleagues. One wrote with grudging admiration that 'we've got a rabbit here from the Antipodes and he's burrowing mighty deep'.

When news of Röntgen's X-rays reached Cambridge, a greatly excited J. J. Thomson obtained one of the very first X-ray photographs and urged Rutherford to study the phenomenon. He progressively weaned Rutherford away from radiowaves, leaving the field of commercial radio development to Guglielmo Marconi, whose work at this time was not as advanced as Rutherford's. Rutherford began replicating Röntgen's experiments. The methodology for producing X-rays struck him as very simple, and by the end of 1896 he classed himself as an authority. He was by then working closely with Thomson on explaining how X-rays made gases capable of conducting electricity. He was fascinated by the behaviour of the ions – electrically charged atoms – which made this possible. When a colleague cast doubt on their existence he indignantly replied that ions were 'jolly little beggars, you can almost see them'.

Reports of Henri Becquerel's discovery of penetrating rays emitted by uranium salts and of Marie Curie's experiments with uranium ore roused Rutherford's curiosity still further. By wrapping uranium in successively increasing layers of thin aluminium foil and observing how the growing thickness of the foil affected the nature and intensity of the escaping radiation, he realized that the uranium was emitting at least two distinct types of radiation. He named them 'alpha' and 'beta' after the first two letters of the Greek alphabet. Alpha rays could be easily contained, but beta rays, a hundred times more penetrating, could pass through metal barriers. He also believed he had detected the presence of a third and highly penetrative radiation, later called 'gamma rays' by Frenchman Paul Villard who is also sometimes credited with their formal discovery. However, the cause and origin of each of these radiations was, as Rutherford wrote, a mystery which he determined to solve.

At the same time, Rutherford was keen to enjoy Cambridge. With interests far beyond science, he relished the rich texture of university life and, as he wrote to his fiancée, overcame 'my usual shyness or rather self-consciousness'. His vigorous intellect attracted people from all fields, including a Hegelian philosopher who invited him to breakfast. It was not, apparently, a success. Rutherford wrote that 'he gave me a very poor breakfast, worse luck. His philosophy doesn't count for much when brought face to face with two kidneys, a thing I abhor . . .' Rutherford was elected to several exclusive academic clubs and had plenty of friends to holiday with. At a seaside resort he was amused when a policeman asked him to swim further along the beach because the landlady of a boarding house opposite objected to the sight of young men in swimsuits. He wrote to Mary that 'the alarming modesty of the British female is most remarkable – especially the spinster, but I must record to the credit of those who were staying there, that a party of four girls used to regularly do the esplanade at the same hour as we took our dips . . .'

Meanwhile, Rutherford's mentor J. J. Thomson was about to make the most significant scientific find of the late nineteenth century, which would profoundly influence Rutherford's own career. Thomson had been investigating the nature of cathode rays. He was convinced that they were some kind of electrified particles, and to prove his theory he began testing their behaviour in electric or magnetic fields. By measuring both the extent to which such fields deflected them and their electrical charge, he discovered that cathode rays consisted of very small negatively charged particles whose mass was about 1,800 times less than the lightest known substance – the hydrogen atom. They were, in fact, totally different from an atom. He initially named these tiny carriers of electricity 'corpuscles'. Later they would become known as 'electrons'.

The corpuscles were, in fact, the first sub-atomic

particles to be found, but their nature was much debated at the time. Their discovery hinted that the atom was not indivisible. Thomson himself admitted that 'the assumption of a state of matter more finely subdivided than the atom is a somewhat startling one'. A colleague later told him he thought Thomson had been 'pulling their legs'. Thomson's work suggested an alternative vision – the instability of matter – to that of the indivisible atom. It was revolutionary stuff. Since the seventeenth and eighteenth centuries most leading scientists, including Newton, had believed the atom to be the smallest unit of matter. Some of the ancient Greeks had shared this view: the word 'atom' comes from the Greek *atomos*, meaning 'indivisible'.* In the early nineteenth century the English Quaker scientist John Dalton had defined the atomic theory that in J. J. Thomson's day remained the orthodox view. This stated that atoms were the basic and smallest units of matter. Each chemical element consisted of huge quantities of identical atoms; what differentiated the respective elements was only the atoms' weight and chemical activity. Dalton's vision of atoms was the Newtonian one of hard, indestructible billiard balls whose arrangement determined the characteristics of chemical compounds.

While the scientific world mulled over the implications of Thomson's discovery, the ambitious Rutherford was preparing to move on after just three years at the Cavendish. In August 1898, helped by a testimonial from

* The ancient Greeks had two theories about the nature of matter. Some, like Aristotle, believed matter was infinite and continuous and so could be infinitely subdivided. Others, like Democritus and Epicurus, thought that matter consisted of minute and indivisible particles.

Thomson praising his originality of mind, the twenty-seven-year-old New Zealander was appointed Professor of Physics at McGill University in Montreal. Tobacco magnate William MacDonald – a man who hated smoking – wished to use his wealth to fund a world-class physics laboratory. Rutherford's task, as he wrote enthusiastically to Mary, would be 'to do a lot of original work and to form a research school to knock the shine out of the Yankees!' It was the perfect outlet for his ambitions. As early as 1896, as he weighed up the significance of Röntgen's X-rays, he had written to Mary that the challenge was 'to find the theory of matter', in other words to discover what matter consisted of 'before anyone else, for nearly every professor in Europe is now on the warpath . . .' It was a race in which, in his view, 'the best sprinters' were the Curies and Henri Becquerel, but he believed that he, too, had a chance.

Although Rutherford was stirred personally by the spirit of competition, the early twentieth century was still a time when scientific results were shared internationally and scientists met one another on friendly terms. However, the world in which they operated was highly nationalistic and competitively imperialist. Even the United States was busy putting down a guerrilla insurgency in its new colony of the Philippines. Britain was involved in the long struggle with the Boers of South Africa. The cause was partly for foreigners' rights in the Boer republics, but also partly about control of the Rand diamond fields. When the British won, *Life* magazine concluded, 'A small boy with diamonds is no match for a large burglar with experience.'

Japan was still largely unknown to the West, but she had been modernizing rapidly since the Meiji Restoration in 1868. Her defeat of China in 1894–5 had shocked the world and prompted the German Kaiser to coin the expression 'die gelbe Gefahr' – 'the Yellow Peril'. Western guidebooks praised the port city of Hiroshima for its lacquer work, bronzes, exquisite landscaped gardens and succulent oysters. The latter were cultivated on bamboo stakes driven into the sea-bed and regularly exposed at low tide. But during the Sino-Japanese war it became the most important military base in western Japan. Hiroshima's sixteenth-century founder, the warlord Mori Terumoto, had named the city after its striking and strategic waterside setting – 'Hiroshima' means 'wide islands'. The delta of the River Otagawa breaks into six channels as it flows down from the mountains to the north through the city to the silver waters of the Inland Sea, producing a series of finger-like, sandy peninsulas that were then criss-crossed from east to west by more than seventy bridges. At the southern tip of the easternmost peninsula sat the newly constructed Ujina port, built partly on reclaimed land and connected to the main city railway station by a four-mile spur built in just over two weeks.

In 1894, after making this short rail journey from barracks in the city, troops had embarked for China from the harbour. Lighters carried men and supplies out to the larger transport ships which lay at anchor side by side with the navy's grey warships. The Emperor moved his Imperial Headquarters from Tokyo into the sixteenth-century Hiroshima castle. Imperial officials chatted in the city's bustling tea houses and formal gardens landscaped with maple and cherry trees. The Emperor ordered the construction of a new building to house meetings of the Japanese Parliament, known as the Provisional Diet, and himself came to Hiroshima to attend its meetings.

Hiroshima for a period assumed the status of a temporary capital. In 1900, Hiroshima's port was busy once more as Japanese troops sailed to China to help Western forces suppress the Boxer Rebellion. With the support of the formidable Empress Dowager of China, the Boxers – a peasant sect opposed to the increasing territorial and commercial exploitation of China by the West and Japan – had risen up, murdering the Japanese and German envoys and imprisoning the Western ambassadors for fifty-five days in their legations in Beijing. Japanese troops made up roughly half of the international relief force and impressed Western observers with their discipline and courage.

They would be even more impressed in 1904 when Russia and Japan went to war over their conflicting commercial and territorial aspirations in Korea and Manchuria. Hiroshima would again become a major port of embarkation. Its citizens cheered the departing troops and nursed the returning wounded. Kimono-clad members of the Shinshu Aki Women's Association met in Hiroshima's Honganji Temple where, kneeling decorously back on their heels, they rolled more than ninety thousand bandages to bind the soldiers' wounds. They rejoiced at news of Japanese success.

The Russian Baltic Fleet sailed round the world to ignominious destruction at the battle of Tsushima by the Japanese fleet commanded by Admiral Togo. On land, Japanese troops won many victories and occupied the Russian island of Sakhalin. American President Theodore Roosevelt brokered a peace conference – a pioneering move onto the world stage by the United States. Under the terms of the peace treaty, Port Arthur and the southern half of Sakhalin were leased to Japan, Korea became a

Japanese dependency and Manchuria returned to Chinese sovereignty. Many Japanese thought the terms too generous to Russia and protested with considerable civil disturbances. Admiral Togo's flagship was sunk in Tokyo harbour and a fire in a major army storehouse in Hiroshima was rumoured to be the work of arsonists opposed to the treaty. To the rest of the world, Japan's victory meant that she had become a major power and a considerable naval presence in the northern Pacific.

Ernest Rutherford, the young scientist from the southern Pacific, settled in happily at McGill. He enjoyed his first winter, breathing in the glacial air, walking on the frozen St Lawrence River and watching huge chunks of ice being cut and stored, ready for sale when summer came. In 1900, the year of the Boxer Rebellion, he was able finally to go to New Zealand and wed Mary. They set up house in Montreal. A piece of student doggerel, 'Ernie R-th-rf-rd, though he's no fool, / In his lectures can never keep cool', suggests that Rutherford did not find dealing with less gifted undergraduates always easy. Nevertheless, he and Mary welcomed research students to tea. It was a friendly atmosphere; Rutherford talked and blew clouds of smoke from the ubiquitous pipe which Mary reluctantly but indulgently allowed him to smoke. As a letter to her from Rutherford in 1896 shows, she had initially been strongly opposed to the habit. Rutherford pleaded,

> A good long time ago, I gave you a promise I would not smoke . . . but I am now seriously considering whether I ought not, for my own sake, to take to tobacco in a mild degree. You know what a restless individual I am, and I

believe I am getting worse. When I come home from researching I can't keep quiet for a minute, and generally get in a rather nervous state from pure fidgeting. If I took to smoking occasionally, it would keep me anchored a bit and generally make me keep quieter ... Every scientific man ought to smoke, as he has to have the patience of a dozen Jobs in research work.

There was, however, no whisky or wine. One young man recalled regretfully that 'in the Rutherford household alcohol was regarded with suspicion'.

1900 was also the year that Rutherford made the first in a chain of discoveries that would challenge the accepted laws of chemistry and establish his reputation. While investigating the properties of the heavy element thorium, he identified a mysterious discharge or 'emanation' whose radioactivity reduced 'in a geometrical progression with time'. In this case it declined to half its original value in sixty seconds and by half of that half-value in the next sixty seconds, so that after two minutes only a quarter of the original activity remained and after three minutes only one eighth. By inspired but careful experimentation he had uncovered a phenomenon at the very core of radio-activity – the *half-life*.

The timely arrival at McGill of English chemist Frederick Soddy gave Rutherford a partner to help analyse the chemical significance of his findings. Initially the two young men sparred. At a meeting of the Physical Society chaired by Rutherford the subject for debate was 'The existence of bodies smaller than an atom'. Soddy's paper, 'Chemical evidence of the indivisibility of the atom', lambasted physicists such as J. J. Thomson for unjustifiably attacking classical atomic theory. Soddy's passion surprised Rutherford but, impressed by his intellect, he invited him to collaborate on examining the mysterious thorium emanation. Soddy agreed, recognizing in Rutherford 'an indefatigable investigator guided by an unerring instinct for the relevant and important'.

They began work in October 1901 and soon proved that the emanation was not merely the result of some disturbance of the air caused by the radioactivity in thorium. The emanation was an inert gas – one without active chemical properties – which would not react or combine with anything. The evidence suggested it was another element, and this moment of discovery was awesome. Soddy, 'standing there transfixed as though stunned by the colossal import of the thing', turned to his companion and said, 'Rutherford, this is transmutation: the thorium is disintegrating and transmuting itself into an argon gas.' Rutherford replied, ' "For Mike's sake, Soddy, don't call it transmutation. They'll have our heads off as alchemists. You know what they are." After which he went waltzing round the laboratory, his huge voice booming, "Onward Christian So-ho-hojers" which was more recognizable by the words than by the tune.' Rutherford urged Soddy to call their discovery not 'transmutation' but 'transformation'. They checked and re-checked, but their results held good. Their discovery, which was indeed akin to alchemy, suggested that radioactive elements disintegrate spontaneously and unstoppably, forming different 'daughter' elements in the process. They contain unstable atoms which decay over time, shedding radiation in the form of alpha or beta particles in an attempt to reach stability.

However logical it might have seemed in the laboratory, Rutherford and Soddy knew that their 'disintegration theory' contradicted another basic law – the immutability and indestructibility of chemical elements. As they expected, their work provoked scepticism and hostility. Alarmed colleagues warned they would bring discredit on McGill University and urged them to delay publishing their findings. British chemist Henry Edward Armstrong demanded to know why atoms should indulge in an 'incurable suicide mania'. But Rutherford and Soddy refused to be brow-beaten, facing down their opponents with confidence and hard evidence.

They were helped by J. J. Thomson in England, who steered them through these potentially damaging and difficult times, ensuring early publication of their papers and lending his authority to their findings. By 1903 they had published a series of papers they considered con-clusive. The final paragraph of their final paper stated, 'All these considerations point to the conclusion that the energy latent in the atom must be enormous . . .' Around this time Rutherford made a 'playful suggestion' that if a proper detonator could be found, it was just conceivable that 'a wave of atomic disintegration might be started through matter, which would indeed make this old world vanish in smoke'.

The Curies were among the sceptics. In the generous, collaborative spirit of the time, they had loaned Rutherford a sufficiently powerful radioactive source to allow him to conduct his research and they were keenly interested in the findings. As early as 1900 Marie Curie had written that the idea of some kind of transformation was very seductive and explained the phenomena of radioactivity very well, but, despite her belief that radioactivity was an atomic phenomenon, she had shied away. Transformation seemed too revolutionary, too alien to the laws of chemistry. The Curies wondered whether Rutherford and Soddy were rushing to unjustified conclusions based too narrowly on findings from thorium. They also worried that the transmutation theory threatened the status of their discoveries, radium and polonium, by redefining them as transitional entities rather than new elements.

In fact, as the theory developed, the reverse would prove true. The theory would explain where radium and polonium fitted in despite their instability. Uranium slowly but inexorably decays, transmuting through a series of radioactive elements, all present in uranium ores. The chain ends when uranium finally transforms into stable, unradioactive lead. Radium is the fifth element in the chain descending from uranium to lead, and polonium is the penultimate link in the chain before lead. The fact that uranium is still present in the Earth's crust, created some 4.5 billion years ago, shows just how slowly uranium decays.

The Curies' perplexity was heightened by Pierre's discovery in 1903 that radium released an astonishing amount of heat. Just 1 gram of radium could heat around 1.3 grams of water from freezing point (0°C) to boiling (100°C) in an hour. These seemingly bizarre findings contradicted the nineteenth-century law of conservation of energy which stated that while energy might change from one form to another, for example from heat to motion, it could not be conjured out of nowhere. The Curies speculated whether some sort of external energy might be responsible; others wondered whether gravitational energy might have something to do with it. Nevertheless, the Curies were uncomfortably aware that the transformation theory offered an explanation –

that the energy was being conjured from within the atom. Eventually they would come to accept it.

Rutherford's knowledge of the Curies' work had made him keen to meet them. In 1903, the opportunity came. While visiting England from McGill to defend his heretical transformation theory, Rutherford, accompanied by Mary, took a trip to the Continent. Reaching Paris on a hot June day, he was alerted by a postcard from Soddy that Marie Curie wished him to call. He hastened to her ramshackle workplace to find it locked. It was, in fact, the very day she was being examined on her triumphal doctoral thesis 'Researches on Radioactive Substances', reporting her work on isolating radium. However, he managed to track down Paul Langevin, whom he had met during his Cavendish days, and Langevin invited the Rutherfords to the celebration that night at which Pierre Curie brandished his tube of glowing radium in his damaged hands.

It was, by all accounts, a lively evening, unmarred by any differences of opinion. Rutherford admired Marie Curie's intellect, 'no-nonsense' style and directness. She, in turn, appreciated that he treated her as an equal. This was to be the first of many meetings between them, but, sadly, it was the one and only time he would talk with Pierre Curie. Just three years later on a wet, windy, overcast Paris afternoon, Pierre absent-mindedly stepped out in front of a horse-drawn wagon in the Rue Dauphine. Too late he tried to scramble out of the way, slipped and fell. The wagon's iron-rimmed rear left wheel crushed his skull, bloodily spilling his brains on the wet boulevard. He was only forty-six.

Marie was left a widow at thirty-eight, with Irène but also with her second daughter, Eve, born in 1904, to care for. The University of Paris decided to maintain their Chair of Physics created for Pierre two years earlier and invited Marie to assume his duties, but did not award her the professorship. It was, nevertheless, the first time in France that such an appointment had been given to a woman, and she accepted. Her first lecture, delivered fifteen years to the day since she had first entered the Sorbonne to register as a student, was a highlight of the social calendar. The fashionable and curious craned their necks for a good look at the first woman to lecture at the Sorbonne. She entered the lecture room quietly with downcast eyes and commenced her course at the exact point at which death had halted Pierre's. Newspapers hailed her performance as 'a victory for feminism'.

Marie rejected a government proposal to build her a laboratory. Pierre had been haunted by the lack of proper facilities and she was bitter that it had taken his death to induce the authorities to provide them. Single-mindedly, at times obsessively, she immersed herself in her work, shunning celebrity. Her greatest dread, as Eve Curie later recalled, remained the 'crushing, mortal boredom which dragged her down when people rambled on about her discovery and her genius'. Her response, repeated like a mantra over the years to come, was 'in science we must be interested in things, not in persons'. Rutherford would prove one of her greatest allies during some difficult personal times ahead.

Rutherford's findings on radioactivity had established his international reputation as one of the leading experimental physicists of the day. Universities courted him eagerly, and in May 1907 he returned to England as Professor of Physics and director of the Manchester

University Laboratory. The laboratory was only seven years old and, unlike the Cavendish with its 'sealing wax and string', was magnificently equipped. The only drawback was that it possessed almost no radioactive materials. Since Rutherford's primary interest was to follow up his work with Soddy and unravel the sequence of elements generated through radioactive decay, this deficiency had to be remedied. A generous loan of some five hundred milligrams of radium bromide from Professor Stefan Meyer of the Radium Institute in Vienna, who had access to the same Bohemian mines which had furnished Marie Curie's pitchblende, solved the problem.

In 1908, the year in which Kenneth Grahame wrote *Wind in the Willows* and Jack Johnson became the first black man to win the world heavyweight boxing championship, Rutherford received the Nobel Prize for Chemistry for his investigations into the disintegration of the elements, and the chemistry of radioactive substances. He was amused that the prize was for chemistry not physics, joking about his instantaneous transmutation from physicist to chemist. Students from around the world flocked to Manchester to study under the Nobel laureate. They found Rutherford an inspirational but taxing taskmaster with a facility to concentrate on a problem for long periods at a stretch without getting tired or bored. A young Japanese scientist named Kinoshita from Tokyo Imperial University, who studied briefly under Rutherford in 1909, wrote wistfully from Japan that 'I wish I could go back again to your lab so that I shall be able to do some decent work'. Visiting Japanese Minister of Education Baron Kikuchi was so impressed by Rutherford's vitality as well as his intellect that he remarked, no doubt tongue in cheek, that he must be the son of the famous Professor Rutherford.

The matter now absorbing Rutherford, and which would lead to the dissection of the atom, was the nature and behaviour of alpha rays – the least penetrating form of radiation. While still in Montreal he had begun to think that helium found in the atmosphere was probably the product of radioactive decay. Studies by Soddy – by then in London and working with chemist Sir William Ramsay, the discoverer of the inert gases – suggested he was right. Soddy demonstrated that, as it disintegrated, radium emitted streams of helium atoms, travelling at tremendous velocity. Rutherford suspected that these were the same as the alpha rays or particles emitted by radioactive materials, and began investigating them.

Together with one of his research students, the German Hans Geiger, Rutherford invented an electrical instrument capable of counting individual alpha particles.* However, Rutherford abandoned this method in favour of one capable of actually making alpha particles visible, using a plate coated with zinc sulphide. When the plate was hit or 'bombarded' with alpha particles, tiny flashes of light occurred at each impact.† The method, called 'scintillation' from the Greek word for 'spark', was time-consuming and hard on the eyes, straining to count every flash, but reliable. Hans Geiger recalled the atmosphere: 'I see the gloomy cellar in which he had fitted up his delicate apparatus for the study of the alpha rays. Rutherford

---

* Hans Geiger would later develop this device into the Geiger counter, still used in radiation laboratories.

† Scientists used the military term 'bombard' to describe how they placed a source of radioactivity near an experimental subject – for which they again used a military term, 'the target' – to determine the effect of the radioactivity released upon the subject.

loved this room. One went down two steps and then heard from the darkness Rutherford's voice . . . Then finally in the feeble light one saw the great man himself seated at his apparatus . . .'

Rutherford's next eureka moment resulted from a routine experiment which he had instructed Geiger and another researcher, Ernest Marsden – by his own account a callow youth from Blackburn – to conduct using the scintillation method. Their task was to see what happened when alpha particles were fired at metal foils, so they positioned a source of alpha particles near a thin gold foil. Most of the particles passed through with little deflection, as they expected, given the particles' weight and velocity. However, a few – one in eight thousand – came bouncing straight back. To Rutherford, this was 'almost as incredible as if you had fired a 15-inch shell at a piece of tissue paper and it came back and hit you'. It suggested the presence of incredibly strong forces in the atoms of gold.

Rutherford mused over these results, which he simply could not understand. He followed his own advice to his students, 'Go home and think, my boy', and over a period of eighteen months, by logic and intuition, found an explanation for his experimental findings and so solved the puzzle. In December 1910 Rutherford, 'obviously in the best of spirits', burst into Geiger's room and, as Geiger recalled, excitedly announced that 'he now knew what the atom looked like'. He had worked out that it was not the solid structure studded with electrons like plums in a pudding as suggested by J. J. Thomson and others. The atom Rutherford visualized was almost empty. Nearly all its mass was concentrated in a powerfully charged but tiny nucleus the size, comparatively, of a pin's head in St Paul's Cathedral. The reason why most of Geiger's and Marsden's alpha particles had barely been knocked off their trajectory as they passed through the gold atoms was that, like ships skimming a great, empty ocean with no other vessels for thousands of miles, they had passed too far from the tiny nucleus to be affected. However, occasionally and randomly, a particle had skimmed close enough to the nucleus to be violently repulsed by an electrical force so enormous that it had virtually been flung back on itself.

Rutherford's interpretation of what had happened was revolutionary. Not only had he established the planetary model of the atom where electrons orbit a tiny nucleus, he had also changed for ever the way in which people regarded the world around them. He had revealed that the stability and solidity of everyday objects – tables, cups, spoons – are an illusion. At the most minute level, human beings and everything around them consist almost entirely of voids with insubstantial boundaries defined by whirling particles.

Rutherford conducted a final suite of alpha-particle scattering experiments to check his hypotheses and then in early 1911 announced to his startled colleagues his discovery of the atomic nucleus. It was, as one later recalled, a 'most shattering' revelation.

CHAPTER THREE

FORCES OF NATURE

IF 1911 WAS A TRIUMPHANT YEAR FOR RUTHERFORD, IT WAS an *annus horribilis* for Marie Curie. Since her husband's death in 1906 she had scored two notable coups. In 1908 she was finally given the full rank of Professor of Physics at the Sorbonne. That same year, she coaxed and bullied the university and the Pasteur Institute into co-founding a Radium Institute to comprise two parts: a laboratory of radioactivity, under her direction, and a laboratory of biological research and 'Curietherapy' – the use of radium to treat cancer and other diseases.

Yet she remained a retiring individual who flinched from the limelight. When she learned that the International Congress on Radiology was to meet in Brussels in the autumn of 1910 to establish an International Radium Standard – a physical benchmark specimen against which radium to be used in industry, medicine and research could be measured – she was reluctant to go. She consulted Rutherford, who sensibly advised that, as the figurehead for radium, she had to be there.

The congress endorsed Marie's unique authority by agreeing that she should prepare the standard and that the unit in which measurements were to be made against the standard should be named the 'curie'. However, arguments broke out over the definition of the unit. An angry Marie believed that she, and she alone, should decide the parameters. A female Swedish scientist had been correct in observing that Marie Curie regarded radioactivity as her 'child' which she had 'nourished and educated'. She resented the interference of others. When Marie failed to get her way, she claimed she was too unwell to continue debating and withdrew. Finally she prevailed, but her stubbornness had roused considerable and lasting resentment. Rutherford, who considered her genuinely frail and 'very wan and tired and much older than her age . . . a very pathetic figure', was one of her few defenders.

Rutherford would meet Marie Curie again the following year when the Belgian industrialist and entrepreneur Ernest Solvay invited thirty leading physicists to the first Solvay Conference, held in Brussels. The conference's primary purpose was to debate a revolutionary scientific idea, quantum theory.

The theory's rather apologetic creator was the German physicist Max Planck. This melancholy-eyed scientist had been investigating how hot solids radiate heat since 1897. He realized that he could only make sense of his experimental findings if he assumed that heat was emitted in 'energy parcels', or separate 'quanta' as he called them, from the Latin meaning 'how much'. The conservative Planck cautiously called his findings a 'hypothesis' rather than a 'theory' when he first published them in 1900. His problem was that, while on the one hand his hypothesis worked, on the other it conflicted with the established laws of physics which decreed that energy was emitted in an uninterrupted flow, not in discrete packets. Planck was in the paradoxical but not unique position of having discovered something intuitively that he did not understand fully in logic.

Albert Einstein had the visionary brilliance to grasp what Planck could not. Challenging, analysing and stepping outside the conventional bounds of life and thought came naturally to him. Brought up in a secular, free-thinking Jewish family in Germany, the son of an engineer, he had quickly rejected what he considered the militaristic character of German education where children were marched and drilled like small soldiers. He completed his education at the Zurich Polytechnic Institute where he studied mathematics and natural sciences. With his thick dark hair and shining dark brown eyes he exuded both energy and a potent sensuality. In 1903 he married Mileva Maric, a Serbian also studying at the institute. She was four years older and apparently walked with a limp. A daughter, Lieserl, born to them the previous year and whose existence only came to light in 1987, either died in infancy or was adopted.

Having failed to find a permanent academic post, in 1905 Einstein took a job as a patent examiner in the Swiss Patent Office in Berne. In his spare time he read Planck's work and found it a revelation. 'It was', he later wrote, 'as if the ground was pulled from under one.' Realizing that quantum theory explained some hitherto inexplicable phenomena, he worked to confirm and extend it. In particular, he applied the theory to the 'photoelectric effect' – the way in which light colliding with certain metals expelled a shower of electrons. Just as Planck had found with heat, Einstein realized that his experimental findings could be explained if he assumed that light was not a smooth, wavelike phenomenon as previously thought but was emitted in tiny, discrete 'energy quanta', separate packages more akin to tiny bullets.*

1905 was a fertile year for the twenty-six-year-old Einstein in other ways. His facility for thinking the unthinkable, which had led him to uphold Planck's quantum theory, also led him to the discoveries for which he is best known. Since the days of Galileo and Newton, scientists had believed that objects at rest and objects moving straight and at constant speed behaved in the same way. However, James Clerk Maxwell's theories suggested that light was an exception to this principle so that measurements of the velocity of light would vary depending on the effects of motion. Einstein, however, believed intuitively that the velocity of light did not vary. One morning he awoke feeling as if a tempest was raging in his mind but that somewhere in the maelstrom were the answers he had been seeking. As he later put it, 'The solution came to me suddenly . . .' It was nothing less than a revolutionary analysis of space and time.

Einstein described his theory in one of five remarkable papers he published that year in the leading German physics journal the *Annalen der Physik*. It was called 'On the Electrodynamics of Moving Bodies'. He postulated how light travelled from place to place with the same velocity regardless both of direction and of whether the source of light was moving relative to the person observing it. This was Einstein's 'special relativity theory',

---

* His discoveries about the properties of light would one day lead to television.

which, as C. P. Snow wrote, 'quietly amalgamated space, time and matter into one fundamental unity'. It was the first step on the path to his 'general' theory of relativity.

Einstein's three-page supplement to the paper, added as an afterthought, argued that if a body emits energy, then the mass of that body must decrease proportionately – in other words, that light transfers mass. He articulated the ideas that he would soon express in the world's most famous equation, $E = mc^2$ – energy is equal to mass times the speed of light squared. Einstein's groundbreaking insight was that energy and mass were not separate phenomena but interchangeable. Each could be converted into the other, and the speed of light was the conversion factor. Implicit in $E = mc^2$ was the potential for enormous amounts of energy to be squeezed from tiny amounts of mass, given the enormous size of the conversion factor.* However, more than thirty years would pass before scientists would finally grasp how to access that energy.

Einstein, who privately nicknamed the 1911 Solvay Conference 'a witches' sabbath', found it more enjoyable than he had anticipated. He wrote to a friend that he spent 'much time' with Marie Curie and Paul Langevin. He was 'just delighted with these people' and praised Marie's 'passionateness' and 'sparkling intelligence'. As was about to emerge in a thundercloud of scandal, one reason for Marie's animation was that she and Langevin

---

* The speed of light is 670 million miles per hour, and the huge factor obtained by squaring this means that just a single pound of matter, if wholly converted to energy, would be equivalent to burning over a million tons of coal.

were in love. This did not, however, soften her insistence at the conference that the International Radium Standard she had prepared should remain 'chez moi' – in other words, in her personal laboratory and under her sole control. When others argued that this was unacceptable, she retreated to her room, once again claiming nervous exhaustion and headaches. Critics claimed her ailments were psychosomatic, and even Rutherford's patience was wearing thin. He wrote, 'Madame Curie is rather a difficult person to deal with. She has the advantages and at the same time the disadvantages of being a woman.' He told her firmly that an international standard should not be 'in the hands of a private person'. Marie would later back down, personally sealing the radium standard in a glass tube and depositing it at the International Bureau of Weights and Measures at Sèvres, near Paris.

At the conference, though, such squabbles were pushed aside as the sensational 'Affaire Langevin' broke in the press. Paris newspaper *Le Journal* reported that Paul Langevin's wife Jeanne was accusing him of having an affair with the forty-three-year-old Marie Curie and intended to divorce him. Newspapermen ambushed Marie in Brussels, thrusting copies of *Le Journal* at her. At first she refused to comment; then, in a handwritten note to the Brussels correspondent of the Paris *Le Temps*, she rebutted the accusations as 'pure fantasy'. However, other papers enthusiastically took up the story. *Le Petit Journal* titillated its readers with a story headed 'A Laboratory Romance – The Adventure of Mme. Curie and M. Langevin'. It included an interview with Jeanne Langevin in which she claimed that the affair had been going on for several years. She had kept quiet about it, hoping for a reconciliation, but her husband's recent behaviour, including slapping her face for spoiling a fruit compote, had forced her to speak out.

The story broadened. Some suggested that the affair might have started before Pierre Curie's death, even that it had prompted him to commit suicide. One journalist used the scandal to attack not just Marie's morals but her credibility as a scientist, querying whether women were capable of creative, independent research. He quoted an eminent but conveniently unnamed scientist who claimed she was a mere 'plodder' and that a woman could only shine in science when 'working under the guidance and inspiration of a profoundly imaginative man' with whom she was in love.

Returning to Paris, Marie Curie continued to deny the affair, seeking refuge from the press with friends. However, the allegations were almost certainly true. In mid-July 1910 Langevin is known to have rented an apartment near the Sorbonne under an assumed name. He and Marie were observed meeting there almost daily. In early 1911, friends had noticed how Marie had suddenly appeared dressed in white with a rose at her waist, rather than in her usual sombre hues. One wrote that 'something signified her resurrection like the spring, following a frozen winter'. Paul Langevin was five years her junior, handsome, charismatic and an acknowledged ladies' man. He would later father a child by one of Marie Curie's pupils. He had married very young and the relationship soured early. He had turned for advice and solace to Marie. An old friend, she considered Langevin a genius, but weak and in need of affection. She feared his wife would force him to desert science in favour of going into industry to make money. Moreover, letters between Marie and Paul were stolen, probably by Langevin's brother-in-law,

Henry Bourgeois, who prised open a drawer in Langevin's marital home. There is evidence that Langevin paid blackmail money, given him by Marie, to try to prevent the letters' disclosure. Marie lent Langevin a total of five thousand francs – more than a tenth of her salary – over this period and Langevin made 'loans' never recorded in writing to his brother-in-law. Marie's friend Jean Perrin wrote angrily of 'odious blackmail'.

In November 1911, while the scandal still raged, came news that Marie Curie had been awarded a second Nobel Prize, this time for chemistry, for her original isolation of pure radium. It was an unprecedented honour, but the press attacks continued. Some contained darker undercurrents than mere simulated moral outrage. Only five years after the end of the Dreyfus Affair,* they reminded readers that Marie was a foreigner and suggested incorrectly that she was very probably a Jew. They demanded she resign from the Sorbonne and return to Poland. Matters finally came to a head when Gustave Téry, editor of the weekly *L'Oeuvre*, published extracts from the Curie–Langevin letters and derided 'the Vestal Virgin of radium' as 'an ambitious Pole who had ridden to glory on Curie's coat-tails and was now trying to latch onto Langevin's'.

Langevin challenged Téry to a duel. He told a friend, 'It's idiotic, but I must do it.' It proved more farcical than dramatic. Dressed in black and wearing bowler hats, the duellists met at the Parc des Princes Bicycle Stadium. Téry,

---

* The Dreyfus Affair was a notorious French miscarriage of justice in which anti-Semitism played a major part. Jewish army officer Alfred Dreyfus was wrongly convicted of passing military secrets to Germany and imprisoned on Devil's Island. His conviction was eventually quashed after a long campaign led by the writer Émile Zola.

as the man who had been challenged, was entitled to raise his weapon first but kept his gun pointed at the ground while he gazed up at the sky. Unable to shoot a man who had not discharged his weapon, Langevin also lowered his. They left the field, honour satisfied. Téry wrote piously, 'The defence of Mme. Langevin does not oblige me . . . to kill her husband . . . I could not deprive French science of so precious a brain.' With this ridiculous encounter, public interest waned, although the Affaire Langevin provoked at least four further duels between defenders and detractors of Madame Curie.

A subdued and frail Marie Curie went to Stockholm to claim her Nobel Prize. She collapsed on her return to Paris with fever and kidney problems, but her health picked up when she learned that Madame Langevin's writ formally seeking separation from her husband did not name her. However, her relationship with Langevin could henceforth, sensibly, be only professional. Einstein, who had remarked on Marie's passion in 1911, observed a change while hiking with her in 1913. He wrote, 'Madame Curie is highly intelligent but has the soul of a herring, which means that she is poor when it comes to the art of either joy or pain. Almost the only time she shows emotion is when she's grumbling about things she doesn't like.'

Rutherford loyally supported Marie Curie throughout the brouhaha. He was by then deeply involved in further attempts to dissect the atom, in the aftermath of his finding of the nucleus. Shortly after his return from the Solvay Conference, a twenty-six-year-old Danish physicist had joined his team at Manchester. Niels Bohr was about to bring quantum theory to the heart of the understanding of the atom. Bohr was an athletic, strong-jawed, huge-handed man with an enormous domed forehead. He spoke in long, complex sentences studded with sub-clauses in a voice that was usually soft and trailed off into a whisper when he reached a crucial point. He came from a distinguished family: his father was Professor of Physiology at the University of Copenhagen. Like Rutherford, Bohr showed an early interest in understand-ing how things worked, and one of his boyhood pleasures was repairing clocks. Also like Rutherford, he was a lateral thinker, quick to spot connections. He was gentle but intellectually tenacious and unafraid to challenge any-one, however high their reputation.

Bohr studied at Copenhagen University, where physics became his passion. He was intrigued by the new dis-coveries – Röntgen's X-rays, Becquerel's rays, the detection of radioactivity, Thomson's electron and Rutherford's identification of alpha and beta radiation. For his doctoral thesis he explored the behaviour of electrons in metals. His findings were so new and unusual that, as with Marie Curie when she was examined on her thesis, no-one was equipped to question them. Bohr then decided he wished to study with J. J. Thomson at the Cavendish Laboratory and arrived in Cambridge in the autumn of 1911. However, shortly before Christmas he heard Rutherford speak at the annual Cavendish dinner about his discovery of the nucleus. Bohr was mesmerized, and the following April he moved to Manchester University.

Bohr found the atmosphere there exhilarating. Rutherford encouraged his young scientists to gather every afternoon for tea. Perched on a stool, his great voice booming out, he urged everyone to speak up, provided they 'made sense' and avoided 'pompous talk'. One of the subjects most eagerly debated was the structure of the atom. Bohr accepted Rutherford's model of the atom as a miniature solar system, with electrons orbiting around the nucleus like planets around the sun, but recognized an inherent flaw. According to Newtonian physics, which saw the world in mechanical terms, the whirling, negatively charged electrons should have gradually dissipated their energy through their movement. As a result, they should have collapsed into the positively charged nucleus in the heart of the atom that was pulling them to their doom, gradually shrinking anything and everything. Yet clearly this did not happen. It was a mystery because, as Rutherford acknowledged, not enough was yet known about either the orbiting electrons and their paths or the nucleus.

Bohr reasoned that, if Rutherford's model was correct, some kind of stabilizing or balancing effect must be at work within the atom. Over the next eighteen months he set out to prove this, turning to the quantum theories of Planck and Einstein. Unlike Planck, who was at the time developing his theory further and even coming round to believing in it himself, Bohr did not worry that the theory could not be properly explained. What mattered was applying it. His guiding principles were that science needed paradoxes to progress, and that, provided they were well founded, seemingly contradictory ideas should not be changed but reconciled. A story frequently related by Bohr exemplified his mental flexibility. A visitor, surprised to see a horseshoe above the entrance to Bohr's house, asked whether Bohr really believed it would bring good luck. 'Of course not,' Bohr replied, 'but I am told it works, even if you don't believe in it.'

Bohr instinctively accepted the existence of quanta and

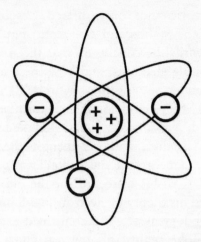

looked for ways to fit a theoretical structure to observed experience of atomic behaviour. By late June 1912, fewer than three months after arriving in Manchester, he had developed an initial version of what would become known as the 'Rutherford-Bohr' model of the atom, and which, once accepted, would be used by scientists ever after. Over the ensuing months, during which he returned to Denmark and married, Bohr refined and developed his ideas further for publication in a trilogy of papers on the 'Constitution of Atoms and Molecules'. He applied quantum theory to matter as well as energy. The heart of Bohr's insight was that the orbits in which electrons travel around the nucleus are specified by quantum rules which provide each orbit with a defined level of energy. While orbiting, an electron suffers no energy loss. Building on this, Bohr envisioned successive layers of electrons 'binding' into a structure around the nucleus until a stabilizing electrical neutrality was achieved. By a 'quantum leap', electrons could switch orbits within an atom, emitting or absorbing energy in bursts.

Bohr's theories not only offered a solution to the problem of the stability of the atom, but he was also nudging towards the conclusion that the structure of the rings of orbiting electrons, and how these built up, held the key to understanding the hierarchy of elements and how and why they could combine to form new ones.

Rutherford, who initially found Bohr's ideas ingenious if hard to visualize, was his mentor throughout. He regarded the Dane as 'the most intelligent chap I've ever met' and admired his disregard for the old orthodoxies. He welcomed his theory of electrons, without yet giving it his formal endorsement. As a confirmed experimentalist, he warned Bohr against placing too much credence on theory alone. He also warned him not to be long-winded when he published his findings, writing, 'it is the custom in England to put things shortly and tersely in contrast to the Germanic method where it appears to be a virtue to be as long-winded as possible'. Bohr dug in his heels. When Rutherford offered to edit Bohr's work for publication, the Dane hurried to Manchester to defend his work, not just paragraph by paragraph but right down to the complex structure of his extensive sentences which, he insisted, were essential to the detailed logic of his case, even if initially confusing. It was one of the few battles Rutherford ever lost. He submitted with good grace, telling his protégé he never thought he would prove so obstinate.

The scientific community responded to Bohr's theories with everything from enthusiasm to incredulity. According to a letter from the Hungarian scientist Georg Hevesy to Rutherford, when Einstein learned of them his 'big eyes . . . looked bigger still, and he told me, "Then it is one of the greatest discoveries" '. Others were openly sceptical, including J. J. Thomson, who was developing his own, different model of the atom. In Germany a number of physicists swore 'to give up physics if that nonsense was true'. Yet, supporting evidence was emerging all the time. Some of this was provided by another of Rutherford's students, the obsessively hard-working old Etonian Harry Moseley, who had arrived in Manchester in September 1910.

Moseley was using X-rays, the penetrating radiation discovered by Röntgen about whose nature scientists were still arguing, to explore variations between elements. To do this, he built an ingenious piece of equipment resembling a toy train with a number of wagons. On each of these he placed a specimen of the element he wanted to examine, and then, by winding silk cords on brass bobbins, moved his 'train' along a pair of rails inserted inside an X-ray tube so that each of his elements in turn was bombarded by cathode rays. When he examined the spectra his specimens produced, Moseley found that they differed according to a regular pattern. The difference between elements seemed to depend on a 'something' which Moseley interpreted as a difference of one unit charge on the nucleus – in other words, a difference of one in the number of electrons possessed by the atom. He knew this would support Bohr's theory of the atom and the Dane's intuition that it was the number of electrons that determined the chemical and physical characteristics of matter.

In late 1913, Moseley left for Oxford University to continue his work there but kept Rutherford and Bohr abreast of his findings. He worked through the naturally occurring elements, from the lightest, hydrogen, to the heaviest, uranium, arranging them in the light of his experimental findings in a revised Periodic Table. Until this time elements had been ranked by their atomic weight. This went back to the days of scientist John Dalton who in the early nineteenth century had developed a theory attaching experimentally determined weights to chemical elements. The idea of a Periodic Table had first been introduced in 1869 by the Russian Dmitry Mendeleyev, who had noticed that when the elements were arranged in order of their atomic weights they could be grouped according to their chemical behaviour. However, no simple relationship governed differences between atomic weights in Mendeleyev's table, whereas Moseley's new classification – 'the law of Moseley', as Rutherford later called it – provided a ladder with ninety-two regular rungs. It was beautifully simple and has provided the basis for physical and chemical analysis of atomic structure ever since.* At the end of his work, Moseley had no remaining doubts that his findings supported Bohr's theories, and said so firmly in the papers he published.

By identifying that there were gaps in his table, Mendeleyev had turned it into a tool for the prediction of new elements. By 1886, three with the chemical properties he had identified – scandium, gallium and germanium – had been discovered. Moseley's 'law' suggested that between hydrogen at number one and uranium at ninety-two there were still seven elements (whose characteristics were predicted) as yet undiscovered. Moreover, Moseley's classification placed several element-pairs in their correct

* Hydrogen, the smallest atom with its one orbiting electron and a charge of one on the nucleus, occupies the first place; helium, with its doubly charged nucleus and two orbiting electrons, is in place number two; and so on until uranium with its ninety-two whizzing electrons.

order in the Periodic Table whereas Mendeleyev, in order to get the chemical properties to fit, had had to place them out of sequence in his ranking by atomic weights.

At the same time, however, there was a difficulty. Moseley's tabulation left no room at the upper, heavier end of the range for the recently identified products resulting from radioactive decay, such as some discharges from radium and thorium. While working at McGill, Rutherford and Soddy had argued that such products were elements in their own right. If so, it had to be possible to fit them into the table.

The anomaly was resolved by Frederick Soddy, who identified the 'Law of Radioactive Displacements' which revealed the existence of 'isotopes'. Soddy deduced that elements could exist in several forms, identical in their chemical and most of their physical properties but differing in their atomic weight. To name them, he borrowed two words from ancient Greek, *isos* meaning 'the same' and *topos* meaning 'place', to signify that isotopes of the same element occupied the same place in the table of chemical elements. Others had also been moving towards these same conclusions, which were an integral part of the jigsaw puzzle of the atom being assembled with such rapidity.

In the spring of 1914, Rutherford, convinced by the accumulating evidence, put his own considerable weight firmly behind the 'Rutherford-Bohr' model of the atom. This was also the year when, on 12 February, the forty-two-year-old Rutherford was knighted by the King. 'Sir Ernest' reacted to the honour with due modesty but was plainly delighted, revelling in his costume of velvet breeches, cocked hat, sword and silver buckles. Former pupils from around the world wrote to congratulate him. One of these was the German chemist Otto Hahn, who had studied under Rutherford at McGill and would one day play a critical part in the discovery of nuclear fission.

Hahn was born in Frankfurt in 1879, the son of a prosperous artisan. Rejecting his father's suggestion that he become an architect, he instead studied organic chemistry. He was, by his own admission, a 'slightly superficial, easy-going' young man, not a hard worker. In his final school report, two of his three top marks were for gymnastics and singing. At Marburg University, he enjoyed 'beery days' and once duelled with sabres. However, in 1904, a chance event changed his life. As preparation for working in industry, Hahn went to London to learn English. By sheer good fortune, he managed to get a place at University College, in the laboratory of Sir William Ramsay.

Hahn at this time knew nothing of radioactive substances, but Ramsay set him to work extracting radium from barium salt. Somewhat to Hahn's surprise, this task led him to the discovery of a new radioactive substance, radiothorium. He watched the material glowing in his darkroom where he was sometimes distracted by a female assistant who found excuses to join the personable young man in the gloom, though, as he later wrote, 'I never dared to kiss her'. He was very fond of women but his English sometimes let him down in the chase. Once, while dancing the fashionable two-step at a university ball, he whispered conversationally in his partner's ear, 'You, here in England, you dance on the carpet. We in our country prefer to dance on the naked bottom.' The girl left the dance floor.

Fascinated by his new area of work, Hahn abandoned

thoughts of industry. Instead, he wrote to Rutherford, then in Montreal, believing him to be 'the only person who had real grasp' of the new science. Rutherford agreed to take Hahn for six months. He enjoyed life in the 'New World', although the discovery that the Rutherford household was teetotal was a shock. He sought solace in his pipe, lending his 'much-chewed specimens' to Rutherford, who frequently mislaid his own. Hahn admired Rutherford's directness, even his simple way of dressing. When a photographer arrived to take Rutherford's photograph, Hahn had to lend him some detachable cuffs because he had not bothered to put any on. More than anything, though, Hahn had found his vocation.

He returned to Germany in 1906 to the Institute of Chemistry in Berlin and began working on the sample of radiothorium Ramsay had given him as a parting gift. He was joined the following year by a slight, dark-haired theoretical physicist from Vienna, Lise Meitner, who would earn from Einstein the accolade 'the German Marie Curie'. She had arrived in Berlin to research under Max Planck and been immediately drawn to the confident, energetic, easy-going Hahn. They decided to work together on radiation experiments, but the institute's director, Emil Fischer, had barred women from the premises. His pretext, after an incident involving a wild-haired Russian student and a Bunsen burner, was that he feared they would set their hair alight. However, he allowed Meitner to work with Hahn in a room which had formerly been the carpenter's workshop and had its own entrance from the street. When she needed the lavatory she had to visit a nearby restaurant.

Lise Meitner's difficulties reveal how extraordinary Marie Curie's achievements had been and the scale of the problems then facing women scientists. Meitner was one of just thirty women working in the new field of radio-activity between 1900 and 1910. She was such a rarity that even Rutherford, who encouraged women in his own laboratories, committed a gaffe. Passing through Berlin in 1908 after receiving his Nobel Prize, he was introduced to the thirty-year-old Lise Meitner. He had seen her name in publications, but even 'Lise' had failed to alert him. He exclaimed, 'in great astonishment, "Oh, I thought you were a man!" '

In the period leading up to the First World War, Rutherford's ability and personality had made him the hub of the international scientific community. When hostilities began in the summer of 1914 he was shocked and depressed. Believing that science should know no boundaries, he did his best to maintain contacts with colleagues overseas. He also worried what would happen to his 'boys', as he called his current and former students, whether foreign like Hans Geiger, by then back in Germany, or British, like James Chadwick, whom the outbreak of war left stranded in Berlin.

Chadwick had arrived in Rutherford's physics department at the age of eighteen, having won a scholarship to Manchester University. He was from a poor working-class background, shy and, as he later confessed, 'very definitely afraid' of Rutherford who did not immediately take to the tall, thin, nervous, bird-like young man. However, he was soon convinced of Chadwick's rare gifts and backed his nomination for an 1851 Exhibition Science Research Scholarship – the same award that had enabled him to fling down his spade in New Zealand and renounce digging potatoes for ever.

Chadwick had arrived in Berlin in 1913 to take up his scholarship, working with Hans Geiger. When war came the following year Chadwick and a German friend were denounced and thrown into prison for, in Chadwick's words, 'having said something we hadn't said'. Chadwick was held for ten days on a diet of coffee and mouldy bread and then released, but not for long. Several weeks later he was rounded up and interned with four thousand others including 'an Earl ... musicians, painters, a few race-horse trainers, a few jockeys' and around a thousand merchant seamen in an improvised prison camp at the racecourse at Ruhleben, near Spandau. He was barely twenty-three and remembered the experience as the time 'when I really began to grow up'.

To preserve his sanity and distract him from the miser-able living conditions, like rations of 'kriegswurst' – 'war sausage made from bread soaked in blood and fat', from which his digestion would never fully recover – and 'the agony when my feet began to thaw out about 11 o'clock in the morning' in unheated stables in the winter, Chadwick gave lectures. He also set up a makeshift physics laboratory in a condemned barracks. Geiger and other German scientists supplied him with bits and pieces of spare equipment. Chadwick also managed guilefully to acquire some radioactive material. Hoping to cash in on the public's passion for radium, the Berlin Auer company was manufacturing toothpaste containing thorium, promising its customers that it would whiten their teeth and give them a radiant smile. Chadwick used it as a radioactive source in experiments. He also acquired a copy of a new paper by Einstein published in Germany in November 1915 expanding his work on relativity into a new theory which he called 'general relativity'. And so, as

Chadwick later described, he became 'probably one of the first English people to know about it'. He could not follow the mathematics but found another internee who could explain it to him.

While Chadwick tried to make the best of things, Rutherford's other star protégé, twenty-seven-year-old Harry Moseley, lost his life. A patriot from a patrician family, he had seen it as his duty to enlist at once. He was killed in hand-to-hand fighting with the Turks on 10 August 1915 in the battle for Gallipoli, where he was serving as brigade signal officer. Rutherford, who had tried hard behind the scenes to have Moseley reassigned to scientific work, wrote sadly that 'his services would have been far more useful to his country in one of the numerous fields of scientific enquiry rendered necessary by the war than by exposure to the chances of a Turkish bullet'.

The field 'rendered necessary by the war' to which Rutherford turned his own talents was anti-submarine tactics. In early 1915, Germany, in an effort to break the deadlock on the Western Front, had declared unrestricted submarine warfare under which, contrary to international law, merchant shipping could be torpedoed on sight, without first being stopped and searched. On 7 May 1915, the German submarine *U-20* torpedoed the Cunard passenger liner *Lusitania* off the coast of Ireland with the loss of 1,200 lives including 128 citizens of the then neutral United States. The Admiralty realized that Britain needed better ways of locating and destroying U-boats and Rutherford threw himself with his natural energy into a programme for developing underwater listening devices. The result was an early forerunner of sonar, known by the acronym ASDIC (Anti-Submarine Detection Investigation Committee).

Marie Curie also plunged herself into war work. She scoured laboratories and hospitals for X-ray equipment, solving the problem of how to move it to where it was most needed by converting vehicles into 'radiological cars'. French aristocrats put their limousines at her disposal and she equipped twenty vehicles, nicknamed 'little Curies'. The X-ray machines themselves were driven by dynamos powered by the car engines. Her own 'radiological car' was a flat-nosed Renault, painted regulation grey with a red cross on the side, in which she dashed from place to place just behind the front lines. She found it distressing work, later writing that 'To hate the very idea of war, it ought to be sufficient to see once what I have seen so many times . . . men and boys . . . in a mixture of mud and blood . . .' As the war progressed she was joined by her elder daughter Irène. Marie also set up two hundred radiological units in field hospitals and trained hundreds of technicians to man them. Over the course of the war, the units assisted in the treatment of more than a million wounded.

First, though, on the instructions of the French government, she had taken steps to protect her precious gram of radium. In the opening weeks of the conflict, when it seemed that the Germans would soon be in Paris, she took the radium, packed into tiny tubes shielded by lead in a case weighing twenty kilos, by train to Bordeaux where she deposited it in a bank vault. The following year, 1915, when things seemed safer, she retrieved it and began 'milking' its radioactive emanation for use in radiotherapy to treat cancers and other diseases.

Elsewhere, science and technology were being applied as never before to the art of war. In November 1911, fewer than eight years after the first flight by Orville

Wright, during his country's colonial war in Libya the Italian lieutenant Giulio Gavotti had dropped the first aerial bombs from his flimsy Etrich monoplane. Less than a month after the sinking of the *Lusitania*, a German Zeppelin dropped the first bombs on London, bringing home to its inhabitants that neither Britain's status as an island nor their own as civilians any longer provided protection.

On the evening of 22 April 1915, Germany launched the world's first poison gas attack, releasing 168 tons of chlorine over the French and Canadian lines on the Western Front. German Jewish chemist Fritz Haber had, from the early stages of the war, been pioneering chemical warfare – the use of poison gases, starting with chlorine – to kill the enemy or to drive them from their trenches. Otto Hahn was summoned to join Haber's unit, together with fellow scientists such as physicist James Franck. After discharging gas over Russian trenches, Hahn came across some of the victims. They were lying or crouching 'in a pitiable position'. The sight left him 'profoundly ashamed and perturbed', but as the war progressed he and his colleagues became 'so numbed that we no longer had any scruples about the whole thing'. As Hahn recalled, Fritz Haber justified the use of gas by stating 'it was a way of saving countless lives, if it meant that the war could be brought to an end sooner'. Even after the war, Haber argued that the use of gas was 'a higher form of killing', the use of which would be essential in future wars. Haber's wife Clara, also a chemist, did not agree. After pleading unsuccessfully with her husband to give up his work, she killed herself in despair the very night in 1915 he returned to the front to prepare for further attacks.

Although Britain, France and America initially

condemned gas attacks, by the Armistice Allied production of chemical weapons outstripped Germany's. The
First World War exposed, as never before, the conflicts
and ambiguities between expediency and morality in
warfare. At its end, the British Air Ministry opposed the
trial as war criminals of German bomber pilots such as
those of the Gotha bombers who had killed 162 civilians
in air raids on London in June 1917, including eighteen
children whose school took a direct hit. The officials'
reasoning was that 'to do so would be placing a noose
round the necks of our airmen in future wars'. They were
reluctant to deny Britain the possibility of carrying out
bombing acts which, when undertaken by others, they
called war crimes. Indeed, in 1920, in Mesopotamia, as
Iraq was then known, Britain would become the first
power to attempt 'to control without occupation' a
country from the air.*

* The British Army would withdraw, leaving the task to the Royal Air
Force. However, the use of airpower alone would, in this instance, fail,
with many civilians killed in ill-directed bombing raids or machine-
gunned when mistaken for hostile forces, thus promoting increased
resistance.

CHAPTER FOUR

'MAKE PHYSICS BOOM'

THE WORRIES AND DISTRACTIONS OF WAR DID NOT DIVERT
Rutherford from yet another major discovery – how to
split the atom. In 1914, Ernest Marsden had been bombarding hydrogen gas with alpha particles. To his surprise
he found that this produced far more 'H-particles' – the
fast-moving nuclei of hydrogen atoms – than he could
account for. His departure to become Professor of Physics
at Victoria College in Wellington, New Zealand, prevented him from investigating further, leaving the
anomaly for Rutherford. Systematically eliminating all
other possibilities, such as the contamination of
Marsden's equipment by hydrogen, Rutherford proved
that the mysteriously prolific H-particles were fragments
chipped off the nuclei of nitrogen atoms in the air
surrounding the experiment. He showed that the
bombarding alpha particles had forced the nitrogen atoms
in the atmosphere to release hydrogen nuclei – the
simplest, lightest nuclei consisting solely of what
Rutherford would soon term 'protons'.

This was the first time that human action had split the

atom. Rutherford had sensed all along that he was on the brink of something major. He defended his absence from a submarine warfare meeting with the statement, 'If, as I have reason to believe, I have disintegrated the nucleus of the atom, this is of greater significance than the war.' By early 1919 his paper announcing the splitting of the atom was on its way to the printers. He had shown that humans could deliberately manipulate and transmute the elements and that, as C. P. Snow put it, 'man could get inside the atomic nucleus and play with it if he could find the right projectiles'. The only snag was that, although it was a simple matter to aim alpha particles at nitrogen nuclei, there was no certainty of hitting them. In fact, most missed, passing by like spent bullets. It was, as Einstein characteristically put it, 'like shooting sparrows in the dark'.

That same year, Rutherford left Manchester for Cambridge to replace an ageing J. J. Thomson as head of the Cavendish Laboratory – the most prestigious scientific academic post in Britain. Thomson wished to step aside to focus on his own research. The Rutherfords installed their modest possessions in Newnham Cottage, despite its name a comfortable house on the banks of the Granta with a large garden which became Lady Rutherford's passion. It was also useful in ensuring that student guests had no opportunity to outstay their welcome. Rutherford would hospitably invite his students to tea on Sunday afternoons. They arrived at 2.30 p.m. in 'best suits and dresses', as the young Australian Mark Oliphant recalled, and sat in a semi-circle. Rutherford kept up lively conversation while his short, plump, down-to-earth wife poured the tea. She would loudly remind her husband, 'Ern, you're dribbling,' if while trying to talk, eat and drink at the same time he spilled tea or food from his mouth in the excitement of the moment. After an hour or so Lady Rutherford, who called everyone 'Mister' regardless of status, would ask her guests whether they would like to see the garden. It was a command rather than an invitation. After a stroll 'we were led firmly to the door in the outer wall where we shook hands and departed'.

Rutherford's first task at the Cavendish was to re-organize the laboratory which, with so many men being demobilized, was, in his view, crowded to excess with students and sadly lacking in space and equipment. These returning researchers included physicist Francis Aston, who in 1919 invented the mass spectrograph, an instrument capable of differentiating both elements and isotopes by mass and which helped validate Rutherford's model of the atom. James Chadwick proved a staunch adminstrative ally. He had returned from his long internment in Berlin malnourished, dyspeptic and impoverished, but matured by his experiences. Rutherford brought him to Cambridge, where he not only showed himself a creative and intuitive scientist, helping Rutherford disintegrate further elements, but progressively became Rutherford's lieutenant. A natural administrator, he kept the Cavendish running, watching over both its finances and its researchers.

The 1920s were hectic, even chaotic, years for atomic physics. Scientists were teasing out ever more facts but also seeking theories and systems to make sense of the bewildering, often conflicting mass of new information. Sometimes supplies of data ran ahead of theory. At other times, theories could not be validated for want of satisfactory data. The main centres of atomic science were the British, revolving around Rutherford, the French, centred on Marie Curie's Radium Institute, and the Germans in Berlin. Each school had its pet interests and its own personality. Each believed itself superior. The British view of the French was that 'where we try to find models or analogies, they are quite content with laws'. The French, conversely, considered their own approach a model of synthesis, simplicity and precision and a happy contrast to 'the haphazard fact-finding sorties of the British, who wanted to turn everything into wheels within wheels', or the 'grandiose, woolly theorising and niggling accumulations of useless data' of the Germans.

Atomic science was also becoming well established in Japan, helped by close and enduring links with Western universities. The Japanese had entered the First World War on the Allied side towards the end of August 1914, two weeks after fighting had begun. In doing so they had cited a strict interpretation of their recent alliance with Britain. In reality, they were keen to enhance their strategic position in the Pacific and in China at Germany's expense. Their initial action was to give the Germans six days to surrender Kiaochow, one of the treaty ports they held in China. The Germans refused. The Kaiser sent a telegram to the Governor of Kiaochow proclaiming, 'It would shame me more to surrender Kiaochow to the Japanese than Berlin to the Russians.' However, the Japanese captured the port within three months and also seized several German colonies and other treaty ports, including among the latter Tsintao in northern China, famous for its brewery, where the Japanese took 4,600 POWs. According to one German prisoner, the Japanese 'treated them as guests' and provided plentiful food, including German sausage, and allowed exercise. Among the considerable gains from the Germans which Japan retained as colonies at the end of the war were the Caroline Islands, the Marshall Islands and the Northern Marianas group, including Saipan and Tinian, in the Pacific.*

Once the First World War was over and Japan held a respected place among the victors, Japanese scientists quickly resumed their academic contacts overseas. In 1923, Yoshio Nishina, an urbane thirty-three-year-old who would become the founder of experimental nuclear and cosmic ray research in Japan, arrived in Denmark to study with Niels Bohr. Seven other young Japanese physicists also came to Copenhagen. Another, Nobus Yamada, worked in Paris with the Curies, preparing polonium sources. Their mentor back in Japan was Hantaro Nagaoka, Professor of Physics at Tokyo University, who had studied in Germany in the 1890s and later visited Ernest Rutherford in Manchester. He once wrote to Rutherford of his admiration for 'the simpleness of the apparatus you employ and the brilliant results you obtain'.

In post-war France, Marie Curie's problem was shortage of money to fund research. Her laboratory had no new equipment and only one gram of radium which was being used to treat cancer. An American benefactress, Mrs William Brown Meloney, editor of the New York magazine *The Delineator* and known as 'Missy', came to the rescue. She raised over $150,000 in the United States

* During the Second World War Saipan and Tinian, once captured by the Americans, would become major air bases for the US assault on Japan.

to purchase a gram of American radium for the scientist she considered 'the greatest woman in the world'. The news so excited the French press that they forgot Marie Curie was the scarlet woman of the Affaire Langevin a decade earlier and eulogized her. At a gala evening at the Paris Opéra, Sarah Bernhardt tremulously declaimed an 'Ode to Madame Curie', hailing her as 'the sister of Prometheus'.

Missy Meloney coaxed an initially reluctant and prematurely frail Marie to cross the Atlantic to receive the radium in person. In 1921, President Warren Harding presented it to her – or at least a symbol of it: the radium in its lead-lined containers was far too precious to be brought to the ceremony. The transatlantic journey was a strain, but the radium enabled Marie Curie to continue her work, helped by her daughter Irène who had become her closest collaborator. Now in her mid-twenties, Irène was tall and sturdily built with a direct, piercing, sometimes disconcerting gaze. Einstein thought she had the characteristics of a grenadier. Other contemporaries recalled her as sometimes haughty and conscious of her status as Marie Curie's daughter and at other times 'very uncouth'. She had little concern for appearances or convention, happily hiking up her skirts to rummage in her petticoat for a handkerchief on which she then noisily blew her nose, and at mealtimes throwing unwanted bread over her shoulder.

In 1926 Irène married Frédéric Joliot, three years her junior. He was athletic, high-spirited, ambitious and, as her own father Pierre had been, the son of a Paris Communard. Joliot had joined Marie Curie's institute the year before, feeling very nervous of 'La Patronne' ('the owner'), as Marie was known, as well as of her daughter.

Potentially dangerous radium products: above, a 1919 advertisement for watches with radioactively luminous dials; below, a 1934 newspaper advert for hair tonic

On his joining the laboratory, a colleague quickly told him that Irène was 'a cow' but, nevertheless, he soon won her affections. Marie Curie introduced him to visiting dignitaries as 'the young chap who has married Irène' while otherwise paying him little attention.

At the age of nearly sixty, Marie could not visualize life without her laboratory. However, cataract problems, which she concealed for a long time, and increasing frailty were hampering her. In 1926, Hungarian organic chemist Elizabeth Rona, working alongside her, was horrified by Marie's clumsy and ill-advised attempts to open a flask containing a solution of radium salt. The contents were highly volatile. As she approached a naked flame with the flask 'a violent explosion scattered glass all over'. It was a miracle neither woman was badly injured. Marie did not associate her physical decline with radiation and her approach was characteristic of the casual attitude at the Curie Institute towards handling radioactive material. A student once watched Irène 'shaking the radioactivity out of her hair and clothing'. Even fifty years after her death Marie Curie's home cookbooks remained radioactively contaminated by contact with her.

Neither was the general public yet alert to the risks. Radioactivity was still regarded as the great panacea, and there was a ready market for associated products. Greedy manufacturers offered the public 'Curie Hair Tonic' which supposedly prevented hair loss and restored its original colour, and a cream guaranteed to confer eternal youth. Gullible purchasers were assured that Marie Curie 'promises miracles'. Other radioactive products included bath salts, suppositories and chocolates.

But danger signs were emerging around the world. In France, several radiologists and researchers died of leukaemia and severe anaemia. A newspaper published their photographs, together with gruesome accounts of amputations, lost eyesight and dreadful suffering. It posed the question, 'Can one be protected against the murderous rays?' In Japan, scientist Nobus Yamada, who had worked in the Curie laboratory preparing polonium sources, sickened and died within two years of returning to Japan. In America in 1925, a young woman working as a painter of luminous watch dials in New Jersey sued her employer for putting her at risk. Her work required her to moisten her brush, dipped in a luminous paint containing radium, with her lips. Nine co-workers had already died; others were suffering from 'radium necrosis', severe anaemia and damage to their jaws. An investigation concluded that radiation was to blame. By 1928, fifteen watch-painters had died.

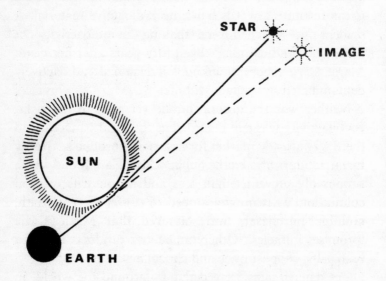

The deflection of starlight by gravity

An American journalist asked Marie whether she had any advice that might help the dial-painters. She was sympathetic, but her only suggestion was that they should eat calves' liver as a source of iron and take plenty of exercise in the fresh air – her universal remedy for radiation-related sickness. Irène's view was that anybody who worried about radiation hazards was not committed to science.

In bleak, post-war Germany, 'the stronghold of physics' was Berlin. Nobel laureates Max Planck and Max von Laue were teaching at the university. So was Albert Einstein, who in the spring of 1914 had accepted a professorship there and membership of the Prussian Academy of Sciences. Separated from Mileva, who had returned to Switzerland with their two sons, and living alone, he had been extending his ideas on relativity. He was concerned, in particular, that his 'special theory', published in 1905, did not give due weight to gravitational forces. In November 1915 he had published his new 'general theory' – read by the interned James Chadwick – postulating that light was bent by gravity at twice the value predicted by Newton. If correct, this meant that space was not flat but curved.

J. J. Thomson, the discoverer of the electron, hailed Einstein's theory as one of the greatest achievements in the history of human thought and the greatest discovery in connection with gravity since Newton. Many, though, remained sceptical, until on 29 May 1919 English astronomer Arthur Eddington took advantage of a solar eclipse in West Africa to photograph beams of starlight. Eddington's image showed the deflection of starlight by gravity to be exactly as Einstein had predicted. The *New*

York Times declared that stars were 'not where they seemed or were calculated to be' but added reassuringly that 'nobody need worry'. The report in the London *Times* was headlined 'New Theory of the Universe – Newtonian Ideas Overthrown'. However, it was for his work on the photoelectric effect and light quanta, not relativity, that Einstein received the Nobel Prize for Physics in 1921.

That year Mileva had divorced him for adultery committed with his cousin, Elsa Einstein, whom he had subsequently married. Einstein was now so famous that a little girl wrote to him asking whether he really existed. However, his celebrity had made him the focus of virulent attack from parts of the German media and academia angry that this much-lauded international figure was a Jew. They also resented his determined and outspoken pacifism during the war. Einstein received death threats and was warned 'it would be dangerous for him to appear anywhere in public in Germany'. He and Elsa departed on a trip to Japan and the Far East until the mood calmed.

Otto Hahn and Lise Meitner were working together at the prestigious Kaiser Wilhelm Institute for Chemistry in the Dahlem suburb of Berlin. The institute – sponsored jointly by government and industry and one of a network of such bodies set up in Germany across the scientific disciplines, including one for physics under Einstein's directorship – had opened back in 1912 in a blaze of celebration led by the Kaiser in a white-plumed hat. That year Meitner had for the first time begun to receive a salary.

Like Rutherford, she and Hahn had continued their research sporadically, despite their war work. Meitner had volunteered as an X-ray nurse with the Austro-Hungarian army but had returned to the institute in 1916 to continue a task started two years earlier – the tracking down of a new element. She consulted Hahn, engaged in gas warfare research, by letter. He replied when he could and occasionally visited her in Berlin. The work was often overshadowed by wartime tragedies, such as the news that one of Max Planck's two sons had been killed in France in 1916. However, in March 1918 she and Hahn announced that they had found the new element – proactinium. Meitner had done most of the work but the paper was in their joint names. Later that year she worked briefly with Einstein, and their admiration was mutual. Shortly afterwards she was given the title of professor at the Kaiser Wilhelm Institute. It was some compensation for the difficult and uncertain times in which she was living. A brief visit to friends in Sweden provided an opportunity to eat things which were just a memory in Germany – 'eggs, butter, bacon, puddings, in short everything good'.

Germany's defeat and the Kaiser's abdication in November 1918 had produced revolution, mutiny, street fighting and strikes throughout the country. Discharged soldiers and sailors joined rival 'red' and 'white' militias supporting the socialist or conservative factions. Civil order disintegrated and living conditions deteriorated as workers quit their posts for the barricades. In Berlin, Hahn was among those volunteering to keep the local power station going, raking the hot cinders so that the coal burned well. The establishment of the Weimar Republic – named after its seat of government, Weimar, in eastern Germany – brought some stability, but life remained very tough. In 1922, terrifying inflation took hold, reducing the mark's worth to almost nothing. The professors brought rucksacks and suitcases to collect salaries that were now paid daily in bundles of increasingly worthless paper. Hahn's wife Edith, whom he had married in 1913, met him every day to pick up his wages and then cycled frantically off to the grocer's hoping to be in time 'to do her shopping at the previous day's prices'.

In November 1923, the height of the economic mayhem, food riots broke out and Adolf Hitler failed in his attempted putsch in Munich. Against this background, work was a welcome refuge for the scientists of Berlin. Hahn wrote that 'while we were busy in the laboratory we simply forgot all our worries about food and food-coupons'. Paradoxically, despite the political and economic turmoil that launched them, they would remember the 1920s as a period of enthusiasm, openness, generosity, collaboration and achievement in German science. They had stumbled on 'the secrets of nature' and 'whole new processes of thought, beyond all the previous notions in physics, would be needed to resolve the contradictions'.

The University of Göttingen, founded in 1737, played a leading role in reconciling these contradictions. Göttingen was an ancient city on the slopes of the Hain mountain in Lower Saxony, some sixty miles south-east of Hanover. Its professors, living in creeper-clad villas, seemed like demi-gods. One of the most highly esteemed was the theoretical physicist Max Born, who had found solace from his war-work with Einstein. Together they played violin sonatas and discussed relativity. Born had been attached to an army research unit whose task was 'sound ranging' – calculating the position of enemy guns by measuring the arrival times of their reports at various listening posts. His experiences convinced him that 'henceforth not heroism but technology would become decisive in war'.

The intellectual atmosphere in Göttingen was highly charged and at times surreal. Young scientists argued and debated in cafés, improvising mathematical formulae on tablecloths. Reputedly they roamed the streets at night unable to sleep and impatient for the doors of their laboratories to open. In 1922, Göttingen hosted a Bohr Festival. Niels Bohr was by then a major international figure. He had persuaded the University of Copenhagen to open a theoretical physics institute and, reluctantly declining Rutherford's invitation to come to England and 'make physics boom', had become its director. Later that year, he would be awarded the Nobel Prize for Physics for his quantized model of the atom. The chance to hear Bohr attracted a fit, blond, boyish twenty-year-old student, Werner Heisenberg, from Munich.

Heisenberg's adolescence had been traumatic. As he later recalled, the war had burst open 'the cocoon in which home and school protect the young in more peaceful periods'. In 1919 he had witnessed street fighting between the communists of the Munich Soviet Republic and government troops. With his family close to starvation, he had dodged through the lines to fetch bread, butter and bacon. While serving in an anti-communist militia he had seen a friend shoot himself in the stomach by accident and die in agony before his eyes. Disintegration, chaos and civil war had awakened a desire to seek new certainties in a world untainted by politics – that of science. However, they had also left him with an enduring fear of communism and a patriotic recognition of the need for stronger government structures if Germany were to prosper once more.

While recovering from a serious illness, Heisenberg read about Einstein's theories of relativity. The mathematical arguments and the abstract thoughts under-lying them both excited and disturbed him. He enrolled at Munich University to study theoretical physics under Professor Arnold Sommerfeld, whose contributions in the fields of quantum theory and relativity and brilliance as a teacher were legendary. Another of Sommerfeld's students was the sharply clever Wolfgang Pauli. Pauli and Heisenberg became close friends, though their habits were diametrically opposed. Heisenberg loved rambling and camping expeditions and became a leader in one of the many movements then springing up with the aim of renewing the spiritual and physical vigour of German youth. Pauli was a night-owl, happiest in smoky cafés. He worked through the night and would not rise until noon. He teased the fresh-faced Heisenberg for being a 'prophet of nature'. In 1925, Pauli would propose his famous 'exclusion principle' suggesting, on the basis of his experimental observations of how electrons behaved when subjected to magnetic fields, that no more than two electrons could inhabit the same orbit around a nucleus. This resolved a hitherto puzzling anomaly and earned him the nickname the 'Atomic Housing Officer'.

It was Sommerfeld who brought Heisenberg with him from Munich to hear Niels Bohr at Göttingen. The lecture hall was crammed. Heisenberg was excited not only by what the Dane had to say, but also, as he later recalled, by how he said it: 'each one of his carefully formulated sentences revealed a long chain of underlying thoughts, of philosophical reflections, hinted at but never fully expressed'. At the end of Bohr's third lecture Heisenberg summoned enough courage to voice a critical remark. Bohr listened gravely and at the end of the lecture invited

Einstein in 1924

Heisenberg for a walk over the Hain mountain. It obviously went well because during it, Bohr asked him to visit Copenhagen. Heisenberg later wrote that 'my real scientific career began only that afternoon'.

Also that year, 1922, Sommerfeld suggested that Heisenberg attend a scientific congress in Leipzig where Einstein was speaking. As Heisenberg entered the lecture hall, a young man pressed a red handbill into his hand. It attacked Einstein and derided relativity as wild, danger-ous speculation alien to German culture and put about by the Jewish press. The lecture went ahead, but Heisenberg was too distracted by the eruption into science of such 'twisted political passions' to concentrate. He recalled that he had no heart, at the end, to seek an introduction to Einstein. It was his first but by no means last experience of what he termed 'the dangerous no-man's land between science and politics'.

After completing his doctorate at Munich, Heisenberg moved to Göttingen as Max Born's assistant. He also made frequent visits to Bohr in Copenhagen. During further long walks he and the Dane became good friends while debating quantum theory. Heisenberg was becoming increasingly troubled by the theory's reliance on the unobservable and hence the unmeasurable. Hypothesizing about what was happening within the atom and about orbiting electrons was, he felt, all very well, but he yearned for proof of what was actually occurring. He therefore decided to focus on what could be observed – the frequencies and amplitudes of light emitted from inside the atom – and to seek mathematical correlations between them.

It was a complex task, but in 1925 Heisenberg had something akin to a vision. A severe bout of hayfever sent him to the bracingly windy, relatively pollen-free North Sea island of Heligoland. He arrived with a face so swollen his landlady thought he had been in a fight. He worked late in his room, churning out reams of calculations until he felt that 'through the surface of atomic phenomena, I was looking at a strangely beautiful interior, and felt almost giddy at the thought that I now had to probe this wealth of mathematical structures nature had so generously spread out before me'. He was so exhilarated that instead of going to bed he went out and climbed a jutting sliver of rock and waited for the sun to rise.

Down from 'the mountain' and back at Göttingen, Heisenberg was sufficiently sure of himself to parade his thoughts to Max Born and his colleagues. Together they evolved what Heisenberg called 'a coherent mathematical framework ... that promised to embrace all the multifarious aspects of atomic physics'. This new approach was the earliest version of 'quantum mechanics' – a tool using experimental evidence to predict physical phenomena. It was based on matrix algebra, a species of mathematics originally developed in the 1850s, and later refined, as a means of analysing large amounts of numbers using a system of grids. In keeping with his original aim, Heisenberg's quantum mechanics focused on what could be observed, such as radiation emitted from an atom, and otherwise involved only the use of fundamental constants. In contrast with the Rutherford-Bohr model, Heisenberg's abstract mathematics provided nothing in the way of a picture of atomic structure, but its predictions proved remarkably accurate.

Heisenberg's approach had a competitor – 'wave mechanics', outlined just a few weeks later by an urbane Austrian physicist, Erwin Schrödinger. Building on an idea of the Frenchman Louis de Broglie that particles such as electrons behave like waves, Schrödinger invented a neat equation capable of embracing those wave-like characteristics. An important feature was the incorporation in the calculation of a likelihood of occurrence – a probability – which meant, for example, that the location of an electron was not predicted as a point but rather as a smear of probability whose density gave the likelihood of the electron being found at any point. At first, Schrödinger's different approach appeared to threaten Heisenberg's quantum mechanics, and the respective proponents indulged in vigorous debate. Heisenberg wrote crossly to Wolfgang Pauli, 'The more I think about the physical portion of Schrödinger's theory, the more repulsive I find it ... What

Schrödinger writes about the visualizability of his theory is probably not quite right, in other words it's crap.' However, Schrödinger proved that his 'wave equation', as it became known, provided results mathematically equivalent to Heisenberg's formulae and that the two theories complemented each other, rather than conflicted. Schrödinger's waves and Heisenberg's matrices were analogous.

Heisenberg's next step, in 1927, was his renowned 'uncertainty principle'. It grew out of an intellectual pummelling from Bohr over whether apparent ambiguities in atomic physics could be reconciled. A cold walk under a star-lit sky in Copenhagen led Heisenberg to a conclusion that some uncertainties were unavoidable. Given the atom's tiny dimensions, the scientist's ability to measure events must be inherently limited. The more accurately one aspect was measured, the more uncertain another must become. Although it was possible accurately to observe either the speed or the position of a nuclear particle, doing both simultaneously was impossible. 'The more precisely the position is determined', he wrote, 'the less precisely its momentum is known and vice versa.' In the mechanical world of Newtonian physics, future behaviour could be predicted with certainty, just as what had happened in the past could be accurately determined. Under Heisenberg's principle, while past behaviour could be known accurately and future behaviour could generally be predicted using a series of approximations based on probability, the future behaviour of an individual atom was subject to inherent uncertainty.

Heisenberg's ideas at first provoked a fierce reaction from Bohr, who taxed him with flying in the face of previous interpretations and reduced him to tears with his vehemence. When both had cooled off, they agreed that their approaches could, after all, be reconciled. Bohr incorporated Heisenberg's uncertainty principle into a broader thesis of his own – 'complementarity'. He argued that conflicting or ambiguous findings should be placed side by side to build a comprehensive picture – the particle and wave nature of matter should be accepted – and each aspect should recognize 'the impossibility of any sharp separation between the behaviour of atomic objects and the interaction with measuring instruments'. He borrowed the word 'complementarity' from the Latin *complementum*, meaning 'that which completes'.

Bohr's and Heisenberg's friendship emerged unscathed from their confrontation. However, Heisenberg's uncertainty principle sparked a famous row with Einstein, who argued that probability was far too vague a tool for assessing the physical world. 'It seems hard to sneak a look at God's cards. But that He plays dice and uses telepathic methods . . . is something that I cannot believe for a single moment.' Neither did he or any other scientist yet believe that this rash of new intellectual tools would be used to predict how atoms could be split to release their latent energy explosively.

That same year, 1927, Heisenberg was appointed professor at Leipzig at just twenty-six. His youth, lack of formality and skill at ping-pong endeared him to his students, one of whom was the young Hungarian Edward Teller, later to be known as the 'father' of the H-bomb. Science was Teller's earliest passion. He had gained his first respect for technology from a ride in his grand-parents' car. But the end of the First World War and the collapse of the Austro-Hungarian Empire, when Teller was ten, had destroyed his comfortable, middle-class world, just as Heisenberg's had disintegrated. Many of his games consisted of playing with numbers, finding security in the patterns they created. In the newly independent Hungary a communist take-over was followed by hunger and uncertainty. Soldiers were billeted on the Tellers in their Budapest home and Edward had perforce to learn to sing the 'Internationale' at school. Many of the communist leaders were Jews, and when the communist regime collapsed it triggered a vicious anti-Semitic back-lash against Jewish families like the Tellers. In 1919 the new right-wing 'white' Hungarian government under Admiral Horthy conducted a purge. Over five thousand people, many of them Jewish, were executed and thousands more fled. Anti-Semitism became so open and pervasive that even as a youngster Teller worried whether 'being a Jew really was synonymous with being an undesirably different kind of person'.

During his final years at school, knowing that science was his great love, Teller sought the company of three young scientists, all from Budapest's Jewish community and all of whom were studying in Germany. The theoretical physicist Eugene Wigner, winner of the Nobel Prize for Physics in 1963, and the mathematician John von Neumann, the designer and builder of some of the first modern computers in the late 1940s, were in their early twenties. The third man, the eccentric Leo Szilard, was a little older. Listening to their discussion, occasion-ally daring to ask questions, Teller decided to study mathematics but knew that it would be hard to climb the academic ladder in Hungary where Jews were subject to a quota system. His father urged him to go to Germany, which in the 1920s, according to Teller, appeared to be free of anti-Semitism. He also urged his son to study something more practical than mathematics, and they compromised on chemistry.

In 1926, Teller's protective parents accompanied the seventeen-year-old onto an express train to Karlsruhe where he enrolled in the Technical Institute. However, within two years Teller had abandoned chemistry and was studying physics and mathematics with Arnold Sommerfeld at Munich. He did not achieve the rapport that Heisenberg had enjoyed with his brilliant teacher. Teller wrote of Sommerfeld that he was 'very correct, very systematic, and very competent. I disliked him.' However, he found his new field, particularly the new science of quantum mechanics, deeply exciting.

Lost in thought on his way to meet friends for a hike in the Bavarian Alps in 1928, Teller absent-mindedly slipped while dismounting from a trolley bus and was caught by its wheels. Unlike Pierre Curie, he survived, but the bus severed his right foot. What Teller remembered most about his recuperation was the sudden disappearance of a Dr von Lossow, who had been treating him. He later worked out that the doctor was a relative of the General von Lossow who had arrested Hitler after his abortive 1923 Munich beer-hall putsch. By 1928, public dissatisfaction with the weak Weimar Republic and the weak economy over which it presided was growing, and conflicts between the extreme right and left were beginning again. As Hitler's Nazis re-emerged as a political and street-fighting force, Dr von Lossow had probably realized that Germany held no future for him.

Teller, however, still caught up in the heady atmosphere of new ideas, did not allow the sinister undercurrents to worry him. Having been released from hospital, and having learned that Sommerfeld had gone abroad for a year, he headed happily for Leipzig and Heisenberg. He was eager to study under the man he revered not only for giving mathematical expression to quantum mechanics but also for giving it philosophical expression through his uncertainty principle.

CHAPTER FIVE

DAYS OF ALCHEMY

ATOMIC PHYSICISTS, LOOKING BACK FROM A LESS INNOCENT age, would recall the 1920s as 'a heroic time . . . a time of creation'. Such an intoxicating atmosphere exactly suited a charismatic young Russian by the name of Peter Kapitza, who arrived at the Cavendish Laboratory to become Rutherford's star pupil. The son of a Tsarist general, Kapitza had in 1921 left a Russia riven by civil war and famine as a member of a Soviet mission sent to renew scientific relations with other countries. The mission's leader, Abram Joffé, a sympathetic individual as well as one of Russia's foremost physicists, had brought Kapitza to help him overcome a devastating trauma. Kapitza had recently lost his two-year-old son to scarlet fever followed, within a month, by the loss of his wife, baby daughter and father to the Spanish flu epidemic sweeping through Europe.

Liking what he saw in Cambridge, Kapitza asked Rutherford to take him on as a research student. Rutherford, fearing that Kapitza might be a left-wing agitator, consulted James Chadwick, who advised that the

Russian would be a good acquisition provided he agreed not to talk politics. Kapitza accepted the condition and soon formed an unlikely friendship with the quiet, retiring Chadwick, allowing the Englishman to pilot his motorbike and, by misjudging the bends, to send them both flying. When Chadwick married Aileen Stewart-Brown, daughter of a prominent Liverpool stockbroker, in 1925, Kapitza was his best man in a borrowed top hat.

Kapitza's enthusiasm attracted other students and a lucky thirty were invited to the 'Kapitza Club', which met in his rooms every Tuesday evening for milky coffee and boisterous debate. Above all, Kapitza came to idolize Rutherford, calling him 'the crocodile', for 'in Russia the crocodile is the symbol for the father of the family and is also regarded with awe and admiration because it has a stiff neck and cannot turn back. It just goes straight forward with gaping jaws – like science, like Rutherford.' He could twist Rutherford around his finger, winning concessions that others would not even have dared to seek. Kapitza's great interest was creating magnetic fields of greater and greater power, and in 1928 he was put in charge of the Cavendish's new Department of Magnetic Research.

Rutherford had become convinced that using subatomic particles naturally emitted by radioactive substances as projectiles to smash atoms was too limiting. The particles lacked the energy to barge through the electrical defences of the nucleus. Under Rutherford's guidance and with industrial help, two of the Cavendish team, John Cockcroft and Ernest Walton, began to develop machines, today known as 'accelerators', that would use high voltages to hurl particles at sufficient speed to penetrate the nuclei of the target.

Elsewhere, others were having similar ideas. In America, at MIT, Robert van de Graaff was building a huge electrostatic device, while at the University of California at Berkeley, Ernest Lawrence, a young experimental physicist from South Dakota, was planning the world's first 'cyclotron' – a machine combining electric and magnetic fields to send particles spiralling away at high speed. He was determined to invade the nucleus sitting snug behind its protective screen of electrons like, as he put it, 'a fly inside a cathedral'.

Lawrence was an extrovert of overpowering drive and energy, much like Rutherford as a young man. He also had some of Rutherford's intuition, and this had helped him conceive the cyclotron. In 1929, the year of the Wall Street crash, Lawrence came across an article by Rolf Wideroe, a Norwegian engineer working in Germany, describing a linear device that would accelerate charged particles down a straight tube – similar to the approach being pursued at the Cavendish Laboratory. Lawrence's German was not good enough for him to understand everything Wideroe had written, but as he studied the accompanying diagram an inspirational thought struck him. If he could confine particles with electromagnets within a circular track, rather than push them along a straight line, he could accelerate them indefinitely, causing them to whizz faster after each burst of voltage. It would, in his words, be a 'proton merry-go-round'. He told his friends confidently, and accurately as it turned out, 'I'm going to bombard and break up atoms! I'm going to be famous.'

Lawrence's first machine was 'a four-inch pillbox sprouting arms like an octopus'. When he demonstrated it to the US National Academy of Sciences, he secured it in place on a kitchen chair with a clothes hanger. Despite its absurd appearance, its potential caused a sensation. Newspapers hailed the invention of a device 'to break up atoms', and they were right. So good was his progress that by the end of the 1930s Lawrence would build a cyclotron with a magnet weighing 200 tons. Inspired by the desire to explore one of the tiniest things in existence, the nucleus of the atom, big science was coming.

While the creators of the new atom-smashing machines honed their early designs, quantum mechanics continued to forge bridges between Europe and the United States. Just as young Americans eager to understand the new theories were flocking to Arnold Sommerfeld in Munich, Max Born in Göttingen, Werner Heisenberg in Leipzig and Niels Bohr in Copenhagen, European scientists were touring America to spread the word. The big names like Einstein were eagerly sought, but so too were younger

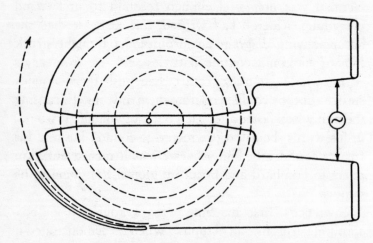

Acceleration of a particle in a cyclotron

scientists. Hungarians John von Neumann and Eugene Wigner were invited as guest lecturers. Their task, in Wigner's words, was 'to modernise' America's 'scientific spirit'. They saw themselves as 'pioneers who break new ground', their mission to make quantum mechanics and relativity theory a reality to people to whom it was still 'an abstraction'.

The experimenters of the Cavendish Laboratory were less immediately impressed by the deluge of fresh ideas. James Chadwick recalled that 'It took quite a time to absorb the meaning of the new quantum mechanics. It was rather slow . . . there was no immediate application to the structure of the nucleus, which was what we were interested in.' Rutherford was frankly sceptical of the complex new mathematical theories, preferring to scent new discoveries in some unexpected experimental result rather than indulge in abstract theorizing. Only in the late 1920s did he concede somewhat grudgingly that wave mechanics might aid the understanding of the nucleus. In the meantime, his laboratory remained the greatest centre of experimental physics in the world. His only rivals were Lise Meitner and Otto Hahn in Berlin and Marie Curie and Irène and Frédéric Joliot-Curie – as the pair chose to be known to emphasize their close collaboration – in Paris.* All the other major players were theorists.

Rutherford had been convinced for many years that an undetected particle at the heart of the nucleus, the 'neutron' as he called it, was the great unclaimed prize. As early as June 1920 he had talked to the Royal Society of

* Frédéric Joliot was sensitive to suggestions that their choice, highly unusual at the time, reflected a desire to retain the fame of the Curie name, or any subordinate status for him.

Ernest Rutherford in 1925

the possible existence of such a particle. His discovery, the year before, of the positively charged proton, residing in the nucleus of every atom, had provided tantalizing clues. For example, the simplest, lightest atom, hydrogen, had one single, positively charged proton counterbalanced by one external, negatively charged electron. The next heaviest atom, helium, had two protons and two orbiting electrons. However, its mass, or atomic weight, was not, as might have been expected, double that of hydrogen, it was quadruple. This could only mean that it had to have one or more electrically neutral particles, equivalent in mass to and complementing the two protons. Rutherford speculated intuitively that the missing piece of the jigsaw, his 'neutron', consisted of electrons and protons parcelled together.

Although Rutherford continued to think about the neutron throughout the 1920s and undertook experiments when he could, he was frequently distracted by other pressures, including the work of university administration and serving on national public committees. His ennoblement in 1931 by King George V as Baron Rutherford only added to the commitments of a man who was still considerably shaken by the sudden death in 1930 from a blood clot of his only child, his daughter Eileen. She had left four children, to whom Rutherford was deeply attached, from her marriage to a Cavendish mathematician.

Realizing that domestic concerns and public duties would continue to hamper his search for the neutron, Rutherford entrusted more and more of the hunt to James Chadwick, who had already been working on the topic for him since the mid-1920s and who, in his own words, 'just kept on pegging away' and 'did quite a number of quite silly experiments' just in case they turned something up. In fact, he worked obsessively. His efforts attracted affectionate satire from junior colleagues who staged a show raucously lampooning the hunt for the elusive 'Fewtron'.

Chadwick made his breakthrough in January 1932, precipitated by a paper by the Joliot-Curies in the French *Comptes Rendus*. This described how, building on work by German scientist Walther Bothe, they had bombarded the light element beryllium – a hard, silvery, toxic metal – with an intense source of polonium, causing an unusually penetrating radiation to stream out of the beryllium. The Joliot-Curies experimented with various substances, including wax, to see whether they could halt the rays from the beryllium, but the rays not only passed through the barriers but appeared to get stronger. The puzzled Joliot-Curies concluded in their paper that the radiation had to consist of some particularly powerful form of gamma ray, the most penetrating of the three types of radiation emitted by radioactive substances. Rutherford read their conclusions and roared, 'I don't believe it.' Chadwick, too, 'knew in his bones' that they were wrong. Their description of the pattern and path of the radiation they had observed convinced him that it consisted of uncharged or neutral particles knocked out of the nuclei of the beryllium – in other words, neutrons.

Chadwick rushed to replicate their experiments. Applying the classic 'sealing wax and string' principles of the Cavendish to make his equipment the simplest fit for the purpose, an excited but careful Chadwick worked day and night. He violated Rutherford's rule that all work in the laboratory should cease by six p.m., partly through irrepressible enthusiasm but also so that his sensitive counting equipment would not be affected by other work going on in the laboratory. After three weeks he had shown that radiation from bombarded beryllium was powerful enough to knock particles out of hydrogen, helium, lithium, beryllium, carbon and argon. The particles expelled from the hydrogen were clearly protons and the others were whole nuclei of the target substance. His measurements of their penetrating power and velocity proved that gamma rays could never have caused the ejection of particles of such energy. The only viable conclusion was that the radiation flowing so powerfully from the bombarded beryllium consisted of 'particles of mass 1 and charge 0' – neutrons.

Chadwick chose the Kapitza Club as the forum for revealing his findings. There was an air of keen anticipation as Chadwick, grey-faced from lack of sleep but plainly exhilarated, addressed his audience. Mark Oliphant captured the moment in the restrained language of the day: 'Kapitza had taken him to dine in Trinity [College, Cambridge] beforehand, and he was in a very relaxed mood. His talk was extremely lucid and convincing, and the ovation he received from the select audience was spontaneous and warm. All enjoyed the story of a long quest, carried through with persistence and vision . . .' At the end, the exhausted Chadwick asked 'to be chloroformed and put to bed for a fortnight'. In fact, he was up again the next morning writing to Niels Bohr and, a month after first reading the Joliot-Curies' paper, sending a letter to *Nature* cautiously headed 'The possible existence of the neutron'. His entry in the notebook recording presentations to the Kapitza Club was similarly guarded: it read 'Neutron?' Chadwick was instinctively cautious, yet however he hedged his findings he knew in his heart he was right.

Chadwick was not, as he freely acknowledged, the first to produce neutrons. Walther Bothe had done so in Germany in the 1920s; so had the Joliot-Curies, following in Bothe's wake. However, none of them had interpreted their experiments correctly and established the existence of the neutron. Chadwick's achievement, in the words of the distinguished Italian physicist Emilio Segrè, was 'immediately, clearly and convincingly' to recognize neutrons for what they were – the true hallmark 'of a great experimental physicist'. Chadwick put it more modestly and prosaically: 'The reason that I found the neutron was that I had looked, on and off, since about 1923 or 4. I was convinced that it must be a constituent of the nucleus.'

The discovery was a blow to Frédéric Joliot-Curie, who wrote privately of his frustration: 'It is annoying to be overtaken by other laboratories which immediately take up one's experiments.' However, in public he was gracious and generous. It was 'natural and just' that the final steps of the journey towards the neutron were undertaken at the Cavendish, since 'old laboratories with long traditions have . . . hidden riches'.

Chadwick's achievement marked a watershed. Nuclear physics (the study of the atom's nucleus) as opposed to atomic physics (the study of atoms) had been in the doldrums. Scientists had faced difficulties of inter-pretation that arose far more swiftly than they could be resolved. Chadwick's discovery provided the all-important clue to many unresolved problems. For example, the neutron added to the understanding of isotopes first discovered in 1913 by Frederick Soddy. Until then, no-one had known exactly what differentiated isotopes from their 'sister' element. The suspicion was that the difference lay in the nucleus, but it took Chadwick's findings to prove that suspicion correct: what made isotopes different was the number of neutrons in their nuclei. But most exciting of all was the realization that, since the neutron carried no electrical charge, it would not be deflected by the positive nuclear charge. It was the ideal missile with which to bombard and probe elements as it could hurtle on until it penetrated the nucleus of the atom.

Across Europe, scientists took note. In Germany, the physicist Hans Bethe, later head of theoretical physics at Los Alamos and an architect of the atomic bomb, decided that the discovery of the neutron made nuclear physics the field in which to work. In Rome, Italian scientist Enrico Fermi – yet another of the fraternity who had studied under Max Born in Göttingen in the 1920s, and till then a theoretical physicist – plunged into experimental nuclear physics, setting up a small group to explore the inter-actions of neutrons 'with any elements he could get hold of'.

What none of them yet knew was that the neutron was also the catalyst for achieving an explosive nuclear chain reaction. Curiously, though, that very year, 1932, Harold Nicolson published a novel, *Public Faces*, about a catastrophically destructive new weapon made from a powerful raw material. This substance could transmute itself with such violence that it could cause an explosion 'that would destroy all matter within a considerable range and send out waves that would exterminate all life over an indefinite area'. 'The experts', Nicolson wrote in his novel, 'had begun to whisper the words . . . "atomic bomb".' They claimed it could 'destroy New York'.

Neutrons were by no means the only reason 1932 would be recalled as a spectacular year in the history of science. In January, just a few weeks before Chadwick's coup, American chemist Harold Urey made another discovery that Rutherford had long predicted. Working at Columbia University, he found that natural hydrogen consisted of 99.985 per cent ordinary hydrogen but also of 0.015 per cent 'heavy hydrogen' – an isotope given the name 'deuterium' – which also existed naturally in combination with oxygen in water. This so-called 'heavy water', which appeared to the naked eye identical to ordinary water, boiled and froze at different temperatures and was 10 per cent heavier. A decade later it would become a substance much sought after by the Nazis, and people would die to deny it to them.

But in 1932 Urey thought of deuterium as a 'delightful plaything for physicists' to use in bombarding other more complex atoms so that they could better understand nuclear structure. He speculated whether heavy water itself might be 'valuable in understanding more of living processes', perhaps even in the study of cancer since some initial research showed that yeast cells, which had some similarities to cancer cells, multiplied less quickly in heavy water than in ordinary. This proved impracticable. Nevertheless, heavy water caught the American public's attention. In a 1935 crime novel, the victim died after entering a swimming pool filled with heavy water which the author described as 'lethal'.* In a review, a scientist wrote, 'it is the most expensive murder on record . . . at the present cost that pool of heavy water would have cost about $200 million'.

On 21 April 1932, Rutherford reported another Cavendish triumph, writing exuberantly to Bohr 'it never rains but it pours'. John Cockcroft and Ernest Walton had just become the first scientists to split the atom using a man-made machine, an accelerator – the device Rutherford had asked them to develop some time earlier. They had created it lovingly and carefully, smoothing plasticine – an innovative new material which had replaced the sealing wax previously used for this purpose – over the joints to create a vacuum. Fearing that rivals might overtake them, Rutherford had urged them to stop perfecting it and 'do what he'd told them to do months ago' – start experimenting. His bullying paid off.

---

* Drinking a few glasses of heavy water would not be lethal, but the replacement of more than one third of the hydrogen in the human body's fluids by deuterium would be fatal.

Cockcroft and Walton bombarded lithium with accelerated protons and succeeded in disintegrating the lithium nucleus into two helium nuclei. According to one of his colleagues, Cockcroft, 'normally about as much given to emotional display as the Duke of Wellington', ran through Cambridge shouting, 'We've split the atom! We've split the atom!' An additional excitement was that the energies of the particles measured by Cockcroft and Walton provided the first experimental confirmation of the validity of Einstein's proposal that $E = mc^2$.

Rutherford asked Cockcroft and Walton to temper their jubilation in favour of discretion to allow them time to exploit their discovery without alerting rivals. However, with a media increasingly hungry for further revelations about nuclear physics following Chadwick's discovery of the neutron a few weeks earlier, soon it seemed only sensible to court press attention. The team chose the Marxist science correspondent of the *Manchester Guardian* to announce their achievement.

Rutherford had been right to fear competition. The Cavendish might easily have been upstaged by Ernest Lawrence at Berkeley. While Cockcroft and Walton had been busily massaging plasticine over the joints of their accelerator, Lawrence had been developing a successor to his small octopus-armed pillbox. His new cyclotron was an eleven-inch version. In August 1931, his assistant Stanley Livingston achieved an energy of over one million electron volts with the new machine – surely enough to accelerate particles to split atoms. Livingston asked Lawrence's secretary to send him a telegram which read, 'Dr. Livingston has asked me to advise you that he has obtained 1,100,000 volt protons. He also suggested that I add "Whoopee!".' When he received it, Lawrence 'literally danced around the room', pale blue eyes shining with excitement and already planning bigger, more powerful devices.

It was therefore a shock to Lawrence, honeymooning happily in Connecticut in the summer of 1932, to learn that Cockcroft and Walton's linear accelerator had become the first device to disintegrate the nucleus with accelerated particles. He sent agitated telegraphic orders to Berkeley: 'Get lithium from chemistry department and start preparations to repeat with cyclotron. Will be back shortly.' Success was not far off. A few weeks later, the president of the university despatched a jubilant message to the Governor of California: 'In September of 1932 artificial disintegration was first accomplished outside of Europe in the Laboratory of Professor Ernest O. Lawrence. This laboratory has taken the lead, in all the world, in the disintegration of the elements.'

*   *   *

Meet the genius who built the cyclotron, the fantastic machine that has ushered in a new era

Ernest Lawrence

ATOM SMASHER

Magazine article lauding Lawrence's achievement

Lawrence had been joined at Berkeley in the autumn of 1929 by a young scientist who shared his ambition to help the United States take 'the lead, in all the world', the twenty-five-year-old Robert Oppenheimer. Slenderly built, with intensely blue eyes, friends thought him 'both subtly wise and terribly innocent'. He was also sensitive, conceited, often neurotic, but charismatically engaging. Though passionate about physics, he was a Renaissance man with obsessions ranging from Hindu philosophy to Dante's *Inferno*.

Oppenheimer had grown up in New York, the product of a wealthy, cultured Jewish family whose Riverside Drive apartment was hung with paintings by Impressionist masters. He had been, in his own words, 'an abnormally, repulsively good little boy'. After attending New York's exclusive Ethical Culture School, he went on to Harvard. Like many contemporaries in continental Europe, Oppenheimer's early years were not free of anti-Semitism, albeit differently expressed. He arrived at Harvard shortly after its president had recommended a quota for Jewish undergraduates. When he applied to go and study under Rutherford at the Cavendish, his Harvard professor's letter of recommendation concluded, in character with the times, 'As appears from his name, Oppenheimer is a Jew, but entirely without the usual qualifications of his race. He is a tall, well set-up young man, with a rather engaging diffidence of manner, and I think you need have no hesitation . . . in considering his application.'

Rutherford, who would never have dreamt of being influenced by matters of race and had a deep contempt for racists, accepted Oppenheimer but was unimpressed by his abilities as an experimentalist. Bohr, while visiting the

Cavendish, asked an obviously unhappy Oppenheimer how his work was going. Oppenheimer replied that he was having difficulties. When Bohr asked whether his problems were mathematical or physical, he despairingly said that he didn't know. Bohr replied with devastating if unhelpful honesty: 'That's bad.' Oppenheimer spent tortured days standing by a blackboard, chalk in hand, unable to write anything. He could hear himself saying, over and over, 'The point is. The point is. The point is . . .' Such were Oppenheimer's inner frustration and turmoil that during a reunion with a friend, Francis Fergusson, in Paris he became so enraged by something Fergusson said that he leapt on him and tried to strangle him, forcing the more powerfully built Fergusson to fend him off. Back in Cambridge a contrite Oppenheimer wrote seeking forgiveness for his bizarre behaviour and explaining how his failure to live up to 'the awful fact of excellence' was tormenting him.

He remained troubled, depressed and occasionally deluded. On one occasion he insisted that he had left a poisoned apple on the desk of a colleague at the Cavendish. For a while a psychiatrist treated him for dementia praecox. There are conflicting stories about why the treatment ended in 1926. According to one, the psychiatrist warned that continuing would do more harm than good; according to the other – and this sounds more likely – Oppenheimer decided he understood more about his condition than his doctor and cancelled further sessions. When Max Born visited the Cavendish in 1926 and invited him to Göttingen, Oppenheimer accepted with gratitude but little confidence in his own abilities.

However, Oppenheimer shook off the worst of his depression and mood swings and flourished at Göttingen.

More than at either Harvard or Cambridge he felt, in his words, 'part of a little community of people who had some common interests and tastes and many common interests in physics'. His passion was theoretical physics, and Göttingen was the focus of the theoretical physics world with all of its leaders teaching there or regularly visiting. Oppenheimer wrote to a friend, 'they are working very hard here, and combining a fantastically impregnable metaphysical disingenuousness with the go-getting habits of a wall paper manufacturer. The result is that the work done here has an almost demonic lack of plausibility to it and is highly successful.'

After sampling other leading centres of European theoretical research, Oppenheimer had come home at last. Ten American universities were eager to secure him and he eventually signed concurrent contracts with two of them: the eight-year-old California Institute of Technology at Pasadena, Caltech, where he agreed to teach in the summer; and Berkeley, where he was to teach in autumn and winter. The twenty-five-year-old Oppenheimer loved fast cars but was, he confessed, 'a vile driver' who could 'scare friends out of all sanity by wheeling corners at seventy'. Unsurprisingly, therefore, when he reached Pasadena after a marathon journey across the States he had his arm in a sling and his clothes were stained with battery acid – the results of a car accident en route.

Oppenheimer had chosen Caltech because he believed its blend of theorists and experimentalists would be good for him ('I would learn, there would be criticism . . .'). His reasons for selecting Berkeley were a little different. Despite possessing Lawrence's unrivalled experimental facilities, the faculty was weak on the theoretical side, with no-one versed in quantum mechanics. Oppenheimer intended to do most of his teaching at Berkeley to remedy these deficiencies and to establish a theoretical and inter-pretative group to complement Lawrence's work. In the autumn, Oppenheimer arrived at Berkeley ready to begin teaching, fresh from holidaying at the ranch he had just leased in the Sangre de Cristo Mountains of New Mexico. He had named it Perro Caliente at the suggestion of a female friend. The words were the Spanish translation of the raucous cry of joy, 'Hot dog!', he had uttered when he learned the ranch was available. The red, raw beauty of the desert stirred him. He often told friends that 'I have two loves, physics and the desert. It troubles me that I don't see any way to bring them together.'

Oppenheimer hit it off at once with Lawrence, just three years his senior, admiring his 'unbelievable vitality and love of life'. They socialized and womanized together, drinking Oppenheimer's famous frozen martinis from glasses rimed with lime juice and honey, and eating his speciality, the spicy Indonesian dish *nasi goreng*, soon nicknamed 'nasty gory' by Oppenheimer's Berkeley friends. They also went riding. Photographs of the two men show Lawrence, tall, sturdy, smiling; Oppenheimer, with a frizz of dark hair, his slighter frame clad in heeled Mexican boots and tight jeans, and a quizzical yet dreamy expression, resembles a young Bob Dylan.

Lawrence the experimentalist and Oppenheimer the theoretician got on well intellectually as well as socially. They attended weekly seminars for theoreticians and experimentalists where Oppenheimer amazed everyone with his ability to assimilate new ideas, his extraordinary memory and the fact that he 'knew more experimental physics than even the experimental physicists did'. He relished the new horizons opened up by the neutron and the development of powerful machines to probe the nucleus. In 1932 he wrote to his brother Frank, 'We are busy studying nuclei and neutrons and disintegrations; trying to make some peace between the inadequate theory and the absurd revolutionary experiments.'

Just as Oppenheimer had hoped, atomic physics was no longer Europe's exclusive preserve. On a visit to Berkeley in 1933, John Cockcroft was startled to find it run more like a factory than a laboratory. 'The experimenters were divided into shifts: maintenance shifts and experimenters. When a leak or fault developed in the cyclotron the maintenance crew rushed forward to plug the leaks ... and fixed the fault when the operating shifts rushed in again.' It was far removed from the small-scale, expense-conscious academic world of the Cavendish, and a warning that the Cavendish might soon be outclassed.

The discoveries of 1932 also gave a fillip to Russian atomic physics. Abram Joffé, who had brought Peter Kapitza to England in 1921, had continued to keep abreast of developments in the West. By the early 1930s he was presiding over the Leningrad Physicotechnical Institute, known as 'Fiztekh' and the crucible of Soviet physics. Joffé also encouraged Western scientists to study and lecture in Russia. However, until 1932 the only serious nuclear work had been research into cosmic rays. This soon changed. Before the year was out, Soviet scientists had replicated Cockcroft and Walton's experiments. Also in that year, inspired by reports of Lawrence's work, the Radium Institute in Leningrad began building Europe's first cyclotron, while Joffé set up a dedicated nuclear physics group. He soon had thirty scientists working in four laboratories. Igor Kurchatov, who would later direct the Soviet nuclear programme, was sufficiently excited by the new science to divert from his study of the behaviour of crystals in magnetic fields to head the new group.

With this surge of interest, Peter Kapitza's absence was increasingly noted, and regretted, by the Soviet authorities. He had retained his Soviet citizenship and made annual visits home at the invitation of the Kremlin. He was a Russian patriot and happy to advise the Soviet government on science and technology in pursuit of Stalin's goal of 'catching up and overtaking the technology of the developed advanced capitalist countries'. But he had no inclination to return to a place where living conditions were so tough. He would not have enjoyed the conditions faced by one young scientist at Joffé's institute, who found himself sharing a freezing dormitory with eight others, with rats trying to chew at his ears. Kapitza wrote to his mother that life without gas, electricity, water and apparatus would be simply impossible. Furthermore, in 1930 Rutherford had persuaded the Royal Society and others to give £30,000 to fund a new laboratory for Kapitza to run.

Kapitza bridged two worlds and took a sly pleasure in doing so. On one occasion he is said to have invited senior Soviet politician Nikolai Bukharin to dinner with Rutherford solely for the pleasure of being able to make the introduction, 'Comrade Bukharin – Lord Rutherford.' Kapitza was elected a member of the Royal Society – a highly unusual honour for a foreigner – and was apparently interested in other prizes of the British establishment. After Rutherford was ennobled, he enquired whether a foreigner could be given a peerage. As it turned out, however, Stalin had other plans for him.

In fact, life was about to change for many members of the international scientific community. Much that had been taken for granted for so long – openness, the freedom to travel and exchange ideas, the right to pursue science without a thought of politics – was about to come under attack. Robert Oppenheimer was one of the few to sense the challenges ahead, writing bleakly but perceptively that 'the world in which we shall live these next thirty years will be a pretty restless and tormented place. I do not think there will be much of a compromise possible between being of it, and being not of it.'

The restlessness divined by Oppenheimer was already evident in Japan, which during the 1920s had seen increasing prosperity, much of it prompted by technological change. Developments in Hiroshima were typical. Though many citizens had continued to make their livings in traditional ways – harvesting and drying sardines; cultivating *nori*, the seaweed which they dried in sheets and used to wrap their sticky rice; growing hemp for ropes and fishing nets; and making *geta*, wooden sandals secured to the foot with thongs – new industries had grown rapidly centring on activities such as manufacturing rayon, rolling tobacco for cigars and canning food – especially beef boiled in soy sauce for the military commissariats based in the city.* The pick of Hiroshima's manufactured goods were displayed in the green-domed Prefectural Products Exhibition Hall, which was one of

* Hiroshima was also a leading producer of the hair extensions used by many women to create the traditional and luxuriant *bunkin takashimada* hairstyle. By 1922, 70 per cent of Japan's hair extensions were manufactured in Hiroshima.

the city's favourite landmarks. Constructed in 1915 to the design of the Czech architect Jan Letzel, it fronted the river near the Aioi Bridge.

Rising wealth had brought many benefits. Hiroshima had become an academic centre with one of the only two higher schools of education in Japan. It was also known for sport: baseball, rowing and track events flourished, and Japan's first Olympic gold medallist, Mikio Oda, who triumphed in the triple jump at the 1928 Games in Amsterdam, came from the city. A new entertainment district – Shintenchi, meaning, literally, 'New World' – was built which by its peak in the late 1920s had more than 120 shops, music halls, theatres and cinemas. Visitors could attend performances ranging from musical comedy to silent samurai movies. Sunday was the day for going to the cinema and shortages of daytime electricity did not spoil the fun: the projectionist simply hand-cranked the film past a large gas lamp. 'Modern boys and girls', as those who espoused Western dress and habits were called, played billiards or had their pictures taken, posed cigarette in hand and dressed in the latest Western fashions, in one of the many photographic studios, or just sat and chatted in cafés. Huge advertising hoardings gaudily promoted everything from Lion brand toothpaste to scented hair oil. By night, elegant electric lanterns fashioned to resemble lilies-of-the-valley cast a glamorous glow. Photographs of the period reveal a relaxed and prosperous ambience in Hiroshima and other leading cities.

However, the political mood was changing. On Christmas Day 1926, a new emperor, the twenty-five-year-old Hirohito, succeeded to the Japanese imperial throne. The name he chose for his reign was Showa –

'illustrious peace' – but the reality would be different. His year-long enthronement festivities were celebrated with enthusiasm in Hiroshima as well as in the rest of the country. Reverence for the Emperor and a desire to separate his divine person from human contact led to the disinfection of cars and trains in which he was to travel, and to the requirement for his people not to look at him but to cast down their eyes in his presence. The celebrations reinforced a growing cult of the emperor.

The early 1930s brought an economic downturn. Turmoil among politicians led to the increasing involvement of the military in the running of all aspects of Japanese life and, at their behest, a further emphasis on the Emperor both as a divine religious figure and as head of a strong and united nation requiring and receiving his subjects' unquestioning loyalty and obedience. In September 1931, the Japanese military fabricated a crisis, 'the Manchurian incident', as a pretext for their occupation of that much disputed Chinese province in which Japan had substantial commercial interests and from which it obtained many scarce primary resources. They installed the last Emperor of China, Pu Yi, as the puppet emperor of their client state, which they named Manchukuo. When the League of Nations condemned their actions, the Japanese left the League.

In 1932, right-wing officers murdered both the Japanese prime minister and finance minister because they would not follow sufficiently militaristic policies. In the wake of the murders Japan abandoned any kind of party system and the military's influence simply increased. The cinemas in Hiroshima's Shintenchi district showed a film called *Japan in the National Emergency*. The script underlined a growing policy against Westernization and for a return to the old values: 'In the past we have just followed the Western trend without thinking about it ... as a result, Japanese pride has faded away ... Today we are lucky to see the revival of the Japanese spirit throughout the nation.' The film disparagingly depicted two Westernized young people, in particular a 'modern girl' who smokes, dances and dares to ask a dignified middle-aged gentleman who accidentally steps on her toe in the street to apologize. He refuses, snorting, 'This is Japan.' The message was strong: women should return home and forget Western fashions and behaviour, and the Japanese public in general should reject Western mores and glory in Japan's unique superiority.

CHAPTER SIX

PERSECUTION AND PURGE

ONE NIGHT, WERNER HEISENBERG HAD A HALF-WAKING 'vision'. as he recalled in his memoirs, he saw a Munich street 'bathed in a reddish, increasingly intense and uncanny glow. Crowds of people with scarlet and black-red-and-white flags were streaming from the Victory Gate toward the university fountains and the air was filled with noise and uproar. Suddenly, just in front of me a machine gun began to cough. I tried to jump to safety and woke up . . .' It was an amalgam of the anarchic scenes he had witnessed as a boy in the Munich of 1919 and of the new, organized National Socialist violence erupting onto Germany's streets. By 1930, even once moderate and conservative German newspapers were hailing Adolf Hitler as the saviour of a Germany deep in financial stagnation. Radical groups of the right and left were fighting in the slums and breaking up each other's meetings.

Perhaps to forget such things, in January 1933 Heisenberg invited some old friends on a skiing holiday in Bavaria which, he recalled, was 'long remembered by all of us as a beautiful but painful farewell to the "golden

age" of atomic physics'. The friends included Niels Bohr
and Bohr's son Christian as well as Carl-Friedrich von
Weizsäcker, whom Heisenberg had known since the latter
was fourteen. They had met in 1927 in Copenhagen,
where von Weizsäcker's father, later the second most
senior official in Hitler's Foreign Office, was Germany's
representative in Denmark. The young Carl had read
articles by Heisenberg and engineered a meeting with him.
Heisenberg, at that time studying under Bohr, was kind to
the quiet, academic, awestruck boy and inspired him
to become a physicist. Von Weizsäcker became not only
Heisenberg's assistant but one of his closest confidants.

Anti-Semitic cartoon from *Der Stürmer*, March 1933

The Bohrs arrived at the local railway station after dark, so Heisenberg and von Weizsäcker went to meet them. Guiding his guests back up the mountain to the sleeping hut, Heisenberg was peering ahead in the lantern-light when he noticed that the snow seemed unusually powdery. Then 'something very odd happened – I suddenly had the feeling that I was swimming. I completely lost control of my movements, and then something pressed on me so violently from all sides that, for a moment, I stopped breathing.' The avalanche had not covered his head and he managed to free his arms. Looking around, he realized he was the only one to have been swept away and had been lucky to survive. Heisenberg and his friends spent their days skiing and talking physics, trying to forget the 'world full of political trouble' below the snow-line.

One of the first signs of those troubles was the racial laws passed on 7 April 1933, soon after Hitler's installation as Germany's new chancellor. The 'Law for the Restoration of the Professional Civil Service' banned 'non-Aryans' – anyone with at least one Jewish grand-parent – from working for the state, and it included the universities as government institutions. There were a few exceptions – Jewish people appointed before the First World War or who had fought or lost fathers or sons at the front. Nobel laureate James Franck had served in the war but refused the 'privilege' of remaining in his post. On 17 April he resigned his position at Göttingen, protest-ing that 'We Germans of Jewish descent are being treated as aliens and enemies of the Fatherland.' Max Born, who could also have claimed exemption, departed quietly but bitterly, writing, 'All I had built up in Göttingen during twelve years' hard work was shattered.' He went for a walk in the woods 'in despair, brooding on how to save my family . . .' Fritz Haber, the man who, as Franck's wartime boss, had masterminded Germany's chemical warfare strategy in the First World War and was a German patriot through and through, also refused his exemption, resigning after being ordered to purge his institute of other 'non-Aryans'.

Viewed as a Jewish stronghold, physics attracted special virulence. Nobel Prize-winning physicists Johannes Stark and Philipp Lenard spearheaded the attack. As early as the 1920s they had set themselves up as figureheads of true 'German physics', denouncing the 'Jewish Physics' of Einstein. Stark had been sacked from the University of Würzburg for breaching the rules of the Nobel foundation by using his prize money to buy himself a china factory, but had convinced himself that Jews were responsible for his fall.

Stark and Lenard also savaged the 'Jewish-minded' Aryans who took their inspiration from quantum mechanics and relativity. In particular they launched a very personal crusade against Heisenberg for his espousal of 'Jewish science' and for being a 'Jewish pawn'. In November 1933, when news broke that he had won the Nobel Prize for Physics, Nazi thugs threatened to disrupt his lecture the following day. In 1935, when it seemed likely that Heisenberg would replace his former teacher at Munich University Arnold Sommerfeld, who was retiring, Stark objected. He denounced Heisenberg as the 'spirit of Einstein's spirit', deploring that he was 'to be rewarded with a call to a chair'.

Two years later, Stark used the much feared official weekly SS journal *Das Schwarze Korps* to brand Heisenberg 'a White Jew', one of the 'representatives of

Judaism in German spiritual life who must all be eliminated just as the Jews themselves'. With the SS taking an ever closer interest in him, Heisenberg's mother, who had known Heinrich Himmler's mother since childhood, begged Frau Himmler to intercede. Somewhat grudgingly, she agreed. However, Heisenberg remained under investigation and was summoned several times to the Gestapo's notorious headquarters in Prinz Albrechtstrasse in Berlin for questioning. He was interrogated in a cellar with, as he recalled, an 'ugly inscription' painted on one of the walls: 'Breathe deeply and quietly'. Finally, in July 1938, Himmler wrote to Heisenberg that there would be no more attacks. On the same day Himmler also wrote to Reinhard Heydrich, chief of the Gestapo, that Heisenberg was too valuable to liquidate. Notwithstanding Himmler's apparent blessing, Heisenberg was still not appointed to Munich University. Instead, a former assistant of Stark's was given Sommerfeld's physics chair. He was, in Sommerfeld's view, a 'complete idiot'.

Einstein severed his links with Germany early and for ever. He was about to sail back to Europe from California when Hitler came to power, and roundly denounced the land of his birth for turning its back on 'civil liberty, tolerance, and equality of all citizens before the law'. A few days later, in Antwerp, he announced his resignation from the Prussian Academy of Sciences, thereby infuriating the Prussian Minister for Education, Bernhard Rust, who had hoped to mark the national boycott of Jewish businesses called for 1 April 1933 by expelling him.

As enraged Nazis ransacked Einstein's house and the authorities confiscated his bank account, Germany's most famous scientist crossed the Channel to England with his wife Elsa, protected by a British naval commander and MP who had had the singular experience of having once been invited to kill Rasputin. Einstein was safe, but he confessed to Max Born that 'My heart aches when I think of the young ones.' He also told him that he had never thought highly of 'the Germans' but the degree of their brutality and cowardice had surprised even him.

In the autumn of 1933, finding England too formal and preferring a life with 'No butlers. No evening dress', Einstein accepted a post at the Institute for Advanced Study at Princeton. Paul Langevin, watching events from Paris, thought his emigration highly significant, remarking only half in jest that 'It's as important an event as would be the transfer of the Vatican from Rome to the New World. The Pope of Physics has moved and the United States will now become the centre of the natural sciences.'

Einstein rounded on German intellectuals for behaving 'no better than the rabble'. Certainly some prevaricated while books by 'undesirables' were tossed on fires and professors sympathetic to the new order donned brown shirts to lecture on such absurdities as 'Aryan mathematics'. A number hoped that the expulsion of so many scholars would further their own careers. However, many were troubled, and a few, including Max Planck, had the courage to try to help their Jewish colleagues. Planck was given an audience with Hitler on 16 May 1933, but, according to Planck, the Führer 'whipped himself into such a frenzy' that Planck could only listen in appalled silence, then leave. Heisenberg also considered protesting, despite his fragile personal position. He visited a tired-looking Planck, whose 'finely chiseled face', he

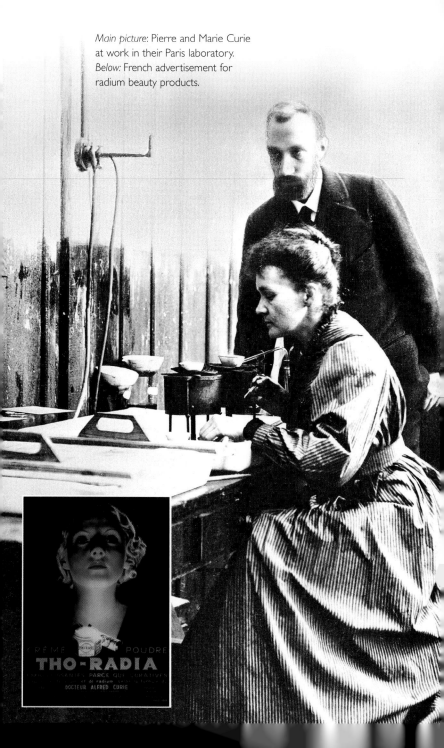

Main picture: Pierre and Marie Curie at work in their Paris laboratory.
Below: French advertisement for radium beauty products.

CRÈME   POUDRE
THO-RADIA
DOCTEUR ALFRED CURIE

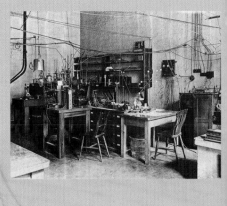

PORT UJINA, JAPAN.

（一其）港品宇

Hiroshima's port of Ujina at the turn of the twentieth century.

Shintenchi – Hiroshima's entertainment district.

Lantern parade in Hiroshima to celebrate the Japanese capture of Nanking in 1937.

Soldiers carrying the ashes of comrades killed in China through the streets of Hiroshima in 1938.

Children farm pigs in their Hiroshima schoolyard to help the war effort in 1944.

Fire-fighting drills in Hiroshima.

Above: The 1933 Solvay Conference.

Below, left: A section of the nuclear pile constructed by Enrico Fermi and his team at Chicago University, in which the world's first self-sustaining chain reaction was achieved on 2 December 1942.

Below, right: Niels Bohr (right) chats to Yoshio Nishina (left) and his colleague Seishi Kikuchi during Bohr's visit to Japan in 1937.

Above: Rudolf Peierls and his Russian-born physicist wife, Genia.

Below, left: Fritz Strassmann as a young man.
Below, right: Leo Szilard.

Above, left: Joachim Ronneberg, leader of the Gunnerside team of Norwegian commandos which attacked the Rjukan heavy water plant in February 1943.
Above, right: Damage to the Rjukan heavy water production cells after the attack by the Gunnerside commandos.
Below: The Rjukan heavy water plant.

thought, 'had developed deep creases' and whose smile 'seemed tortured'. The initiator of quantum theory, shaken by his encounter with completely irrational forces, convinced Heisenberg that protests would be 'utterly futile'.

Heisenberg took Planck's advice, trying to convince himself that the extremism could not last, even that something good might emerge from the mayhem. But his optimism seemed naive to the point of absurdity to his Jewish friends. He told Born that 'Since . . . only the very least are affected by the law – you and Franck certainly not . . . the political revolution could take place without any damage to Göttingen physics . . . Certainly in the course of time the splendid things will separate from the hateful.' Heisenberg would later justify his position as one of 'inner exile' during which he sought to protect 'the old values' so that something would survive 'after the catastrophe'. Looking back after the war, he even suggested that his Jewish friends had faced easier choices than he had. Forced to leave, 'at least they had been spared the agonising choice of whether or not they ought to stay on'. 'Inner exile' would come to involve many compromises, both conscious and unconscious, for Heisenberg.

Lise Meitner wondered anxiously what would happen to her. She was an Austrian national, not a German. Also, the Kaiser Wilhelm Institute was not directly under government control and its staff were not government servants. Nevertheless, she felt threatened, and on 3 May 1933 she wrote to her long-term friend and collaborator Otto Hahn, then in the United States, begging him to come home. Hahn, who had received equally disturbing letters from other Jewish friends, hurried back to Berlin to see for himself. He was so shocked that he suggested a group of prominent Aryan academics should protest against the treatment of their Jewish colleagues. Yet, just as he had counselled Heisenberg, Max Planck, on the basis of his own protest, warned that it would be pointless: 'If today you assemble 50 such people, then tomorrow 150 others will rise up who want the positions of the former . . .' Planck believed the best way to protect German science was for the present to keep quiet. In an amoral, practical sense he was right. Once the Jewish academics were gone, German science was allowed to proceed largely unmolested.

Hahn, too, followed Planck's advice. Like Heisenberg, he steadfastly refused to join the Nazi Party. He also resigned his lectureship at the University of Berlin to avoid having to participate in Nazi Party meetings. In 1935, on the first anniversary of the death of Fritz Haber – he had died of a heart attack during a visit to Switzerland the year before – Hahn and Max Planck, prompted to action again, organized a memorial service, despite official threats, at which they both spoke. University professors, as government employees, were too nervous to attend in case they were sacked but sent their wives in one of the scientific community's very few concerted gestures of solidarity with those who had been ousted. Planck ended his oration with the words, 'Haber was true to us, we shall be true to him.'

The Solvay Conference of October 1933 in Brussels was a refuge and a distraction from the disturbing happenings in the wider world. Forty experimentalists and theoreticians attended, including Rutherford, Chadwick, Lawrence, Madame Curie, the Joliot-Curies, Langevin, Meitner and

Bohr, to debate the 'Structure and Properties of the Atomic Nucleus'. They argued about whether Chadwick's neutron was a composite of particles or, as experiments would shortly confirm, a particle in its own right. They also discussed the recent finding of another new sub-atomic particle – the positively charged electron, or 'positron' – by Carl Anderson, a physicist at Caltech, Pasadena, researching into cosmic radiation. Anderson had made his discovery using a clever device invented many years earlier by Scotsman Charles Wilson, the cloud chamber, designed to make the invisible path of particles visible. This was achieved by shooting particles through a saturated water vapour created in the chamber, causing them to leave a trail of droplets, like the tail of a meteor. Their track, thus revealed, could be photographed through a window in the side.*

Just like the neutron, the positron had previously been glimpsed but misinterpreted by others. Chadwick had come close to it but, fixated on the neutron, had missed the significance of some observations. The Joliot-Curies had photographed electrons in a magnetic field 'going backwards the wrong way' but had not recognized them as positrons, until they read of Anderson's work.

Piqued by their failure to identify the positron, the Joliot-Curies had launched a series of experiments to dis-cover more about it. Placing a cloud chamber in a strong magnetic field, they bombarded elements with alpha particles. While this caused elements in the middle of the Periodic Table to release protons, they found that light elements such as aluminium sometimes ejected a neutron

---

* The discovery of the positron was the first clear indication that the universe consisted of anti-matter as well as matter.

and a positron instead. This caused them to wonder whether a proton might be a compound of a neutron and a positron. However, their suggestion met fierce opposition at the Solvay Conference, particularly from Lise Meitner. Undeterred by the hostile gaze of Marie Curie, who resented her daughter's work being criticized, she stated that 'My colleagues and I have done similar experiments. We have been unable to uncover a *single* neutron.'

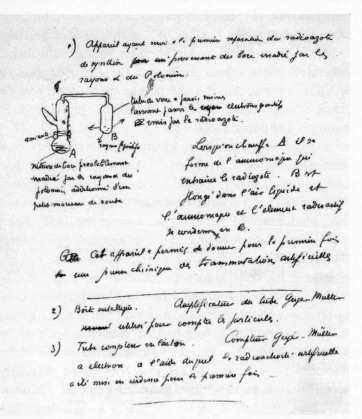

Manuscript notes by Frédéric Joliot-Curie describing his first experiments with radioactivity

Deflated and anxious, the Joliot-Curies hurried back to Paris to re-check their findings but could find no mistakes. The light elements had definitely emitted neutrons. Encouraged, they resumed their experiments and so stumbled upon one of the most significant discoveries so far.

After bombarding ordinary aluminium with alpha particles, Frédéric Joliot-Curie used a Geiger counter to measure the results. To his surprise, when he moved the radioactive source emitting the alpha particles away from the aluminium, the counters, instead of immediately falling silent, continued noisily clicking. He could not believe it. He repeated the manoeuvre, and the results were the same. He fetched Irène, who was equally puzzled. That evening they had to attend a dinner engagement, so they asked a colleague to check that the counters were not faulty. Hastening back to their laboratory the next morning they found his note: the counters were operating perfectly.

Painstakingly, the Joliot-Curies worked out what had happened. Until then, all reactions scientists had produced had occurred immediately and had ceased as soon as the bombarding source was removed. However, on being bombarded with alpha particles, the aluminium had transmuted into an intermediate radioactive isotope of phosphorus which, as it decayed back to its stable state, silicon, continued to emit radioactivity (positrons) for some time after the bombarding source of alpha particles had been removed. They had induced a new phenomenon – 'artificial radioactivity'. An exultant Frédéric told his assistant, 'With the neutron we were too late. With the positron we were too late. Now we are in time.' Previously, physicists had known that by bombarding it with a particle of sufficient energy a nucleus could be dis-integrated and a new, stable one formed. No-one had realized that, in certain circumstances, an unstable element in the process of nuclear decay could be created. In other words, man could force the elements to release their energy in the form of radioactive decay. The Joliot-Curies rushed to publish the news of 'A New Type of Radioactivity'.

The discovery caused consternation and disappoint-ment at Berkeley. As one of Lawrence's team observed, 'We could have made the discovery any time.' If, rather than concentrating on continued improvement of the per-formance parameters of their accelerators, they had only thought to run a Geiger counter over one of their targets, they too would have heard the tell-tale click announcing the creation of a new radioactive element. This had not happened for practical reasons: the laboratory's Geiger counter and the cyclotron worked on the same switch so the team had never had the chance to explore whether the counters kept registering after the cyclotron was switched off. It was an irritating thought. As another man admitted, 'We felt like kicking each other's butts.' They altered the wiring, left the Geiger on after taking the cyclotron down, and sure enough heard the counter's rhythmic tick – 'a sound that none who was there would ever forget', recalled Stanley Livingston.

The discovery brought Marie Curie great satisfaction. Joliot wrote of 'the expression of intense joy which over-took her when Irène and I showed her the first [artificially produced] radioactive element in a little glass tube. I can see her still taking this little tube . . . in her radium-damaged fingers. To verify what we were telling her, she brought the Geiger counter up close to it and she could hear the numerous clicks . . . This was without a doubt the last great satisfaction of her life.' Marie told a friend that 'We're back in the fine days of the old laboratory.'

'La Patronne' was still a powerful presence at the Curie Laboratory. When young chemist Bertrand Goldschmidt arrived for an interview in June 1933 she told him in a strong Polish accent 'for a year or two you will be my slave in chemistry and do everything for me'. He was in awe of this 'rather small old lady with big hairs on her chin' who, dressed entirely in black, looked much older than her sixty-five years. He was also fascinated by the stories still circulating of a once-active love-life. There were, he recalled, 'many rumours', including that Eve Curie, born in 1904 and with her blue eyes and dark hair so different from her older, fairer sister, was not Pierre Curie's daughter 'but André Debierne's', and that Debierne had later been 'succeeded in Madame Curie's heart by Langevin'.

However, any such passions were long spent. Marie's life was ending. She died, aged sixty-six, at dawn on 4 July 1934 in a sanatorium in the mountains. The cause of death was extreme pernicious anaemia 'of rapid, feverish development. The bone marrow did not react, probably because it had been injured by a long accumulation of radiations.' She had insisted on reading her own temperature, holding the thermometer 'in her shaking hand' and recognizing from the sudden fall in her fever that her end was near.

Her coffin was buried above Pierre Curie's. There was no priest and no prayers, as befitted a devout sceptic, but her brother and sister cast a few grains of Polish soil on her coffin.

*   *   *

The following year, the Joliot-Curies were awarded the Nobel Prize for Chemistry for their finding of artificial radioactivity. In his acceptance speech, Frédéric Joliot remarked that 'scientists who can construct and demolish elements at will may also be capable of causing nuclear transformations of an explosive character'. Few paid much public attention to these prophetic words. In 1933, at a meeting of the British Association for the Advancement of Science, Ernest Rutherford had insisted that anyone who believed that atomic energy could be released on a large scale was 'talking moonshine'. Niels Bohr believed that even if a release of explosive power from the nucleus was possible in theory, in practical terms it was unattainable: 'Not only are such energies at present far beyond the reach of experiments, but it does not need to be stressed that such effect would scarcely bring us any nearer to the solution of the much discussed problem of releasing nuclear energy for practical purposes. Indeed the more our knowledge of nuclear reactions advances the remoter this goal seems to become.' To Einstein, the chances of achieving a massive release of energy were like 'a blind man in a dark night hunting ducks by firing a shotgun straight up in the air in a country where there are very few ducks'.

For the present, nuclear science remained an open subject studied for the joy of knowledge.

Rutherford took the Nazi threat to science extremely seriously. He agreed to head up the Academic Assistance Council, created in May 1933 in London to find academic jobs in Britain for Jewish scientific refugees. It was a formidable task. More than 1,500 had been thrown out of work, including at least a quarter of Germany's physicists.

Rutherford chaired a huge gathering in the Albert Hall at which one of the chief speakers was Einstein, unmistakable, according to one of those present, with his shaggy mane of hair, 'great furrowed forehead' and 'enormous bulging chocolate-brown eyes'.

Rutherford helped many personally, including Max Born and his Quaker wife, doing everything from finding the Borns somewhere to live in Cambridge and driving them to visit their quarantined dog to providing Born with temporary work at Cambridge until the offer of the Physics Chair at Edinburgh University came through. Born later wrote, 'A disaster turned out to be a blessing. For there is nothing more wholesome and refreshing for a man than to be uprooted and replanted in completely different surroundings.' Although he supported his right to receive help, Rutherford did not wish to meet Fritz Haber, who in the year prior to his death was living in England. Born recorded how Rutherford 'declined violently' an invitation to his house because Haber was to be present; 'he did not wish to have any contact with the man who had invented chemical warfare with the help of poison-gas'.

Niels Bohr, half-Jewish himself, was also very active on behalf of Germany's displaced scientists, firing off unsolicited invitations to people to come to Copenhagen. The international conference he held there every summer became, in the words of one refugee, 'a sort of labour exchange'. James Franck found sanctuary in Copenhagen before going to America. So did Edward Teller, before going first to England and then on to the United States. Teller was unstinting in his praise of the Academic Assistance Council and of the British, saying, 'They accepted many more scientists than Britain could possibly

use and all of us were welcomed as permanent residents
. . . the English are truly among the most hospitable and
ethical people in the world.'

Bohr also helped Lise Meitner's young nephew Otto
Frisch, who had been sacked from Hamburg University. In
October 1933 Frisch had arrived in England to work
under Patrick Blackett, the left-leaning head of the physics
department at Birkbeck College, London. Learning of the
Joliot-Curies' discoveries, Frisch built himself some
apparatus from a few bits and pieces from Woolworth's
and began investigating artificial radioactivity for himself.
When Bohr came to London to visit Blackett later that
year he offered the ingenious but unknown young man the
chance to come to Copenhagen. Frisch wrote to his
mother that it was as if 'God Almighty himself has taken
me by my waistcoat button and spoken kindly to me'.

In the United States, Hungarian Eugene Wigner, who
had emigrated there before Hitler came to power, joined
forces with another émigré colleague and wrote in
German to a small group of physicists at American
educational establishments. They asked them to set aside
a small percentage of their income for the next two years
to help former colleagues find academic posts in America.
Oppenheimer, who had relatives in Germany and, in his
words, felt 'a continuing, smoldering fury about the treatment of Jews in Germany', was one of the recipients.
However, the originators of the scheme were wary of
launching a general appeal, not only because academic
jobs in the US were in short supply but also because some
universities were known to be reluctant to employ Jews.

In Russia, despite Stalin's desire to push his country to
the forefront of science, the Soviet authorities made few
attempts to attract the talented, jobless people so

anxiously seeking new positions. Growing paranoia about infiltration by foreign agents meant that only those with known communist sympathies were admitted. Until 1937 they were allowed to work in Russia's physics institutes, but many then fell victim to Stalin's purges. Suspected of being spies, they were imprisoned, tortured and exiled.

Just as in Germany, Russia's scientists were becoming progressively caught in an ideological web. Abram Joffé felt he was 'living on top of a volcano'. After the defection of a leading theoretical physicist, George Gamow, at the 1933 Solvay Conference, Joffé was no longer permitted to travel outside Russia. Soviet scientists working abroad also felt vulnerable. Lev Landau, studying in Berlin, implored colleagues never to discuss politics with him in case it got him into trouble.

Peter Kapitza, however, felt perfectly at ease. He was about to take charge of the splendid new Mond Laboratory for magnetic field and low temperature research in Cambridge. It had been financed, at Rutherford's urging, by the Royal Society, drawing on a bequest from the Mond family, co-founders of the industrial giant ICI. The building was a model of art-deco chic, with a crocodile etched on the façade in affectionate tribute to Rutherford. But in 1934 Kapitza made an ill-advised visit to Russia. Just as he was preparing to return to Cambridge, he was informed that the Soviet Union 'could no longer dispense with his services, in view of the danger from Hitler'. His second wife, Anna, whom he had married in 1927, was allowed to return with the news to England, from where she sent her frustrated husband parcels of everything from lavatory paper to trousers, both of which he discovered were, unlike oysters and smoked sturgeon, virtually unobtainable in Russia.

Rutherford was shocked and angry. He tried everything he could to pressurize the Soviet government, but to no avail. In the event, all he could do was to offer to ship Kapitza the brand-new equipment installed for him in the Mond Laboratory, which reached Leningrad jumbled up in a cargo of frozen meat. After much argument, the Russian government paid £30,000 for the apparatus and installed it in a lavish new institute, purpose-built for Kapitza. By 1936, the trapped Russian, for whom there would be no more dinners at high table in Cambridge or thoughts of a peerage, reflected philosophically to Rutherford that 'we are only small particles of floating matter in a stream which we call fate . . . the stream governs us . . .' Rutherford wrote bracingly but sympathetically, 'A reasonable number of fleas is good for a dog – but I expect you feel you have more than the average number.' Anna Kapitza, who had rejoined her husband in Moscow in 1935, believed that Rutherford's staunch support had kept her highly strung husband from suicide, confiding, 'I am absolutely certain I owe to you the life of Kapitza, without his love and gratitude to you, without your invaluable help . . . he would not be alive.'*

Kapitza would survive the savage purges that would reach their peak in 1937–8 when nearly eight million people were arrested including a hundred physicists, many of whom were shot or vanished into the gulags. Abram Joffé would also be spared to sustain Soviet physics through these precarious times. However, fear of the

---

* According to one story, when Kapitza finally returned to Cambridge thirty years later, in the 1960s, he was invited to dine at Trinity College. To his consternation he realized he had no academic gown to wear. The college butler came to the rescue, producing Kapitza's original gown, unworn since 1934.

night-time knock on the door sapped the creativity of Russian scientists at a critical time for nuclear physics.

Kapitza's was not the only departure from the Cavendish. James Chadwick, highly impressed by the performance and potential of Lawrence's cyclotron, was growing restive. As he later wrote, 'It was becoming very difficult to push on without some new equipment . . . we needed a means of accelerating protons or other particles . . . at high energies. But that meant more space, particularly more money, and particularly engineering. It meant complicated equipment, and Rutherford had a horror of complicated equipment.'

And Rutherford would not budge. Lacking Kapitza's shameless ability to charm and manipulate Rutherford, and anxious not to quarrel with his mentor of a quarter of a century, Chadwick thought it best to leave. In 1935, while Rutherford was embroiled in his campaign to persuade the Soviets to release Kapitza, Chadwick was offered the Chair of Physics at Liverpool University and accepted. Rutherford backed him, telling Liverpool they would be fortunate if they could attract Chadwick. He also backed Chadwick for the Nobel Prize for Physics, which he was awarded that year for his discovery of the neutron. It may have been nervousness that made Chadwick drop the cheque handed to him by King Gustav at the presentation ceremony in Stockholm. He remained a quiet, shy man who did not enjoy public occasions, which often prompted a flare-up of his digestive problems. After hearing Frédéric Joliot-Curie deliver his Nobel lecture in Stockholm at the same presentation, Chadwick wrote, perhaps a little enviously, that 'He was a great actor. He liked that kind of thing, and he did it very well.'

*   *   *

In 1937, while Chadwick was struggling to modernize Liverpool University's primitive physics department, experimental atomic physics unexpectedly lost its lynchpin. Unlike Marie Curie, Rutherford's health had not apparently suffered from his cavalier approach to radioactive materials. On a lecture tour in the United States he once happily discarded a paper he had used to funnel uranium salts into a tube. The paper was retrieved by his hosts, who used it as a radioactive source for forty years. However, for some years Rutherford had been suffering from a small umbilical hernia. On 14 October he began to suffer from sharp stomach pains and vomiting. His doctor operated for a routine strangulated hernia, but Rutherford suddenly worsened. On 19 October, Lady Rutherford wrote in despair to Chadwick that her husband was 'only hanging by a thread'. He died that evening aged just sixty-six.

The scientific community was shocked by the unexpected death of a larger-than-life character so full of creative energy, who had roamed his laboratory, dribbling ash from his pipe and encouraging his researchers with raucous cries of 'I feel it in my water!' During his career he had trained a dozen Nobel Prize winners. In the words of one researcher, his death left them feeling 'stupefied rather than miserable'. It simply 'did not seem in the nature of things'. Rutherford was cremated and his ashes buried near the tomb of Sir Isaac Newton in Westminster Abbey. From his exile in the United States, Einstein mourned the passing of 'one of the greatest experimental scientists of all time'. A tearful Niels Bohr, recently returned from a lecture tour around Japan which had included an audience with Emperor Hirohito, remarked that, like Galileo, Rutherford left science 'in quite a different state from that in which he found it'.

Rutherford remained true to his philosophy that 'it was not that the [nuclear] experimenters were searching for a new source of power or the production of rare or costly elements. The real reason lay deeper and was bound up with the urge and fascination of a search into one of the deepest secrets of nature.' Yet even he had come privately to suspect that scientists might not have the luxury of the disinterested pursuit of knowledge for much longer. Despite his public comment about 'moonshine', he had in the early 1930s alerted Sir Maurice Hankey, Secretary of the Committee of Imperial Defence, that the Cavendish Laboratory's nuclear work might one day be crucial to the nation's defence. He had advised the government to 'keep an eye on the matter'.

At the Institute of Physics in Rome, Enrico Fermi had been following up the Joliot-Curies' discovery of artificial radioactivity when he had an inspired thought. What, he wondered, would happen if he used neutrons instead of alpha particles to bombard elements? He reasoned that neutrons should be even more effective in producing artificial radioactivity since, having no charge, they would be more likely to penetrate the nucleus. He set to work with his small team, which included Emilio Segrè. Segrè recalled that, like 'a steamroller that moved slowly but knew no obstacles', Fermi systematically went through the Periodic Table, irradiating each element with neutrons. The first eight produced nothing, but at the ninth, fluorine, the Geiger counter clicked into life, registering artificially produced radioactivity.

As Fermi progressed through the table, some of the

radioactive isotopes he created were so short-lived he had to race down the corridor with them to the Geiger counters before the radioactive emissions ceased as they decayed to stability. In May 1934, he reached the final element, number ninety-two, uranium. He found that bombarding this heaviest of metals appeared to produce one or more new radioactive elements beyond uranium in the Periodic Table – the so-called 'transuranics'. Fermi published his results in a series of reports. In Copenhagen, scientists crowded around Otto Frisch, one of the few able to translate from Italian. The Italian press gleefully hailed Fermi's discoveries as proof that a fascist Italy under Mussolini had resumed her 'ancient role of teacher and vanguard in all fields'.

The chemical complexities of the products formed in uranium by neutron bombardment were, for the present, too great for Fermi to interpret. However, in the process of irradiating elements with neutrons, Fermi made what he regarded as the most important of all his achievements: he discovered that the more slowly neutrons travelled, the more likely they were to penetrate the nucleus of the target. Like many of the great discoveries, it had come about through intuition. Fermi had decided on impulse to see what happened if he filtered the neutrons he was firing at his target through a barrier of paraffin. To his surprise this increased the level of artificial radioactivity produced by a hundredfold, so that 'the counter clicked madly'.

Suspecting that the large amount of hydrogen in paraffin might be a factor, he experimented with another substance also containing large amounts of hydrogen – water. Fermi's assistants brought it in buckets from the goldfish fountain in the garden behind the laboratory, and Fermi channelled neutrons through it. The effect was the same as with the paraffin: the level of artificial radio-activity was enormously enhanced. Fermi deduced that the cause must be the protons in the hydrogen filter. They had a similar mass to the neutrons and, colliding with them, made the neutrons bounce elastically back and forth, absorbing some of their momentum. By the time the neutrons moved on to the target, their speed, ordinarily tens of thousands of kilometres a second, had been sufficiently slowed, or 'moderated', for them to slide more easily into the target nuclei. Fermi and his collaborators quickly filed a patent on the slow-neutron process. Unknown to them, the process would prove critical to the development of the atom bomb.

Across Europe, laboratories began replicating Fermi's bombardment of elements with slow neutrons. Lise Meitner found his idea of transuranics 'so fascinating' that she determined to pursue it. Needing the help of 'an outstanding chemist', she asked Otto Hahn to resume the direct collaboration they had abandoned some years earlier. Together they bombarded uranium with neutrons so that they could compare their findings with Fermi's, and confirmed to their satisfaction that he had indeed created new elements beyond uranium.

They were wrong. Like most scientists of the day, they erroneously believed that all nuclear reactions, even when produced by the neutron, were small. Thus, when a new element was formed, it differed from the original by only a few protons or neutrons. It did not occur to them that they were observing the results of something much more dramatic. Fermi had not created 'transuranics', he had split the nucleus of uranium, thereby creating lighter elements lower down in the Periodic Table.

Only one scientist scented what the new discoveries

might mean. In September 1934 German chemist Ida Noddack, co-discoverer of rhenium, the last naturally occurring element to be identified, published an article challenging Fermi's claim to have created transuranics. She suggested it was 'conceivable that when heavy nuclei are bombarded by neutrons, the nuclei in question might break up into a number of pieces, which would no doubt be isotopes of known elements but not neighbours of the irradiated elements'. In other words, it was perfectly possible for the nucleus of a heavy atom such as uranium to shatter, releasing far more energy than most scientists believed and transmuting into much lighter elements. She was on the brink of revealing nuclear fission.

However, Noddack did not pursue the thought, and neither did anyone else pay any attention. Emilio Segrè wrote in later years of his enduring amazement that no-one had taken Noddack's article seriously: 'It said that fission was observed. Fermi and I read it and we still didn't discover fission. The whole story of our failure is a mystery to me.' Ida Noddack was ignored partly because she was a woman. The respect afforded to Marie Curie and Lise Meitner was the exception. Even they had often been marginalized. Long before the Nazis came to power, editors had refused to publish Meitner's papers on the grounds that she was female. In group photographs of workers at her institute she was often placed at the back or to one side, or even omitted altogether.

Of perhaps greater significance, though, was Noddack's lack of credibility. In 1925, Ida Tacke, as she then was, and her future husband, chemist Walther Noddack, had claimed to have discovered an element, 'masurium', but later could not substantiate it. Emilio Segrè himself considered them worse than incompetent, believing they had been 'plain dishonest'. Lise Meitner had been one of their greatest critics, and felt no inclination a decade later to pay attention to Ida Noddack's work. Otto Hahn was equally dismissive. When Noddack asked him at least to refer in his lectures and papers to her criticisms of Fermi's work he replied that he did not wish to make her look ridiculous. Her 'assumption of the bursting of the uranium nucleus into larger fragments', he said with crushing emphasis, 'was really absurd'.

CHAPTER SEVEN

'WONDERFUL FINDINGS'

WHILE SCIENTISTS BICKERED AND DEBATED, THE INTERNATIONAL situation was worsening. In Spain in 1936, civil war had broken out between the Republican government and General Franco's rebel junta. The Soviet Union supported the Republicans, and fascist Germany and Italy backed the junta, sending arms and men and, in particular, 'volunteers' from their air forces. On Monday, 26 April 1937, German and Italian bombers, including the new German Heinkel 111s and Dornier 17s, attacked the historic Basque capital of Guernica for three hours during the weekly market. Guernica had no air defences so the planes flew low, bombing and machine-gunning with impunity. One survivor remembered 'a sapphire blue light' as incendiaries exploded. Another, a child, recalled, 'You could see the heads of the flyers, see they were German planes . . . the next day the town was still burning in some places and there were corpses in the street.' According to the Basque government, 1,645 people died out of about ten thousand, including three thousand refugees, in the town. The presence of foreign reporters near the town gave the bombing immediate, appalled prominence. Picasso's painting gave it immortal infamy.

In 1936, the Japanese government signed the Anti-Comintern Pact with Nazi Germany, which Italy joined a year later. In so doing Japan allied herself firmly with the rising European dictatorships and against the Western democracies, aiming to create a new order in the East as the dictators did in Europe. She was not long in taking further practical steps to achieve this goal. In July 1937, the Japanese turned their encroachment into China, resumed six years earlier in Manchuria, into a full-scale war of conquest.

Many inhabitants of Hiroshima considered this war 'a Holy Crusade', and Hiroshima's port was once more filled with warships and transports bound for China. Local residents lined the streets to give a rousing send-off to the departing troops of the 5th Division, based in the city's castle. Later they bowed their heads in sombre respect as returning soldiers paraded down silent streets with white boxes around their necks containing the ashes of fallen comrades. In December 1937, the people of Hiroshima celebrated the capture by Japanese troops of Nanking, the capital of Nationalist China, with a massive lantern parade. They then gathered outside the city hall to listen to a military band. Pictures showed numerous women and children among the crowds.

Strict government censorship kept from them that the behaviour of the Japanese troops after the fall of Nanking had been barbarous. Ninety thousand Chinese soldiers, who had surrendered under promise of fair treatment, were killed, many of them bound and used for bayonet practice. A Japanese soldier wrote that among the civilians 'women suffered most . . . we sent out coal trucks to the city streets to seize a lot of women. And then each of them was allocated to 15 to 20 soldiers for sexual intercourse and abuse. After raping we would also kill them.' Some two hundred thousand of the city's civilian population of around half a million are estimated to have been killed, in addition to the ninety thousand surrendered prisoners. The commander of the victorious Japanese forces announced, 'The dawn of the renaissance of the East is appearing.'

Both Britain and the United States protested at the rape of Nanking but took no firm action, nor did they do so after the Japanese bombed Shanghai in September 1937. However, the United States government condemned such bombing in the following words: 'The American Government holds the view that any general bombing of an extensive area where there resides a large population engaged in peaceable pursuits is unwarranted and contrary to principles of law and humanity.'

That same summer, Emperor Hirohito authorized 'special chemical warfare units' to be sent to the Asian mainland. The Japanese were to use poison gas against the Chinese on many occasions. They also used bacteriological weapons, releasing rats infected with the plague and other toxins. When rats were released in the wrong place, 1,600 Japanese troops became infected and died.

At the Curie Laboratory, Irène Joliot-Curie was driving herself and her team unsparingly. After her husband Frédéric was appointed professor at the Collège de France in Paris, where he built a laboratory dedicated to nuclear physics and began work on a thirty-two-inch cyclotron, she worked closely with a Yugoslav colleague, Pavel Savitch. In late 1937 they announced that by bombarding uranium with neutrons they had found a substance that remained radioactive for more than three hours. Over the coming months they published a series of explanations of what this highly active material might be. First they suggested thorium, then actinium, then a transuranic with the chemical properties of lanthanum. Otto Hahn and Lise Meitner, deep in their own investigation of trans-uranics and sceptical of both the Paris team's techniques and their findings, dubbed it 'curiosium'. Meitner thought that Irène Joliot-Curie was 'still relying on the chemical knowledge she received from her famous mother and that knowledge is just a bit out of date today'. Hahn remarked less delicately of some of Irène Joliot-Curie's results, 'This damned woman. Now I will have to waste six months proving that she was wrong.'

However, the long partnership between Lise Meitner and Otto Hahn was about to end. On 12 March 1938, welcomed by rapturous crowds, German troops marched into Austria and annexed it. Austrian citizenship ceased to exist as the country's 'Aryan' population, hailed by Hitler as German racial comrades, became citizens of the Third Reich. Austria's Jews became subject to the Reich's racial laws, enforced by the Austrians with such speed and brutality that even the Germans were startled. Stefan Meyer, Rutherford's 'dear friend', resigned as director of the Radium Institute in Vienna before he could be dismissed.

The day after Austria's annexation, a fervent Nazi member of the Kaiser Wilhelm Institute for Chemistry, Kurt Hess, denounced Vienna-born Lise Meitner, stating, 'The Jewess endangers the institute.' That she had become a Protestant in 1908 was no protection. A friend tipped off Hahn, who hurriedly consulted Heinrich Horlein, treasurer of an organization which sponsored the institute. Horlein's view was unambiguous: Lise Meitner must leave. On 20 March, Hahn told Meitner the news and she wrote in her diary, 'Hahn doesn't want me to come to the institute any more.' She was 'very miserable', feeling that 'he has, basically, thrown me out'. Two days later, Hahn and his wife Edith celebrated their silver wedding anniversary. A depressed Meitner was among the guests at what must have been a subdued occasion with Hahn guiltily aware that 'I too had left her in the lurch'. As he later admitted, 'I lost my nerve.'

Lise Meitner turned for help to Paul Rosbaud, a fellow Austrian who was scientific adviser to the German publishing house Springer Verlag and who had taken over editorial responsibility for the important scientific journal *Naturwissenschaften* after its Jewish editor, Arnold Berliner, was sacked. Rosbaud had a Jewish wife whom he took to England with their daughter for safety in 1938. He loathed the Nazis and would later skilfully exploit his contacts across the universities, the military and industry, as well as within the Nazi Party, as an agent for the British. He would also help many Jews and smuggle food into concentration camps. He took the bewildered and miserable Meitner to visit her lawyer and also to call on several senior members of the hierarchy that governed the institute, who, against all the odds, asked her to remain. Horlein, whose first reaction, like Hahn's, had resulted from panic, had meanwhile changed his mind and was no longer demanding her departure.

Lise Meitner felt paralysed with indecision. No-one had dismissed her but, she wondered, was it safe or sensible to remain? Friends outside Germany perceived her danger with greater clarity. Several wrote, ostensibly inviting her

to come and give lectures or seminars but with the underlying purpose of providing her with an official reason to
be allowed out of Germany. Among them was Niels Bohr,
who asked her to come to Copenhagen, stating that the
Danish Physical Society and Chemistry Association would
pay her expenses and adding that 'you would give my wife
and me special pleasure if you would live with us during
your stay'. The letter was carefully crafted to convey to
the Nazi authorities that Meitner was a scientist of international standing. Experience of helping others get out of
Germany had taught Bohr that this tactic sometimes
helped. He specified no date but urged her to come
quickly.

Still she clung on. The thought of leaving Berlin and her
work and all that was familiar was painful to the fifty-
nine-year-old spinster. She had until now been shielded
from what was happening to Germany's Jews. Cocooned
in the institute, surrounded by friends and colleagues in
the scientific community and until recently an Austrian
citizen, she had not been exposed to the full brunt of the
Reich's anti-Jewish policies. The dismissal of Jewish
academics in 1933 had been only the start of a much
broader campaign to make the lives of Jews intolerable. In
1935 the Nuremberg laws had stripped them of their
citizenship. Jews were progressively denied the right to
make a living, frozen out of contact with 'Aryans', no
longer even allowed to enter public parks. In the first
years of their regime the Nazis had encouraged Jewish
people to emigrate and allowed them to take money and
possessions with them. That was changing. Soon, refugees
lucky enough to get foreign visas would be allowed to
take almost nothing.

Towards the end of April 1938, Meitner learned that

the Ministry of Education was considering her position in the institute. Nervously she tried to find out what was going on while continuing her normal schedule of work. On 23 April she attended Max Planck's eightieth birthday celebrations and presented him with a photograph album that was also, perhaps, a farewell gift. Her diary for those weeks shows that, although trying to live an ordinary life, she was coming to accept that her time in Germany was over. She wrote to James Franck, about to leave the Johns Hopkins University for the University of Chicago, seeking his help. He at once lodged an affidavit on her behalf – the first step in the immigration process – and undertook to support her. Yet going to the United States seemed such a huge step. On 9 May she decided instead that she would prefer to join Niels Bohr's team in Copenhagen. She had admired Bohr since their first meeting in 1920, which she remembered as having 'a magic which was only enhanced' on subsequent occasions. Going to Copenhagen also meant she could be with her nephew, Otto Frisch.

She was shocked when, the next day, officials at the Danish Embassy declared her Austrian passport invalid and refused her a visa. They told her they could only act if she had a German passport. She asked Carl Bosch, president of the Kaiser Wilhelm Society which administered her institute, to lobby on her behalf. He wrote to the Minister of Education, asking that 'the well-known scientist, Professor Lise Meitner' be given a German passport and permitted to leave for a neutral country, but days turned into weeks bringing still no news.

In the midst of Lise Meitner's own anxiety, Otto Hahn's wife Edith – her friend since the earliest days of the marriage – had a complete nervous breakdown, brought on by tension. Then, on 14 June, came chilling news that technical and academic people would not be permitted to leave. On cue, two days later Bosch received a reply to his letter. His request for a passport for Meitner was refused: 'It is considered undesirable that well-known Jews leave Germany to travel abroad where they appear to be representatives of German science.' It concluded, 'This statement represents in particular the view of the Reichsführer-SS and Chief of the German Police in the Reichsministry of the Interior' – Heinrich Himmler.

The thought that she had come to Himmler's personal attention was terrifying. However, the new rules forbidding Jewish scientists to leave were not yet in force, so Meitner's friends worked frantically. Niels Bohr lobbied scientists in countries with more flexible entry requirements than Denmark's. He had particular hopes of Holland and launched an appeal to Dutch physicists. Dirk Coster in Groningen, who had been helping Jewish refugees since 1933, and Adriaan Fokker in Haarlem responded. Both men began trying to raise funds and to find Meitner a job. By late June, Coster sent word that he could offer her a position for a year. Almost simultaneously, Meitner was offered a post at the new Nobel Institute for Experimental Physics in Stockholm, where she would be working under Manne Siegbahn, a scientist she had known for two decades. She decided to go to Sweden if she could. Experimental physics was in its infancy there and would offer her more scope. But a few days later a letter from Bohr brought disturbing news. The Swedish offer was not yet firm after all. In particular, formalities allowing her to enter Sweden had not been completed.

At this worrying moment, she was tipped off by Carl Bosch that the new rules forbidding Jewish scientists to leave were about to come into force. She sent anxious messages to Holland to enquire whether Dirk Coster's offer was still open. She also wrote to her former assistant, Carl-Friedrich von Weizsäcker, asking him to contact his father in the Foreign Office about her request for a German passport. At just that time, the senior von Weizsäcker was overseeing new laws forbidding Jews to transfer any funds out of Germany. Lise was informed that the ministry could not help.

Dirk Coster badgered the Dutch authorities to allow Meitner in without passport or visa. Finally, on 11 July he learned that she would indeed be admitted to Holland. He left immediately for Berlin where only a tiny inner circle knew Meitner was about to flee, among them Otto Hahn, Max von Laue, Paul Rosbaud and Peter Debye, a Dutch physicist who was director of the Kaiser Wilhelm Institute for Physics and had acted as a secret conduit for messages between Meitner and colleagues in Holland. Hahn helped Lise to pack and she spent her last night in Berlin at his house. As he recalled, 'We agreed on a code telegram in which we would be let known whether the journey ended in success or failure. The danger consisted in the SS's repeated passport-control of trains crossing the frontier. People trying to leave Germany were always being arrested on the train and brought back.'

On 13 July, Lise Meitner 'left Germany forever – with 10 marks in my purse' and two small suitcases. She was wearing 'a beautiful diamond ring' which had once belonged to Hahn's mother and which he had given her as they parted because 'I wanted her to be provided for in an emergency'. Paul Rosbaud drove her to the station. She was so frightened that at one stage she begged him to turn back, but she managed to pull herself together. Dirk

Coster was waiting on the train. Greeting each other as if they had met by chance, they sat down together. The journey passed quietly, but as the train approached the Dutch border Coster sensibly suggested that she give him the diamond ring in case it drew attention to her, and he tucked it discreetly into his waistcoat pocket. They crossed the border unhindered and by early evening were in Groningen. Meitner felt in a state of shock, 'uprooted from work, colleagues, income, and language, suspended between a past that was gone and a future that held nothing at all'.

Coster despatched the prearranged telegram reporting that the 'baby' had been safely delivered, and Hahn sent 'heartiest congratulations'. When Meitner's fellow Austrian Wolfgang Pauli – the so-called 'atomic housing officer' – heard the news, he told Coster that 'You have made yourself as famous for the abduction of Lise Meitner as for hafnium', the element Coster had co-discovered in 1922 in Copenhagen and called after the city's Latin name, Hafnia. The jocular remarks masked huge relief. Meitner had left with only hours to spare. Kurt Hess, the man who had been so quick to denounce her, had alerted the authorities that she was planning to flee. Max von Laue later wrote to her recording his thankfulness that 'The shot that was to bring you down at the last minute missed you.'

Manne Siegbahn's offer of a post in Stockholm was clearly Lise Meitner's best hope of security. Dirk Coster, whose invitation to Holland had primarily been a device to help her escape from Germany, urged Siegbahn to complete the outstanding formalities quickly. Meitner finally arrived in Sweden in August 1938 after spending some time with the ever-hospitable Bohrs. She wrote to Coster, 'One dare not look back, one cannot look forward.' At the same time she worried about friends and family still trapped in Germany and Austria.

Despite her eminence, Meitner discovered she was to be paid less than the starting salary of an assistant. Indeed, she was so poor that she could barely pay for a room, meals and small daily expenses. Yet poverty mattered less than the loss of facilities to carry on with her work. She wrote miserably to Hahn that she had 'no position that would entitle me to anything'. She asked him to imagine how he would feel if he had a room at the institute that 'wasn't your own, without any help, without any rights and with the attitude of Siegbahn who loves only big machines and who is very confident and self-assured – and there I am with my inner shyness and embarrassment'.

Hahn, meanwhile, was doing what he could to have her possessions sent on to her, although his personal position was difficult. His name had appeared in a list of dismissed Jewish academics included in a travelling anti-Semitic exhibition, 'The Eternal Jew', designed to disgust the onlooker with examples of supposed 'Jewish' physical characteristics and alleged 'Jewish' moral depravity. Hahn had had to submit fresh affidavits to convince the authorities of his Aryan roots. The more he tried to do on Lise Meitner's behalf, the more he risked attracting further unwelcome attention and suspicion. Yet, stressful though it was, he persisted, arguing with the Education Ministry which at first insisted that everything Meitner owned must stay in Germany. Her clothes finally arrived in October, although she had to pay heavy customs dues. And when her remaining possessions finally reached Stockholm, she found her furniture in splinters, her china smashed and her books ripped apart – a final act of petty malice by the Nazi authorities.

In November 1938, she was at least able to see Otto Hahn again when both were invited to Niels Bohr's institute in Copenhagen. She was waiting on the platform for Hahn as his train drew in, but their reunion was sad rather than joyful for he brought disturbing news. Meitner's brother-in-law, Otto Frisch's father, who was still living in Vienna with his wife, Lise's sister, had just been arrested. He had been rounded up with thousands of Jewish men in the aftermath of the vicious government-inspired attack on Jews and Jewish property on the night of 9 November, remembered ever after as Kristallnacht – 'Crystal Night' – after all the splintered glass left lying in the streets.

The distressed and desperately worried Meitner found some consolation in hearing about Hahn's recent work. He updated her on his ongoing sparring with Irène Joliot-Curie over her mysterious substance – 'curiosium' – and on his continuing efforts to prove her wrong. Hahn's assistant, chemist Fritz Strassmann, had convinced him not to be so dismissive of the French team's work and to replicate some of their experiments. In so doing he and Strassmann had created what they firmly believed to be isotopes of radium. However, when they had attempted to use barium as a 'carrier' to help extract the radium – the use of carrier chemicals to separate substances produced by neutron bombardment had become a standard technique – they had found to their surprise that they could not separate the barium from the radium. Meitner listened carefully and then advised Hahn exactly what experiments to conduct to cross-check his findings.

Hahn returned to Berlin keeping his meeting with Lise

Meitner a closely guarded secret, but Strassmann understood precisely who was setting the agenda. He had a very high regard for the exiled physicist whom he had always seen as the intellectual leader of their team. He also owed her a great deal. He had arrived at the institute in 1929, extremely poor but grateful just to be there, and prepared to work for almost nothing. In 1933, though reportedly so malnourished that he sometimes fainted from weakness, he had turned down a well-paid job in industry because it would have required him to join the Nazi Party, which he abhorred. Meitner had persuaded Hahn to find fifty marks a month out of a contingency fund to keep Strassmann going, and eventually to appoint him as an assistant. Looking back on the events of late 1938, Strassmann wrote, 'she urgently requested that these experiments be scrutinised very carefully and intensively one more time. Fortunately L. Meitner's opinion and judgment carried so much weight with us in Berlin that the necessary control experiments were immediately undertaken.'

Lise Meitner, meanwhile, returned to Stockholm. Having learned that her brother-in-law had been deported to the Dachau concentration camp and that his only chance of release was to emigrate, she tried desperately to get visas for the Frisches to come to Sweden. She felt like 'a mechanical doll', going smilingly through the motions 'but with no real life inside'. She wrote to Hahn of her gratitude that work 'forces me to collect my thoughts, which is not always easy'. In particular, she was keen to know what results Hahn and Strassmann were obtaining.

Over the next few weeks they worked relentlessly, trying to discover more about the material they had created. Yet whatever they did, as Otto Hahn recalled, it behaved like barium – an element in the middle of the Periodic Table and much lighter than uranium. They could not understand how bombarding uranium could possibly have produced such a result. A few days before Christmas Hahn posted his findings to Meitner, hoping that 'perhaps you can put forward some fantastic explanation' for results that were 'physically absurd', adding, 'you will do a good deed if you can find a way out of this'. He knew he needed her expertise; as he later wrote, 'we poor chemists . . . we are so afraid of these physics people'. His letter reached her on 21 December and she immediately replied that his findings were 'very odd. A process in which slow neutrons are used and the product seems to be barium . . . but we've had so many surprises in nuclear physics that one can't very well just say it's impossible.'

Two days later she left Stockholm to spend Christmas with a friend in the small, windswept seaside resort of Kungelv near Gothenburg. Otto Frisch arrived from Copenhagen to join the 'short, dark, and bossy' aunt he was so fond of. With his father still in Dachau and his mother trapped in increasingly desperate conditions in Vienna, he was deeply anxious and, like Meitner, glad to immerse himself in Hahn's scientific dilemma as a distraction. Frisch's initial reaction was that Hahn must have made a mistake, but Meitner insisted that she 'knew the extraordinary chemical knowledge and ability of Hahn and Strassmann too well to doubt for one second the correctness of their unexpected results'.

They 'sort of kept rolling this thing around', as Frisch later recalled, and went out into the snow to think, Frisch on skis and his aunt walking rapidly on foot. The problem confronting them was that, until now, no-one had thought it possible to chip more than tiny pieces off nuclei. Yet the barium nucleus was roughly half the size of a uranium nucleus, which suggested that Hahn had split the uranium in two. Searching her mind for an explanation, Meitner recalled an idea of Niels Bohr's, that the nucleus was like a drop of liquid with nuclear forces playing the part of surface tension and keeping the nuclear 'drop' spherical. She described how 'In the course of our discussions we evolved the following picture: if, in the highly charged uranium nucleus – in which the surface tension is greatly reduced owing to the mutual repulsion of the protons – the collective motion of the nucleus is rendered violent enough by the captured neutron, the nucleus may become drawn out length-wise, forming a sort of "waist", and finally splitting into two more or less equal-sized, lighter nuclei which, because of their mutual repulsion, then fly apart with great force.'

They sat down in the cold on a tree trunk and started to calculate on scraps of paper. 'The charge of a uranium nucleus', they found, 'was indeed large enough to overcome the effect of surface tension almost completely; so the uranium nucleus might indeed resemble a very wobbly, unstable drop, ready to divide itself at the slightest provocation (such as the impact of a single neutron).' Yet, as Frisch recalled, there was a problem:

After separation, the two drops would be driven apart by their mutual electric repulsion and would acquire high speed and hence a very large energy, about 200 MeV [mega-electron volts] in all; where could that energy come from? Fortunately Meitner remembered the empirical formula for computing the masses of nuclei and worked out that the two nuclei formed by the division of a uranium nucleus together would be lighter than the original uranium nucleus, by about one-fifth the mass of a proton. [This was because] whenever mass disappears, energy is created, according to Einstein's formula $E = mc^2$, and one-fifth of a proton mass was just equivalent to 200 MeV. So here was the source for that energy; it all fitted!

Yet, the moment of realization was not entirely comfortable. Frisch felt as if he had 'caught an elephant by the tail' without meaning to and now did not know what to do with it.

On 28 December, Hahn wrote again. Prompted by Meitner's note of 21 December, he too was wondering whether the uranium might have split. He did not understand the full picture, but recognized that, if true, it meant that the transuranics they had studied for four years did not exist. They were, instead, smaller, lighter nuclei, like barium, which formed when uranium was split. He asked her to look at a note he and Fritz Strassmann proposed to publish in *Naturwissenschaften*. On New Year's Eve, Meitner replied cautiously that '*perhaps* it is energetically possible for such a heavy nucleus to break up'. The next day, she returned to Stockholm where she immediately began reviewing the evidence for the transuranics and realized that they could, indeed, be light nuclei. It was a bitter-sweet discovery. She had wasted years of study on transuranics, but on the other hand she had provided the first theoretical interpretation of fission of uranium, showing how it produced radioactive elements and liberated large amounts of energy. On 3 January 1939 she wrote to Hahn that 'I am now almost *certain* that you really do have a splitting to barium and I find that to be a really beautiful result'. They were, she said, 'wonderful findings'.

*   *   *

The most 'wonderful' aspect of all, as Edward Teller
wrote, was that 'the secret of fission had eluded everybody
for all those years'. As Hahn put it, 'none of us realized
that we had done it'. Fermi had failed to see it. So had
Irène Joliot-Curie; her lanthanum-like substance had, in
fact, been lanthanum, created by nuclear fission. Even the
great Rutherford had been deceived. When an excited
Otto Frisch returned to Copenhagen and broke the news
to Niels Bohr, the Dane 'smote his forehead with his
hand', exclaiming, ' "We were all fools." '

Bohr urged Frisch to write a paper with Lise Meitner as
soon as possible and promised to say nothing until it was
published. By dint of long-distance telephone calls, aunt and
nephew drafted a short note to the editor of the British
journal *Nature* describing the splitting of a nucleus and the
theory underlying it. They also found a name for their new
phenomenon. Frisch asked an American biologist, William
A. Arnold, working in Bohr's institute, what he called the
process by which single cells divide into two. He replied,
'Fission.'

However, before submitting the paper, Frisch wanted to
be absolutely certain their conclusions were right. He
conducted some experiments and became the first to provide experimental proof of the fission of a uranium atom
when hit by a neutron. Having finally achieved the results
he wanted, he went to bed at three a.m. on 13 January
1939 but four hours later 'was knocked out of bed by the
postman who brought a telegram to say that my father
had been released from concentration camp'. His
parents had been granted visas to emigrate to Sweden.
His happiness was complete.

Meanwhile, on 6 January Hahn and Strassmann's

report appeared in *Naturwissenschaften*. It made no mention of Meitner's and Frisch's contribution, and could hardly have done so in the political climate in Germany. Neither was there any acknowledgement of Ida Noddack's earlier work. Piqued, she wrote a short article in *Naturwissenschaften* pointing out that five years earlier she had suggested the splitting of the uranium atom. Paul Rosbaud, as editor, asked Hahn to comment but he refused. Rosbaud therefore added a terse note beneath her article stating, 'Otto Hahn and Fritz Strassmann have informed us that they have neither the time nor the desire to answer the preceding note.'

On 16 January, Frisch at last posted his and Meitner's paper to *Nature*, together with a supplementary one reporting his experimental findings. In order to protect their friends, Meitner and Frisch took care to credit Hahn and Strassmann only for work already in the open literature. However, the articles did not attract the attention the authors deserved. Sadly, they were not finally published until 11 February, by which time the world knew all about fission, not only from Hahn and Strassmann but also from Niels Bohr.

On 7 January 1939, Bohr had sailed for America aboard the liner *Drottningholm* together with Belgian physicist Leon Rosenfeld, to whom he confided, 'I have in my pocket a paper that Frisch has given me which contains a tremendous new discovery, but I don't yet understand it. We must look at it.' The two men spent the voyage in Bohr's stateroom going over again and again the theory of fission until Bohr was convinced he had 'got hold of the solution'. As Rosenfeld observed, 'it turned out to be extremely simple'.

A group of scientists was waiting on the quayside to greet Bohr, including the American John Wheeler, who had worked with him in Copenhagen. Wheeler was staggered when, within moments of stepping onto dry land, Bohr murmured, in the low voice he used when imparting information of the highest significance, that the uranium atom had been split. That night Wheeler took Rosenfeld off to Princeton where he addressed the physics club. Unaware of Bohr's promise to keep the news quiet until Frisch and Meitner were in print, Rosenfeld announced the discovery, causing a sensation. A horrified Bohr tried to protect Frisch's and Meitner's primacy, but it was too late. All he could do was refrain from public comment himself. However, in late January the first copies of Hahn's and Strassmann's paper in *Naturwissenschaften* arrived in the United States and Bohr felt free to reveal the physical discovery and theoretical explanation of nuclear fission.

The occasion was a conference at George Washington University on 26 January. Some scientists did not even wait for Bohr to finish before rushing off to try the experiments for themselves. That evening Bohr was invited to watch the Carnegie Institution's accelerator in action. For the first time he saw the uranium atom splitting before his very eyes, the glowing green pulses on the screen of the oscilloscope leaping every time a uranium nucleus fissured. Leon Rosenfeld, by his side, recalled that 'the state of excitement challenged description'. By the end of January over a dozen laboratories worldwide had produced nuclear fission.

At Berkeley, Robert Oppenheimer's initial reaction to the news of fission was 'that's impossible', but within days he had changed his mind and was speculating that this 'could make bombs'.

* * *

The new knowledge could scarcely have been revealed at a worse time. In October 1938 Nazi Germany had been allowed to annex the Sudeten German districts of Czechoslovakia under the Munich Agreement, which an optimistic Neville Chamberlain assured the British people guaranteed 'peace in our time'. Hitler promised once again that this was the end of his territorial ambitions, but many, especially those who had suffered personally at the hands of his regime, doubted this. In the tense political climate, some scientists worried that nuclear fission was far too sensitive to be the subject of cross-border gossip. The old belief in a brotherhood of scientists openly discussing and publicizing their findings from Cambridge to Columbia to California, from Liverpool to Leipzig to Leningrad, now seemed as naive as it was alarming. Colleagues and comrades would soon be competitors.

One of the first to grasp the danger was Hungarian physicist Leo Szilard, an eccentric, conceited man but, in the eyes of many contemporaries, 'sparkling with intelligence and originality'. Szilard had an uncanny prescience. He had been one of the quickest to grasp the peril facing European Jewry, arriving in England in the early months of 1933. Like his fellow Hungarian Edward Teller, he was the product of a liberal, cultured, middle-class Jewish Budapest family. Szilard had developed an early preoccupation with 'saving the world'. After the end of the First World War and the collapse of the Austro-Hungarian Empire, he was swept up in the fervour of Bela Kun's Soviet republic, driving trucks draped with socialist slogans around Budapest. When Kun fled in the summer of 1919, Szilard found, like Teller, that the world had changed. When he tried to enrol at the university other students blocked his way, calling him a Jew. His protestations that he was a Calvinist (he had converted a few weeks earlier, believing it would be prudent) did him no good. They kicked him down the marble stairs.

A shaken Szilard applied for a visa to study abroad. At first the government refused, on the grounds that he had been a socialist agitator, but he applied again and with help from family friends just managed to get out. In Berlin he enrolled at the Technical Institute to study engineering but soon realized that physics was his true interest. In 1920 he boldly sought out Max Planck and announced that he only wanted to know the facts of physics; he would make up the theories himself. Life was hard. Szilard lived in shabby, rented rooms and his family were too poor to send him food parcels. He survived on the most basic of rations and roamed Berlin's streets staring in shop windows at food he could not afford to buy.

But at least the intellectual life was satisfying. In 1921, Szilard asked Max von Laue to supervise his thesis, which von Laue suggested should be on relativity theory. That same year, Szilard persuaded Einstein to tutor him and some friends, including fellow Hungarians John von Neumann and Eugene Wigner. Szilard's particular talent was for intense lateral thinking – teasing out patterns and then seeking ways of uniting them through a theory. His tools were statistics rather than experimental evidence. He applied this approach to a problem in thermodynamics and took his results to Einstein, who listened politely then said, 'That's impossible. This is something that cannot be done.'

'Well, yes,' responded Szilard, 'but I did it.'

After reaching England, Szilard worked to get Jewish academics out of Germany but then, convinced war was coming, he moved on to the United States. He learned of the discovery of fission at Princeton while visiting Eugene Wigner, who was recovering from jaundice in the university infirmary on what Wigner considered a 'miserably' un-Hungarian diet of 'potatoes, beans and everything boiled in water'. Szilard came to see him every day, and the two friends 'discussed fission problems and this and that'. One morning, Szilard said, 'Wigner, now I think there will be a chain reaction.'

As Szilard recognized, the possibility of creating a nuclear bomb depended on whether fission could be used to trigger a self-sustaining chain reaction. In other words, by using neutrons to bombard uranium atoms, was it possible not only to split the uranium nuclei but, in the process, to release enough further neutrons which, if they in turn hit other uranium nuclei, could trigger a self-sustaining chain reaction liberating colossal amounts of energy?

As early as September 1933 he had conceived the idea in theory, sparked by reading a newspaper report of Lord Rutherford's 'moonshine' speech dismissing the idea that energy could be liberated from the atom. Szilard later described the Damascene moment. The article 'sort of set me pondering as I was walking the streets of London, and I remember that I stopped for a red light at the intersection of Southampton Row. As I was waiting for the light to change and as the light changed to green and I crossed the street, it suddenly occurred to me that if we could find an element which is split by neutrons and which would emit *two* neutrons when it absorbed *one* neutron, such an element, if assembled in sufficiently large mass, could sustain a nuclear chain reaction.' So alarmed was Szilard that such a process could be used to create an

explosive device that in the spring of 1934 he applied for
a patent for the process he envisaged. He assigned this
patent to the British Admiralty for safekeeping but did
nothing further.

Now that, four years later, fission had been shown to be
a reality, Szilard wanted urgently to test his theory of
chain reaction. Although he had no formal university
appointment, he secured special permission to conduct
experiments at Columbia University. He borrowed $2,000
from a friend, rented some radium and, using some of the
university's equipment, carefully set up his experiment. As
he later described, 'all we needed to do was to get a gram
of radium, get a block of beryllium, expose a piece of
uranium to the neutrons which come from beryllium'
and then see whether neutrons were emitted in the
process.

On 3 March 1939, 'everything was ready and all we
had to do was to turn an [electrical] switch, lean back,
and watch the screen of a television tube. If flashes of light
appeared on the screen, that would mean that neutrons
were emitted in the fission process of uranium and this in
turn would mean that the large-scale liberation of atomic
energy was just around the corner. We turned the switch
and we saw the flashes.' The pulses of light proved that
bombarding uranium with neutrons could indeed spark a
chain reaction. The spectacle left Szilard with 'very little
doubt in my mind that the world was headed for
grief'. He was the first to perceive that a race would soon
begin.

That night Szilard rang Edward Teller in Washington,
announced tersely in Hungarian, 'I have found the
neutrons,' and hung up. Teller had been contentedly
playing the piano when the phone rang. As he returned to

the instrument, the thought came that 'the world might change in a radical manner. The prospect of harnessing nuclear energy seemed chillingly real.'

CHAPTER EIGHT

'WE MAY SLEEP FAIRLY COMFORTABLY IN OUR BEDS'

ADDRESSING DIGNITARIES AT THE NOBEL PRIZE CEREMONY IN Stockholm in 1905, Pierre Curie had posed a disturbing question: 'One may imagine that in criminal hands radium might become very dangerous ... we may ask ourselves if humanity has anything to gain by learning the secrets of nature.' He had not doubted the answer, adding reassuringly, 'I am among those who think, with Nobel, that humanity will obtain more good than evil from the new discoveries.' Thirty-four years later, scientists had greater knowledge and faced more difficult judgements.

As Leo Szilard had quickly grasped, the fact that nature's 'secrets' might pose a risk to humanity implied new roles and responsibilities for scientists. Even before he had had a chance to conduct his own experiments at Columbia University confirming the viability of a nuclear chain reaction, Szilard launched a campaign to keep 'nature's secrets' secret. The spectre of an explosive atomic device in Nazi hands haunted him. It would be only too easy, he reasoned, for scientists in Nazi Germany to comb through the technical journals and piece together snippets of information. It was a standard technique of intelligence gathering which in the current international climate might prove disastrous. His answer was to persuade scientists in the free world to adopt a policy of self-censorship.

Szilard correctly identified Enrico Fermi as one of the scientists most likely to solve the mysteries of a chain reaction, and targeted him accordingly. Fermi had recently arrived in the United States with his wife Laura and their two children to take up a professorship at Columbia, one of six American universities eager to appoint him. Laura Fermi was the daughter of a Jewish naval officer and they had decided it was not safe for the family to remain in Mussolini's Italy. Their opportunity to flee had come late in 1938 when Fermi was awarded the Nobel Prize for Physics for his identification of new radioactive elements and his discovery of how nuclear reactions were affected by slow neutrons. He was notified of the award the day the Italian authorities announced that Jews were to be deprived of their rights of citizenship and their passports withdrawn.

By this time, the Italian authorities viewed the Nobel Prize with some disfavour. Their German allies had banned their citizens from accepting it after the 1935 Nobel Peace Prize was awarded to a German author and pacifist imprisoned as an enemy of the state. However, no-one prevented the Fermis from travelling to Stockholm for the Nobel ceremony. Here they collected their prize money and never went home. Stepping onto American soil on 2 January 1939, Fermi declared that 'we have founded the American branch of the Fermi family'. Within days his wife was exploring what she called 'the marvels of pudding powders' and of frozen food, just then appearing on the market. The process of Americanization was, she quickly realized, less tangible. It was 'more than learning language and customs and setting one's self to do whatever Americans can do'. It would take time to understand 'New England pride' and 'the long suffering of the South', and even longer, perhaps, to think of Shakespeare before Dante.

When Szilard first explained his concerns to Fermi in February 1939 the Italian was sceptical. He considered the likelihood of a chain reaction to be less than Szilard and regarded his censorship plans as alarmist and against the spirit of science. Fermi had seen intellectual freedom stamped out in fascist Italy and was reluctant to participate in any scheme to suppress knowledge. In the circumstances, as Szilard recalled, 'Fermi thought that the conservative thing was to play down the possibility that this [a chain reaction] may happen, and I thought the conservative thing was to assume that it would happen and take the necessary precautions.' Fermi and Szilard 'had high regard for each other' but were 'extremely different in personality, habits of work, outlook on life, and almost everything else' and 'could scarcely work together on the same experiment', recalled Emilio Segrè. One of the problems was that, as another physicist put it, 'Szilard's way of working on an experiment did not appeal to Fermi. Szilard was not willing to do his share of experimental work, either in the preparation or in the conduct of the measurements. He hired an assistant . . .'

With his hallmark persistence, Szilard continued relentlessly to lobby Fermi with the help of an intermediary, American-born physicist Isidor I. Rabi, one of the first Jewish physicists appointed at Columbia and selected thanks to glowing references from Werner Heisenberg. As

Szilard later recounted, the debate resembled a quick-fire comedy routine rather than a serious debate between scientists.

> I went to see Rabi, and I said to him, 'Did you talk to Fermi?'
>
> Rabi said, 'Yes, I did.'
>
> I said: 'What did Fermi say?'
>
> Rabi said, 'Fermi said "Nuts!"'
>
> So I said, 'Why did he say "Nuts!"?'
>
> And Rabi said, 'Well, I don't know, but he is in and we can ask him.'
>
> So we went over to Fermi's office and Rabi said to Fermi, 'Look, Fermi, I told you what Szilard thought and you said "Nuts!", and Szilard wants to know why you said "Nuts!"'
>
> So Fermi said, 'Well, there is the remote possibility that neutrons may be emitted in the fission of uranium and then of course perhaps a chain reaction can be made.'
>
> Rabi said, 'What do you mean by remote possibility?'
>
> And Fermi said: 'Well, ten per cent.'
>
> Rabi said, 'Ten per cent is not a remote possibility if it means that we may die of it. If I have pneumonia and the doctor tells me that there is a remote possibility that I might die, and that it's ten per cent, I get excited about it.'

Fermi, who according to Emilio Segrè abhorred battles, finally gave in. He agreed not to publish any further findings on fission and neutron research and encouraged his colleagues at Columbia to do likewise.

Leo Szilard also singled out Frédéric Joliot-Curie at the Collège de France as one of those likely to stumble on the chain reaction. The potential of nuclear fission, especially an estimate by Lise Meitner and Otto Frisch that a single fissuring uranium nucleus could release enough energy to make a grain of sand jump visibly, had certainly caught the Frenchman's attention. With two ambitious assistants, Russian Lew Kowarski and Austrian Hans von Halban, he was devoting himself to exploring the phenomenon of uranium fission. Kowarski was 'a gruff . . . enormous brute of a man . . . with the memory of an elephant'. The illegitimate son of a Russian Jewish merchant and a Russian Orthodox opera singer, he had fled to France after the Russian Revolution with his father and had struggled to find enough money to complete his studies. The good-looking von Halban was also partly Jewish but from a much more affluent background in Vienna. He had come to France before Hitler's annexation of Austria, and in contrast to Kowarski was an urbane, cultured charmer.

On 2 February 1939, Szilard wrote to Joliot-Curie begging him to publish nothing openly about neutron research: 'Obviously, if more than one neutron were liberated, a sort of chain reaction would be possible. In certain circumstances this might then lead to the construction of bombs which would be extremely dangerous in general and particularly in the hands of certain governments.' He asked the Frenchman to exercise 'sufficient discretion to prevent a leakage of these ideas' to the press and told him of proposals for a concerted approach to physicists across the United States, Britain and France seeking a moratorium on publicizing work on fission. He also pointed out that Fermi was holding back from publication results achieved by the Columbia team.

Szilard's letter arrived at an unfortunate time. The

French team was poised to publish a paper reporting the results of an intricate set of experiments proving that uranium fission produced neutrons. They were, as von Halban described, excited by their findings: 'we were thoroughly convinced that the conditions for establishing a divergent chain reaction with neutrons could be realised'. Their reaction to Szilard's proposal was negative. In the words of Bertrand Goldschmidt, the young French chemist who had promised to be Madame Curie's 'slave', 'the Szilard proposal was neither completely understood nor accepted at the Collège de France'. Joliot-Curie maintained that self-censorship conflicted with his support for internationalism and the freedom of science, but, according to Goldschmidt, reluctance to forgo the glory was a key factor in his thinking. Joliot-Curie's views were not shared by Paul Langevin, Marie Curie's reputed erstwhile lover. He believed the new discoveries to be more dangerous than Hitler, telling a refugee from Germany, 'Hitler? It won't be long before he breaks his neck like all other tyrants. I'm much more worried about something else. It is something which, if it gets into the wrong hands, can do the world a good deal more damage than that fool who will sooner or later go to the dogs. It is something which – unlike him – we shall never be able to get rid of: I mean the neutron.'

In March 1939, the month when Hitler seized the remnants of Czechoslovakia not ceded to him at Munich, Joliot-Curie and his team rushed their paper to the British journal *Nature*. Despite further pleas from Szilard and like-minded allies such as his fellow Hungarian Eugene Wigner and Viennese physicist Victor Weisskopf, on 7 April – the day Mussolini invaded Albania – they despatched a second paper to *Nature*. In it they estimated the number of secondary neutrons produced through the fission of a single uranium nucleus by a single neutron to be 3.5, each of which could fission another uranium nucleus, releasing more and more energy and further neutrons. The figure of 3.5 would prove an over-estimate – the true figure was, on average, around 2.5 – but from Szilard's perspective the damage was done. The article appeared in *Nature* on 22 April 1939.

Joliot-Curie's team thus became the first to publish results showing that fission produced enough secondary neutrons to have the potential to start a chain reaction. Their principal preoccupation was the ability to use the energy released by fission to produce nuclear power. Anxious to protect France's position, they took out a series of secret patents on the construction and operation of nuclear reactors to contain and exploit chain reactions for the production of nuclear power. However, just as Szilard had feared, their articles were spotted by German scientists whose interest was in weapons, not nuclear energy. In a letter of 24 April 1939, Hamburg professor Paul Harteck, a chemical explosives consultant to the German Army who had spent a year at the Cavendish Laboratory with Ernest Rutherford, alerted Erich Schumann, head of weapons research in the German Army Weapons Office, to the potential military applications of nuclear fission. He wrote that recent developments in nuclear physics might lead to the production of an explosive far more powerful than any yet known and that any country possessing it would have an 'unsurpassable advantage'.

In Bertrand Goldschmidt's view, Joliot-Curie's team 'started the Germans off'. After the war, when Lew Kowarski was asked why they had published such sensitive information at such a sensitive time, he replied, 'Why not secure priority? Hell, as I always say, it's not vanity – it's bread and butter.' The publication of the Joliot-Curie team's work in *Nature* also blasted Szilard's hopes of a general agreement on self-censorship. He continued to argue fiercely and volubly for restraint, but the Joliot-Curie articles had made his position, for a while at least, untenable.

Fermi had meanwhile begun his own experiments on chain reactions at Columbia. They confirmed everything Szilard had warned of. Looking down towards the skyscrapers of Manhattan from his high office window and shaping his hands into a ball, he reflected to a colleague, 'A little bomb like that and it would all disappear.' By mid-March 1939 Fermi was so concerned that he discussed the need to alert the US government with the head of the Columbia physics department, George Pegram. Pegram wrote to Admiral Stanford C. Hooper, technical director for Naval Operations, warning that 'uranium may be able to liberate its large excess of atomic energy, and this might mean that uranium might be used as an explosive that would liberate a million times as much energy per pound as any known explosive'. He added that, in his own view, 'the probabilities are against this' but that 'the bare possibility should not be disregarded'.

Fermi was invited to Washington to present his findings to a group of senior naval officers. Accounts conflict about his reception. According to one, the presentation began unpromisingly when Fermi overheard himself being announced to Admiral Hooper with the words 'There's a wop outside'. Having heard 'the wop's' careful, measured presentation, the navy gave Columbia a meagre $1,500 for fission research. As Emilio Segrè wryly observed, 'although the sum was puny, it indicated goodwill'.

Niels Bohr shared Fermi's instinctive distaste for secrecy, believing that 'openness is the basic condition necessary for science. It should not be tampered with.' However, he also believed, like Szilard, that war was coming. Laura Fermi, who with her husband had been among the group waiting on the New York quayside to welcome Bohr when he stepped off the *Drottningholm* in January, had noticed how tired and stooped he looked. He had aged since the Fermis had visited him in Copenhagen on their way from Stockholm to the United States. She did not, like John Wheeler, catch his whispered announcement of the discovery of nuclear fission, but she heard him mutter a stream of worried comments: 'Europe . . . war . . . Hitler . . . Denmark . . . danger . . . occupation'. In the weeks after his arrival she recalled that Bohr spoke constantly of 'the doom of Europe in increasingly apocalyptic terms and that his face was that of a man haunted by one idea'.

Almost at once Bohr began working on the consequences of fission at the Institute of Advanced Study at Princeton, helped by John Wheeler. In early February, while puzzling over why the rate of uranium fission he was observing was some hundred times less than he would have expected, he had a burst of inspiration. Perhaps, he reasoned, the two isotopes present in uranium – the dominant U-238 (so named for its 92 protons and 146 neutrons) and the much rarer U-235 (with 92 protons but 143 neutrons) – behaved differently when bombarded with neutrons. If only U-235 – constituting less than 0.7 per cent of natural uranium – was splitting, and not the U-238 of which natural uranium was almost entirely composed, this would explain the low rate of fission. It was hard for bombarding neutrons to find a suitable target.

Pondering why U-235 should be more susceptible to fission than U-238, Bohr deduced that the reason related to the number of neutrons and protons and the effect this had on the binding energy of a nucleus when a neutron was added. Adding a neutron to U-235 resulted in an even number of neutrons and a tightly bound U-236 compound nucleus, whereas adding a neutron to U-238 resulted in an odd number of neutrons and a less tightly bound U-239 compound nucleus. The tighter binding of U-236 meant that its formation released significantly more energy than was the case for the formation of U-239, and this further agitated the neutrons and protons of U-236 to the point where they elongated into the wasp waist required for fission in Bohr's liquid drop model. So he concluded that neutrons travelling at any speed would fission U-235. Conversely, in the case of U-238, the energy release was insufficient for fission by slow neutrons.

As Bohr chalked row after row of formulae on his blackboard, his underlying hope was that if he was right and the isotope U-235 was the key to fission, this would make an atomic bomb unviable. A massive industrial effort would be required to separate out sufficient quantities of the isotope from natural uranium. According to Edward Teller, Bohr told a group including Szilard, Wigner and himself, gathered expectantly in his office at Princeton, that 'you would need to turn the entire country into a factory'. On 15 March, Bohr published his initial conclusions in the US journal *Physical Review*.

Nevertheless, if sufficient U-235 could, after all, be obtained, then an atomic bomb remained a possibility.

Bohr conceded as much to a meeting of the American Physical Society in April 1939. Speculating about the results of bombarding a small amount of uranium with neutrons, he admitted it might produce a chain reaction or an atomic explosion. The press picked up his remarks and presented apocalyptic visions to their readers. The science writer of the *New York Times*, William L. Laurence, portrayed uranium as the 'philosopher's stone' and predicted that a tiny quantity could 'blow a hole in the earth 100 miles in diameter. It would wipe out the entire City of New York, leaving a deep crater half way to Philadelphia and a third of the way to Albany and out to Long Island as far as Patchogue.' The *Washington Post*'s headline was 'Physicists Here Debate Whether Experiments Will Blow Up 2 Miles of the Landscape'.

Scientists in the free world not only faced decisions about whether to publicize their work but about how to respond to old friends and acquaintances still working in totalitarian countries. Some were in no doubt what they should do. In February 1939, American physicist Percy Bridgman announced in the journal *Science* that 'I have decided from now on not to show my apparatus to or discuss my experiments with the citizens of any totalitarian state. A citizen of such a state is no longer a free individual, but may be compelled to engage in any activity whatever to advance the purposes of that state ... Cessation of scientific intercourse with totalitarian states serves the double purpose of making more difficult the issues of scientific information by these states and of giving the individual opportunity to express abhorrence of their practices.'

Yet no such embargos affected Werner Heisenberg that

summer of 1939 when he was invited to lecture at universities across America. Some American scientists speculated openly that his real purpose, as he traversed the country, was to gather intelligence on fission. However, his old friends, many of them émigré Jewish scientists such as Hans Bethe with whom personal bonds of trust and affection were still strong, welcomed him. They also urged him again and again to quit Germany. At Ann Arbor, where Heisenberg stayed with Sam Goudsmit, a Dutchman of Jewish extraction who had emigrated to America in the 1920s, he met Enrico Fermi who was attending the annual physics summer school at the University of Michigan. Their friendship too went back a long way – both had attended Max Born's lively seminars in Göttingen – but while bonds remained, their discussions revealed how much their lives and opinions had diverged.

Heisenberg's view was that 'Italy's leading physicist' had chosen to 'ride out the coming storm' in America. He did not seem to recognize that choice had had little to do with it, if Fermi was to protect his wife and children. According to Heisenberg, the two men discussed whether Heisenberg should also emigrate. It would certainly have been easy for him to find a post. George Pegram at Columbia University was one of several only too eager to offer the German Nobel laureate a professorship. Fermi queried why Heisenberg did not stay in America and play his part 'in the great advance of science'. 'Why renounce so much happiness?' Fermi asked. Heisenberg's reply was that he had gathered around him a small circle of young people anxious to ensure that 'uncontaminated science' could make a comeback in post-war Germany and that 'if I abandoned them now, I would feel like a traitor'.

According to a young graduate hired as a bartender at a party attended by Fermi and Heisenberg, and who overheard them, Fermi tried to convince Heisenberg that his belief that he 'could influence even guide the [Nazi] government in more rational channels' was a naive illusion. He argued that the fascists had 'no principles; they will kill anybody who might be a threat . . . You only have the influence they grant you.' Heisenberg's reply was that 'Germany needed him'. According to his own account, Heisenberg also argued that 'Every one of us is born into a certain environment, has a native language and specific thought patterns, and if he has not cut himself off from his environment very early in life, he will feel most at home and do his best work in that environment.' He added, with characteristic insouciance, that 'people must learn to prevent catastrophes, not to run away from them'.

Heisenberg also recorded how Fermi pressed him on the issue of fission – a subject which, according to Heisenberg, he himself was never the first to raise during his visit. Fermi warned that 'there is now a real chance that atom bombs may be built. Once war is declared, both sides will perhaps do their utmost to hasten this development, and atomic scientists will be expected by their respective governments to devote all their energies to building the new weapons.' Heisenberg, however, recalled assuring him that 'the war will be over long before the first atom bomb is built'.

Announcing cheerfully to his friends in America that he had to get back for machine-gun practice with the Mountain Rifle Brigade to which he had been assigned for annual military service, Heisenberg sailed home in a nearly empty ship, the *Europa*, arriving back in Germany in mid-August 1939. He spent the next few weeks helping his wife furnish and prepare the country house he had bought in Urfeld, high in the Bavarian mountains, so that, as he later wrote, she and the children could take refuge from the coming disaster.

In Japan, with the long war against China prospering, ordinary citizens had little sense of impending catastrophe. Down by the Aioi Bridge in the central Hiroshima district of Salugakucho – the name means 'music' or 'Noh theatre' and was bestowed because many Noh artistes lived there – life was, as one inhabitant recalled, 'lively and busy'. In 1939, the Aioi Bridge itself had been rebuilt into a 'T' shape ingeniously connecting three tongues of land. The area was 'a calm, cosy place with many traditional homes and stores'.

In the Kimatsu family's rice shop, rice polishing machines with their funnelled hoppers stood in a neat line. Behind the storeroom, where bags of rice and sacks of charcoal and firewood were piled, lay the garden and living area. There in the summer months, while crickets clicked noisily, the Kimatsus chilled watermelons and beer to ward off the heat. In the autumn they burned olive wood, scenting their lattice-doored house with its pungent aroma.

The wooden buildings lining the busy nearby streets housed sports shops, photo studios, bike shops and stores whose wares ranged from cosmetics, dolls and ice-cream to soy sauce and white *miso*, a much-loved Japanese flavouring for meats and stews, made of malt and boiled soy beans; customers brought small containers to be filled from the barrels in which it was stored. Salugakucho also had many woodworking shops. Their owners welcomed buyers into interiors designed and furnished to show off their exquisite craftsmanship. People also visited the district to buy specialities; the Kadohatsu caterers, for example, were famous in Hiroshima for providing the best wedding feasts. Some came to learn skills such as flower arranging and the time-honoured ritual of the tea ceremony, both of which were taught by the owner of the Iroha Hotel. On the upper floor of the Ise General Store was a dress-making business where seamstresses stitched indigo-dyed kimonos for daily wear, or sometimes the elaborate black bridal kimonos, heavily and colourfully embroidered with plants and blossoms such as *sho* (pine), *chiku* (bamboo), *bai* (plum) and other symbols of good fortune and longevity.

In the last weeks before war, scientists in the United States, Britain, France and Germany were all actively exploring fission. The publication in June 1939 of a detailed account of a uranium chain reaction by German theoretical physicist Siegfried Flügge showed how far German thinking had developed since Hahn and Strassmann had published their tentative conclusions on uranium fission just six months earlier. The article concluded that 'our present knowledge makes it seem possible to build a "uranium device"'. Flügge, who was a colleague of Hahn's at the Kaiser Wilhelm Institute and, like him, no Nazi, had in fact published the article because he believed such sensitive information should be shared with the wider world. In this spirit he had also given an interview to the big-circulation newspaper *Deutsche Allgemeine Zeitung*. However, the net effect was to worry the wider scientific community that the Nazis were advancing down the path towards nuclear weapons.

Leo Szilard recognized the importance of keeping more

than ideas out of the grasp of Nazi Germany. Up to this point no-one had been much concerned about securing the world's stocks of uranium. However, news came that Germany had forbidden any export of uranium ore from the Joachimsthal mines in Czechoslovakia, which the Nazis now controlled and which decades earlier had supplied Marie Curie with her sacks of pitchblende mixed with pine needles. Europe's only other large stockpile of uranium belonged to the Union Minière du Haut-Katanga, a Belgian company which owned rich uranium mines in the Belgian Congo. Szilard was worried that the Germans might try to get hold of the Belgian ore. Recalling that Albert Einstein had long been a friend of Queen Elizabeth of Belgium he decided to ask whether he would be prepared to contact her to seek her help in warning the Belgian government not to export uranium ore to Germany.

On a hot July day, Szilard and his friend Eugene Wigner set out past the New York World Fair, with its extra-ordinary collection of buildings celebrating the theme of constructing the world of tomorrow, to visit Einstein, who was holidaying on Long Island. The sixty-year-old, clad in rolled-up trousers and singlet, led his visitors to his study where, talking in German, they explained their mission as they sipped iced tea. According to Szilard, 'This was the first Einstein had heard about the possibility of a chain reaction. He was very quick to see the implications and perfectly willing to do anything that needed to be done. He was willing to assume responsibility for sounding the alarm even though it was quite possible that the alarm might prove to be a false alarm. The one thing that most scientists are really afraid of is to make a fool of them-selves. Einstein was free from such a fear and this above all is what made his position unique on this occasion.'

Einstein agreed to write a letter but suggested sending it to a member of the Belgian cabinet rather than directly to his friend Queen Elizabeth. Wigner argued that before sending anything to a foreign government it was surely their duty to inform the State Department in Washington of their intentions. The others agreed, and decided that if the State Department was to be involved their appropriate course of action was to prepare a letter to the Belgian ambassador to be shown in draft to the State Department. Einstein dictated a note in German warning of the possibility that explosive bombs of unimaginable power could be made from uranium and urging the necessity of keeping stocks of uranium out of enemy hands – by implication, German ones.

The problem was how to ensure that the US government paid attention to the views of three refugee scientists who, despite Einstein's fame, had little entrée into the inner circles of government. Szilard had been wondering about trying to enlist the help of influential aviator Charles Lindbergh. However, a friend of Szilard's, Viennese refugee economist Gustav Stolper, suggested they approach Wall Street financier Alexander Sachs, a personal friend of the President and one of an intimate group of advisers possessing, as Roosevelt himself had stipulated, 'great ability, physical vitality and a real passion for anonymity'. The forty-six-year-old Sachs had been following the development of nuclear power for a while and, to Szilard's delight, agreed personally to deliver a letter from Einstein to Roosevelt. No doubt with Sachs' advice, Szilard drafted a letter broader in scope than that originally dictated by Einstein. Addressed directly to the President rather than the State Department, it not only dealt with the need to safeguard the stockpile of uranium from the Belgian Congo, but sought support for the funding and acceleration of nuclear research.

Szilard posted the draft to Einstein, and in early August 1939 travelled once more to Long Island to discuss it. Since Szilard had never bothered to learn to drive and Wigner was away, his replacement chauffeur was Edward Teller in his 1935 Plymouth. This time Einstein greeted them in an old dressing-gown and slippers. Teller served as their scribe, writing down a third draft letter at Einstein's and Szilard's dictation. Szilard took this new draft back to New York and used it as the basis for two further texts, one comparatively short and one rather longer, both addressed to the President. He left it to Einstein to decide which he preferred, and he chose the more detailed one.

In later years, when the question arose of whether Einstein, a declared pacifist, had fired the gun which began the American race for the bomb, he would insist that 'I served as a pillar box', nothing more. Sachs, too, would remember Einstein's role as facilitator rather than prime mover, recalling, 'We really only needed Einstein in order to provide Szilard with a halo, as he was then almost unknown in the United States. His entire role was really limited to that.'

Sachs finally presented the document to President Roosevelt on 11 October 1939 after repeated calls to the White House to secure an appointment. To ensure the President's full attention, Sachs read selected highlights from the letter signed by Einstein. It warned Roosevelt that 'the element uranium may be turned into a new and important source of energy in the immediate future'. This new phenomenon could 'lead to the

Albert Einstein
Old Grove Rd.
Nassau Point
Peconic, Long Island

August 2nd, 1939

F.D. Roosevelt,
President of the United States,
White House
Washington, D.C.

Sir:

Some recent work by E.Fermi and L. Szilard, which has been communicated to me in manuscript, leads me to expect that the element uranium may be turned into a new and important source of energy in the immediate future. Certain aspects of the situation which has arisen seem
to call for watchfulness and, if necessary, quick action on the part
of the Administration. I believe therefore that it is my duty to bring
to your attention the following facts and recommendations:

In the course of the last four months it has been made probable -
through the work of Joliot in France as well as Fermi and Szilard in
America - that it may become possible to set up a nuclear chain reaction
in a large mass of uranium,by which vast amounts of power and large quantities of new radium-like elements would be generated. Now it appears
almost certain that this could be achieved in the immediate future.

This new phenomenon would also lead to the construction of bombs,
and it is conceivable - though much less certain - that extremely powerful bombs of a new type may thus be constructed. A single bomb of this
type, carried by boat and exploded in a port, might very well destroy
the whole port together with some of the surrounding territory. However,
such bombs might very well prove to be too heavy for transportation by
air.

-2-

The United States has only very poor ores of uranium in moderate
quantities. There is some good ore in Canada and the former Czechoslovakia,
while the most important source of uranium is Belgian Congo.

In view of this situation you may think it desirable to have some
permanent contact maintained between the Administration and the group
of physicists working on chain reactions in America. One possible way
of achieving this might be for you to entrust with this task a person
who has your confidence and who could perhaps serve in an inofficial
capacity. His task might comprise the following:

a) to approach Government Departments, keep them informed of the
further development, and put forward recommendations for Government action,
giving particular attention to the problem of securing a supply of uranium ore for the United States;

b) to speed up the experimental work,which is at present being carried on within the limits of the budgets of University laboratories, by
providing funds, if such funds be required, through his contacts with
private persons who are willing to make contributions for this cause,
and perhaps also by obtaining the co-operation of industrial laboratories
which have the necessary equipment.

I understand that Germany has actually stopped the sale of uranium
from the Czechoslovakian mines which she has taken over. That she should
have taken such early action might perhaps be understood on the ground
that the son of the German Under-Secretary of State, von Weizsäcker, is
attached to the Kaiser-Wilhelm-Institut in Berlin where some of the
American work on uranium is now being repeated.

Yours very truly,

A. Einstein

(Albert Einstein)

construction of bombs, and it is conceivable – though much less certain – that extremely powerful bombs of a new type may thus be constructed. A single bomb of this type, carried by boat and exploded in a port, might very well destroy the whole port together with some of the surrounding territory. However, such bombs might very well prove to be too heavy for transportation by air . . .' It urged the need for 'watchfulness and, if necessary, quick action on the part of the Administration'. Sachs also read out a detailed note of his own composition and further extracts from a pile of technical papers he had brought with him.

Roosevelt was not, as Sachs had hoped, electrified. He said politely that the subject matter was interesting but that any government intervention would be premature. However, he invited the disappointed Sachs to join him for breakfast the next day. 'That night I didn't sleep a wink,' Sachs later wrote. Instead he paced his suite at the Carlton Hotel and several times went out to walk in a small nearby park, trying to marshal his thoughts. As he sat on a park bench, reflecting that everything was 'already beginning to look practically hopeless', suddenly, 'like an inspiration, the right idea came to me. I returned to the hotel, took a shower and shortly afterwards called once more at the White House.'

Sachs found Roosevelt alone at the breakfast table in his wheelchair. The President greeted him with two wry questions: 'What bright idea have you got now? How much time would you like to explain it?' Sachs replied that it would not take long and briefly recounted the story of the young American inventor Robert Fulton. During the Napoleonic Wars, Fulton had offered to build Napoleon a fleet of steamships to help him overcome his arch enemy, the British. The French emperor, believing Fulton was talking nonsense, impatiently dismissed the visionary young man who subsequently pioneered the world's first steamships. Sachs reminded Roosevelt of the nineteenth-century British historian Lord Acton's comment that 'England was saved by the shortsightedness of an adversary. Had Napoleon shown more imagination and humility . . . the history of the nineteenth century would have taken a very different course.'

The cautionary tale had the desired effect. For several minutes Roosevelt said nothing. Then he scribbled a note and handed it to a servant, who returned bearing a bottle of fine old brandy from Napoleon's time and filled two glasses. Roosevelt raised his, toasted Sachs, and quietly remarked, 'Alex, what you are after is to see that the Nazis don't blow us up?'

Sachs replied, 'Precisely.'

The President summoned his attaché, General 'Pa' Watson, and consigned to him Sachs' documents with the instruction, 'Pa, this requires action!'

Leo Szilard hoped that the British, too, would take notice of the nuclear risk. In January 1939 he had reminded the British Admiralty about the chain-reaction patent he had taken out in 1934 and asked them to maintain its secrecy. To his relief, several months later they agreed to do so. For a while there seemed a real possibility that the British would take the potential military applications of nuclear fission seriously. Frédéric Joliot-Curie's experiments demonstrating the release of secondary neutrons through fission had caught the eye of physicist George Thomson, son of Rutherford's famous mentor at the Cavendish Laboratory, J. J. Thomson. Within hours of the

Joliot-Curie team's publication of their results in *Nature*, Thomson began lobbying the British government about the need to secure all possible stocks of uranium. He also began work at London University's Imperial College, where he was professor of physics, to assess the feasibility of using nuclear fission to create a bomb.

Many, however, remained sceptical about whether such activities were valuable, given competing research priorities such as the study of microwaves, key to the development of radar. Winston Churchill was among the doubters. On 5 August, he wrote to Sir Kingsley Wood, Secretary of State for War, stating that 'the fear that this new discovery has provided the Nazis with some sinister new secret explosive with which to destroy their enemies is clearly without foundation. Dark hints will no doubt be dropped and terrifying whispers will be assiduously circulated, but it is to be hoped that nobody will be taken in by them.'

This view appeared vindicated when on 1 September 1939, the day Nazi troops invaded Poland and two days before Britain declared war on Germany for her violation of Polish sovereignty, Niels Bohr and John Wheeler published their classic paper presenting the theoretical basis for their hypothesis about the scarce isotope U-235. They explained that fission only occurred in U-235, which was extremely hard to separate from the non-fissionable but prolific U-238 of which natural uranium chiefly consisted. Furthermore, if the bombarding neutrons were slowed down to enhance their chances of smashing into atoms of U-235, this would prevent the tremendously fast reaction needed to spark an atomic explosion. This paper, coupled with early discouraging results from Thomson's work at Imperial College, suggested that pursuing fission for military purposes was not, after all, a priority. Official interest waned, and in the early months of the war a relieved minister in the British War Cabinet wrote, 'I gather that we may sleep fairly comfortably in our beds.'

CHAPTER NINE

A COLD ROOM IN BIRMINGHAM

FOR THE SECOND TIME IN HIS LIFE, JAMES CHADWICK WAS IN the wrong place when war broke out. On 3 September 1939 he was holidaying with his wife and twin daughters in a remote region of northern Sweden where the trout-fishing was good. The news was brought to their farmhouse by a local farmer who had heard it on the wireless. The family at once packed and set off for Stockholm, five hundred miles to the south, only to find all flights to London cancelled. While they waited, hoping to find some other way home, Chadwick contacted Lise Meitner, whom he found lonely, depressed and wondering whether she should accept a post she had been offered at Cambridge University. The Chadwicks managed to get flights to Holland and James just had time to scribble a quick note to Meitner before they dashed to the airport. It concluded, 'I am ready to do anything to help you', but they would not meet again until the war was over.

In Holland the Chadwicks once again found themselves stuck. Another Briton at their Amsterdam hotel was in the same plight – H. G. Wells, whose 1913 novel *The World*

Set Free had uncannily predicted not only the discovery of artificial radioactivity but the year, 1933. He had also predicted the destruction of cities by nuclear bombs in the 1950s. The *Times Literary Supplement* had dismissed his thoughts as 'porridge'. The Chadwicks did not approach Wells. Instead, they hovered nervously about the hotel, afraid of missing a message about a ship or plane that might have room for them. At last they found places on a 'stinking, rusty, tramp steamer' which carried them across the North Sea to England. Among those waiting anxiously to greet Chadwick in Liverpool was the impoverished, thirty-year-old Polish-Jewish physicist Joseph Rotblat.

As a child during the First World War, Joseph Rotblat had experienced and 'witnessed great suffering' and had become a scientist 'as a way of bringing relief, of helping

Sketch of Joseph Rotblat by Otto Frisch

a lot of people'. While still in Warsaw, he had read of the discovery of uranium fission and, like Enrico Fermi, Leo Szilard and Frédéric Joliot-Curie, had conducted experiments showing that during fission more neutrons were emitted than absorbed – the conditions for a chain reaction. He had speculated about the potential for an explosive device, but the idea so terrified him that his 'first reflex was to put the whole thing out of my mind, like a person trying to ignore the first symptom of a fatal disease in the hope that it will go away. But the fear gnaws all the same, and my fear was that someone would put the idea into practice.' His particular fear was that that 'someone' would be the Germans: 'I had no doubt that the Nazis would not hesitate to use any device, however inhumane, if it gave their doctrine world domination.'

However, Rotblat's reason for coming to Liverpool had nothing to do with his fears about fission. He had 'great hopes of building up nuclear science in Poland. I knew I needed a big machine, a cyclotron, and Chadwick was building one.' Chadwick was indeed determinedly pursuing his construction of a cyclotron and was in close touch with Ernest Lawrence at Berkeley, his mentor and adviser. He was also in frequent contact with his old friend John Cockcroft who, after Chadwick's departure, had finally persuaded Rutherford that the Cavendish must have a cyclotron. A massive donation from car magnate Sir Herbert Austin had made it possible. Although construction of the Liverpool device was not proceeding as fast as he would have liked – the builders, Metropolitan-Vickers, had become flooded with defence contracts – by 1938 Chadwick had been able to tell the Royal Institution that the machines at Liverpool and Cambridge were nearly ready for use.

Rotblat had arrived a few months later, in the spring of 1939, supported by a small scholarship which was just enough to keep him but insufficient for his wife to come too. Walking out of Lime Street Station and up the hill towards the university, he was shocked to see 'the worst slums you can imagine'. It was 'not very encouraging generally'. Also, his English was poor, 'even with people who spoke the King's English', and the Liverpudlian accent defeated him completely. He found lodgings in a rambling house full of postgraduate students where the landlord skimped on the food and watered the coffee. As he later wrote, he found a remarkable divergence between the England described in the novels of P. G. Wodehouse, which he read to improve his English, and the deprivation and drabness he saw around him in Liverpool.

Rotblat was also dismayed by the primitive conditions of the Liverpool University physics department, which was not the state-of-the-art facility he had anticipated. It was divided into two parts, the teaching side and the research side, which, though 'they were co-habiting in the same building', hardly spoke to each other. Rotblat was amazed when he visited the teaching lab 'and discovered they had no a.c. [alternating current]'. How, he wondered, 'could you teach electricity' in such circumstances? It was 'almost as though you ran a transportation firm and used a cart and horse'.

Bewildered, disappointed and isolated by his lack of English, Rotblat nevertheless quickly settled in, helped by an amicable welcome from other members of the physics department. Chadwick was particularly welcoming, despite the fact that, as Rotblat quickly recognized, he was a shy man and 'very much liked to be left to himself'. The first weekend after his arrival, Rotblat was asked to tea by the Chadwicks and was amazed to learn from other members of the department that he was the only one to be so honoured during Chadwick's four years at Liverpool. Rotblat was often invited to join the family for weekends at their cottage in Wales, too, and to go fishing with Chadwick. He got on well with Chadwick's wife Aileen, discovering a warmth where others merely found snobbery and chill class-consciousness.

Chadwick set Rotblat to work on investigating a very short-lived isotope. The skill, speed and originality with which the Pole completed the task so impressed him that he offered Rotblat the most prestigious fellowship his department had. It was worth £120, exactly the sum Rotblat received from his Polish scholarship, and it had never before been awarded to a foreigner. A delighted Rotblat exclaimed, 'Oh good, this means I shall be able to bring my wife.' Chadwick, unaware till now that his protégé was married, was aghast at the idea of the couple existing on so little, but Rotblat insisted they would manage.

In August 1939, shortly after the Liverpool cyclotron fired its first beam of accelerated particles, Rotblat returned to Warsaw. His thoughts were not only dominated by the chance to bring his wife to England. Like so many others, he had read the article published that summer by German scientist Siegfried Flügge which talked of a uranium device. As he pondered, his ideas crystallized and he 'worked out a rationale for doing research on the feasibility of the bomb'. He also concluded that 'the only way to stop the Germans from using it against us would be if we too had the bomb and threatened to retaliate', but his scenario 'never envisaged that we should use it, not even against the Germans'.

Once in Warsaw, Rotblat sought out his former professor, Ludwig Wertenstein, who was Jewish like himself, had been a pupil of Marie Curie in Paris and had also spent a year at the Cavendish, where he had got to know Chadwick. Rotblat told Wertenstein of his conviction that 'the only way to stop Hitler was to have the bomb ourselves' and asked his advice. The professor replied that 'he couldn't advise ... it was a matter of conscience' – a comment Rotblat took deeply to heart and would remember when he became closely involved with the Allied bomb project.* For the present, though, Rotblat's chief worry was his wife. She had been taken ill with appendicitis and was too sick to travel. An anxious Rotblat waited until, in his own words, 'the last minute', but finally he left, intending that his wife would come later. In the event he caught almost the last train to leave a free Poland for over fifty years. He reached England on 1 September 1939, the day the Nazis marched into Poland. He never saw his wife again.

Back in Liverpool, he faced pressing financial problems. His Polish funding had dried up, his fellowship was not due to commence until October and he had just seven shillings and sixpence in the world. He could not even pay his rent and hitch-hiked to London to seek the help of the Polish Embassy, but 'there was complete chaos and they asked me, could I help them'. However, his hitherto parsimonious landlord proved unexpectedly kind, agreeing he could stay on and pay him back later. Nevertheless, Rotblat was tremendously relieved when Chadwick

---

* Despite Niels Bohr's efforts to save him, Wertenstein would be killed by flying shrapnel as he tried to flee to Hungary across the Danube in 1944.

returned safely to England and immediately threw him a lifeline, appointing him as lecturer in nuclear physics, despite what Rotblat called his 'very, very shaky' English. Rotblat spent the next weeks studying English as hard as he could, not only to enable him to lecture but, as he later recalled, to allow him 'to go back to the problem which worried me the whole time'. Rotblat decided to go to Chadwick and quietly suggest 'that we should start work on the bomb'.

Chadwick's reaction to the discovery of uranium fission had initially been low-key. He was not convinced that there would be 'any interesting consequences from it' and that 'if something could be done with it, it would be a technical development rather than a search for new physical facts'. His views were shared by the majority of the scientific community in Britain, despite sensational articles in the press speculating about an awesome and dreadful new weapon.

In October 1939, Edward Appleton, a former colleague of Chadwick's at the Cavendish and newly appointed secretary of the Department of Scientific and Industrial Research, asked Chadwick privately for his views. Chadwick replied that although a device was theoretically possible the process was complicated and he doubted its feasibility. He had studied Bohr's and Wheeler's conclusions carefully. Their theory that fission by slow neutrons was entirely due to the rare isotope U-235 not only showed that any chain reaction in ordinary uranium would require huge amounts of the metal, perhaps tons, it also implied that this requirement was in itself an obstacle. To create a chain reaction, the neutrons would have to travel long distances, seeking out the sparse U-235 atoms and causing the whole process to unfold too slowly. As energy was released, the uranium would heat up and evaporate before the chain reaction had gone very far. Even if a chain reaction was achievable it was highly unlikely to lead to a bomb. Nevertheless, Chadwick promised Appleton he 'would think about it again'.

In late November, Rotblat summoned up sufficient courage and grasp of English to present in detail to Chadwick his plans for research on the feasibility of an atom bomb. Recognizing that slow neutrons would not cause the immediate and catastrophic conditions required for an explosion, he argued that the chain reaction must be triggered by fast neutrons instead. At the end of his presentation his mentor, who had remained silent, gave a response that was typically Chadwickian. As Rotblat recalled, 'he just grunted', leaving Rotblat discouraged by his reaction. In fact, his views had melded with Chadwick's own evolving opinions. As Chadwick later wrote, 'It was only the direct impact of war which made me put my mind to such questions. I then saw how simple the problem of producing a violent explosion really was, provided that a suitable material existed . . . which would support a chain reaction with fast, not slow neutrons, so that a substantial part would react, and release large amounts of energy, before the system had time to fly apart.'

Chadwick came back to Rotblat and asked, 'What sort of experiments do you want to do?' Rotblat told him, and began exploring such critical but unknown areas as the energies of neutrons generated by fission and the proportion of neutrons that would be absorbed by other nuclei without producing fission. By then, half of Chadwick's team had been seconded to classified radar work from which Rotblat, as a foreigner, was barred, and research assistants were thin on the ground. Chadwick assigned Rotblat a young Quaker called Flanders to help him. As a conscientious objecter, he had been sent to the university instead of being posted to the army. Rotblat wondered whether he should tell his assistant that he was working on research with possible military applications, but Chadwick had ordered him to divulge nothing of the work's true purpose and Rotblat reasoned that the experiments had independent scientific validity. After the war he discovered that Flanders had 'guessed something was going on', so in a sense 'we were both deceiving each other'.

Chadwick meanwhile reported to Appleton his revised conclusions that 'it seems likely' that fission 'could be developed to an explosive process under appropriate conditions'. These disturbing views prompted the government to unite all uranium and fission research under the Air Ministry. This included not only Chadwick's work but also experimental work being conducted by George Thomson at Imperial College in London and at Birmingham University under Rutherford's former pupil, the Australian Mark Oliphant. All information would be reviewed by the Air Defence Research Committee chaired by Sir Henry Tizard, Rector of Imperial College.

By the spring of 1940 Tizard and his commitee were still uncertain whether nuclear research would be valuable to the war effort. However, a note forwarded by Oliphant from two émigré scientists in his Birmingham department changed everything. One of them was Otto Frisch; the other was Berlin-born mathematical physicist Rudolf Peierls. Just like Rotblat, they believed an atom bomb was possible. Also like him, they had arrived at the idea of nuclear deterrence.

*   *   *

Rudolf Peierls had originally come to England in 1933 to spend the second half of a Rockefeller scholarship at the Cavendish Laboratory in Cambridge, having spent the first half with Enrico Fermi in Rome. He and his ebullient Russian physicist wife Genia had adapted quickly to English life, pleased that the rules of polite behaviour were 'much less rigid' than in Germany, although the food was rather a shock. With characteristic humour, Peierls devised 'a theory of the typical English boarding-house food: it would be undemocratic for the cook to impose his or her taste on the guests, so things are boiled until only a neutral matrix remains, to which the guest can give any flavour by adding salt, pepper, horseradish, mustard, ketchup, and so on'.

As Peierls' fellowship drew to an end, he had looked anxiously for a job in England. As he was learning from every letter posted from Germany, Jewish academics were being thrown out of their jobs, and as a Jew himself Peierls knew he no longer had a future there. His wife's pregnancy added to his personal worries, but he still did what he could for others. When he saw a junior post at Cambridge advertised, he applied himself but generously sent a telegram to Hans Bethe in Germany suggesting he also apply. In the event, neither man was appointed. However, Lawrence Bragg, a professor at Manchester University, came to Peierls' rescue with a two-year grant from a fund similar to that set up by the Academic Assistance Council which Rutherford was then spearheading. The grateful Peierlses moved north with their new baby daughter and were soon joined by Hans Bethe when he came to England. They offered him a room in their 'damp and icy' house.

In 1935, Peierls was offered an appointment at the Mond Laboratory in Cambridge, originally built for Peter Kapitza to conduct his magnetism and low temperature experiments. The money allocated for Kapitza's salary was unused and Rutherford had persuaded the Royal Society to award Peierls a research fellowship. Then, in the spring of 1937, Mark Oliphant suggested he apply for a mathematics professorship at Birmingham University. He was successful, and at last had the security of a permanent appointment. The Peierlses celebrated by buying an old car for £25 and learning to drive. Their peripatetic life appeared to have ended.

In 1939, Otto Frisch turned up in Birmingham. The Nazi invasion of what remained of Czechoslovakia in March 1939 had made him uneasy about remaining in Copenhagen and left him, in his own words, 'in a state of complete doldrums', believing war was coming and fearing that nothing he did would be any good. He was also depressed, fighting 'a pretty strong presentiment' that he had only a few months left to live. This prompted him for 'the only time in my life' to take 'some initiative'. When Oliphant visited Copenhagen, Frisch appealed to him, confessing his fears that Denmark 'would soon be overrun by Hitler' and asking 'would there be a chance for me to go to England in time, because I'd rather work for England than do nothing or be compelled in some way or other to work for Hitler or be sent to a concentration camp'. A 'very sympathetic' Oliphant said, 'You just come over in the summer. We'll find you something to do. You can give a few lectures or something.' Frisch arrived in July 1939 with two small suitcases and the Peierlses took him under their wing. Rudolf Peierls particularly admired Frisch's talent 'to ponder until he could present a problem in a form that admitted of a solution, the mark of a real physicist'. Their mutual talents were about to combine.

Most of Oliphant's work at Birmingham was at this time concerned with radar development. Security regulations did not allow aliens born in enemy countries, such as Otto Frisch, and recently naturalized British citizens, such as Rudolf Peierls, to be employed on sensitive war work, and thus both men were excluded from taking part. Indeed, neither was supposed to know anything about the project. However, the secrecy was 'a bit of a charade'. As Frisch recalled, Oliphant would sidle up to Peierls and pose him a 'hypothetical' question to which Peierls would furnish an answer, knowing full well what it would be used for. 'Oliphant knew that Peierls knew, and I think Peierls knew that Oliphant knew that he knew. But neither of them let on.' However, their formal exclusion from the radar work freed the two men to think about uranium fission.

While Frisch had been at Niels Bohr's institute there had been little belief in a 'superweapon' as a practical possibility. Frisch assumed that to be correct until in early 1940 in Birmingham he was invited to contribute an article on fission to the annual report of the British Chemical Society. He was then living in a freezing bed-sit where in winter, even with the gas fire on, the daytime temperature did not rise above 6° Celsius and where at night 'the water froze in the tumbler at my bedside'. Huddled in his overcoat and with his typewriter balanced on his knees, he typed out his article, writing in unfamiliar English 'there are now a number of strong arguments to the effect that the construction of a super bomb would be, if not impossible, prohibitively expensive and that furthermore the bomb would not be so effective as was thought at first'. He posted his article, but writing it had raised nagging doubts about whether he was right.

Frisch also brooded about the possibilities suggested by some studies he had recently begun of a method for separating isotopes known as 'thermal diffusion'. Invented by German scientist Klaus Clusius, it consisted of filling a tube with a gas mixture. If this mixture was heated at one end and cooled at the other, experiments had shown that the lighter isotopes would migrate to the hotter end and the heavier ones to the cooler region, thus suggesting the possibility of separating the lighter, fissionable U-235 from the heavier U-238.

Frisch sought out Peierls and startled him with the question, 'Suppose someone gave you a quantity of pure 235 isotope of uranium – what would happen?' They began to calculate the consequences, using a formula worked out by Peierls for calculating the 'critical mass' – the amount of fissionable material needed to be brought together to release sufficient neutrons to start a self-sustaining chain reaction. As Peierls recalled, 'The work of Bohr and Wheeler seemed to suggest that every neutron that hit a 235 nucleus should produce fission. Since the number of secondary neutrons per fission had been measured approximately, we had all the data to insert in my formula for the critical size.' The result amazed them. Others who had tried to calculate the critical mass 'had tended to come out with tons'; the Joliot-Curie team had estimated it at around forty tons. Frisch's and Peierls' first estimate was 'about a pound', which as Frisch observed was not, after all, 'such a lot'. Frisch calculated that using the Clusius thermal diffusion method of isotopic separation he could produce a pound of reasonably pure U-235 in a matter of weeks.

The two men also calculated whether the chain reaction would last long enough to cause a catastrophic explosion. Scribbling literally on the back of an envelope, they worked out that a substantial amount of the uranium would fissure, releasing energy equivalent to 'thousands of tons of ordinary explosive'. As Peierls recalled, 'We were quite staggered by these results: an atomic bomb was possible at least in principle! As a weapon it would be so devastating that, from a military point of view, it would be worth setting up a plant to separate the isotopes. In a classic understatement, we said to ourselves, "Even if this plant costs as much as a battleship, it would be worth having." ' With further understatement, Frisch said thoughtfully to Peierls, 'Look, shouldn't somebody know about that?'

Together they composed the famous Frisch–Peierls memorandum entitled 'On the construction of a "super-bomb"; based on a nuclear chain reaction in uranium'. The compelling three-page two-part document dealt with scientific, strategic and ethical issues. It suggested that 'one might think of about 1 kg [of uranium] as a suitable size for the bomb'. Their estimates of the critical mass were, in fact, an underestimate. They were unaware that some of the neutrons colliding with U-235 would simply be absorbed or 'captured' rather than causing fission, but, as Peierls later wrote, 'the order of magnitude was right'. They also described how to explode a bomb with a mechanism that would force two pieces of uranium together at tremendous speed to constitute the critical mass.

The memorandum addressed the human consequences not only of the blast, which could probably destroy 'the centre of a big city', but of the subsequent effect of radiation, 'fatal to living beings even a long time after the explosion'. 'Most of it', the note predicted, 'will probably be blown into the air and carried away by the wind. This cloud of radioactive material will kill everybody within a strip estimated to be several miles long. If it rained the danger would be even worse because active material would be carried down to the ground and stick to it . . .' Frisch and Peierls suggested that the probably very high number of civilian casualties 'may make it unsuitable as a weapon for use by this country' but pointed out that, as there was no effective defence other than the threat of retaliation with the same weapon, it would be worth developing as a deterrent, 'even if it is not intended to use the bomb as a means of attack'. They also warned that although 'we have no information that the same idea has also occurred to other scientists . . . all the theoretical data bearing on this problem are published, [and] it is quite conceivable that Germany is, in fact, developing this weapon'.

The Frisch–Peierls memorandum, with its origins in a cold room in Birmingham where a refugee muffled in an overcoat tapped with chilly fingers on a typewriter balanced on his knees, was the first document to demonstrate scientifically the real possibility of creating an atomic weapon, and the first to describe its shocking effects. For security reasons the two scientists typed the note themselves, making only one carbon copy, and gave it to Oliphant who in March 1940 sent it to a startled Sir Henry Tizard. The depiction of a weapon that would be 'practically irresistible' was about to kick-start the faltering British atomic programme.

CHAPTER TEN

MAUD RAY KENT

ON THE MORNING OF 21 JUNE 1940, THE BRITISH COLLIER *Broompark* docked at Falmouth after a tense thirty-six-hour crossing from Bordeaux during which an accompanying vessel had been sunk by a German mine. On board was a motley cargo of twenty-six drums of heavy water, industrial diamonds worth some £4 million and piles of machine tools. The passengers included a bedraggled group of French scientists and their families. This was the conclusion of a mission entrusted to the eccentric thirty-three-year-old Earl of Suffolk, Charles Henry George Howard, known to all as 'Jack'. Barred by his limp from the armed forces, he had been appointed scientific liaison officer at the British Embassy in Paris, where he lived at the Ritz and, according to his contemporaries, 'spent a lot of time drinking kirsch' and carousing with pretty women. With France about to fall, his bosses ordered him to gather some fifty eminent French scientists and engineers they had identified as useful and whisk them off to safety in Britain. In the time available Suffolk had been able to find only about half of them, and the disappointed reaction of the government official meeting the train bringing the party to London was, 'Oh, is that the lot?' However, the little group included Frédéric Joliot-Curie's two right-hand men, Hans von Halban and Lew Kowarski, though Joliot-Curie himself had opted to remain in occupied France.

The French team had spent a difficult few months. On the outbreak of war in September 1939, Joliot-Curie, a captain in the artillery reserves, had been immediately called up but then given special responsibilities for co-ordinating government scientific research. This provided cover for his studies on fission. Kowarski and von Halban had quickly been naturalized and then drafted, but, at Joliot-Curie's request, arrangements were made for them to remain with him on special assignment. Joliot-Curie's hope was to demonstrate a self-sustaining chain reaction in natural uranium using slow neutrons, to convince government that this could provide a potential new source of energy and thereby to win funding to build a nuclear reactor, or 'uranium boiler'. To do this, he not only needed sufficient quantities of natural uranium but also a suitable 'moderator' with which to slow down the bombarding neutrons. Experiments had shown that ordinary water, as used by Enrico Fermi when he had it carried in buckets from the goldfish fountain in the gardens behind his institute in Rome, was not sufficiently effective for his purposes – too many neutrons were lost from the chain reaction. Joliot-Curie decided he must have the denser heavy water.

Heavy water, though, was scarce and expensive. The only supplier in Europe was the Norsk-Hydro-Electric Company in Norway, of which the German industrial giant IG Farben owned 25 per cent. The company's

Vemork plant, near Rjukan, manufactured it as a by-product of synthetic ammonia. It was a painfully slow process. Tons of ordinary water were electrolysed with cheap electricity to release the hydrogen required to manufacture the ammonia. As electrolysis tends to release ordinary hydrogen, this left behind a tiny residue of heavy water. Joliot-Curie briefed the French Minister of Armaments and Lieutenant Jacques Allier of French intelligence about heavy water's special significance in fission research and pleaded that Norsk-Hydro's entire stock be obtained quickly and brought to France. Allier's interest in the substance had already been raised by reports that IG Farben was demanding without explanation that two tons of heavy water be shipped to Germany. The head of the Norwegian plant, Axel Aubert, a Norwegian of French extraction who was suspicious of the company's intentions, was stalling but could not fend off IG Farben much longer.

Consequently, in February 1940, Lieutenant Allier had left Paris secretly by train for Amsterdam, travelling under his mother's maiden name, Freiss. He was carrying a letter from the French President and a credit note for 36,000,000 French francs.* His orders were to bring the entire stock of heavy water to Paris, or, if that proved impossible, to render it unusable by contaminating it with cadmium, of which Joliot-Curie had given him a small phial. However, despite French precautions, Allier's departure had not gone unnoticed by German agents. French intelligence intercepted a telegram reading 'At any price intercept a suspect Frenchman travelling under the

---

* 36,000,000 francs was a substantial sum, then equivalent to £290,000 or $1,400,000.

name of Freiss'. Nevertheless, by 2 March Allier had reached neutral Sweden where he made contact with French intelligence agents before slipping into Norway. In Oslo he had a clandestine meeting with Axel Aubert, who agreed without demur to loan France, free of charge, Norsk-Hydro's entire stock of heavy water – some 185 kilos – for the duration of the war.

At the Vemork plant the heavy water was sealed into twenty-six seven-litre cans, especially made by an Oslo craftsman working secretly at home. On 12 March, in a carefully planned exercise, the cans were flown out from under the noses of German agents via Oslo airport. Two airliners on scheduled flights, one to Perth in Scotland and one to Amsterdam, were waiting on the runway. Allier acted as if he intended to board the plane for Holland, paying no attention as the other plane's propellers began to revolve. Suddenly, a large taxi rushed up to the airfield. Inside were both a French agent, enacting a charade about being late for the Amsterdam flight, and the carefully concealed cans of heavy water. The agitated man's taxi was allowed to drive on to the airfield and it halted between the two planes, out of sight of the terminal building. The cans were then swiftly manhandled onto the Perth plane aboard which Allier had meanwhile slipped. It took off almost at once and reached Scotland safely. The subterfuge had been entirely necessary, for German fighters forced down the Amsterdam plane, which left soon after, at Hamburg, where it was thoroughly searched. Nothing, of course, was found. By mid-March the French had moved their cans of heavy water to Paris, where they were stored in the vaults of Joliot-Curie's Collège de France.

The French mission had been timely. On 9 April 1940,

a month later, German troops invaded Norway and ended the phoney war. But Paris was not for long a safe home for the heavy water either. On 10 May the Germans attacked neutral Belgium and Holland. The latter quickly capitulated after heavy bombing raids on Rotterdam which killed 814 civilians. The German blitzkrieg swiftly overran Belgium too. By the end of May, with the evacuation of British troops from Dunkirk under way, and with German forces advancing on Paris where French government ministries were burning their papers, Joliot-Curie had found a temporary hiding place for the drums of heavy water in the death cell of the central prison at Riom, near Clermont-Ferrand. A few days later, von Halban and Kowarski loaded a truck with scientific equipment and fled south from Paris with their families. The Joliot-Curies joined them soon after.

On 16 June, two days after the fall of Paris to the Germans, Allier arrived with orders for the French scientists to withdraw west to Bordeaux for evacuation to England aboard the *Broompark*. Early the next day, 17 June, the same day that Marshal Pétain broadcast to the French people that he had assumed control of their government and had applied to the Germans for an armistice, von Halban and Kowarski, a refugee once more, loaded the heavy water onto a truck and, with their wives and children, joined the frightened stream of people heading for the coast. Joliot-Curie followed after leaving Irène, who was suffering from a combination of respiratory problems and anaemia, at a sanatorium.

The port was in chaos, under aerial attack and crammed with more than half a million refugees, troops and abandoned military and civilian vehicles. Von Halban and Kowarski managed to find the *Broompark* and embarked with the heavy water. Despite the Earl of Suffolk's assurances that his wife and children would be brought safely to England, Joliot-Curie decided to remain in France – a decision his mother-in-law Marie Curie would have wholeheartedly endorsed. As Irène later told a friend, 'my mother would never have abandoned her laboratory'. However, it meant that Joliot-Curie would have no further contact with von Halban and Kowarski until the war ended.

Joliot-Curie never revealed his motivation for staying. Some friends thought that Irène and her powerful personality, which dominated their marriage, were a major factor. Bertrand Goldschmidt, who worked with Frédéric again after the war, believed that worries about his poor English and the status and facilities the British would accord him may have swayed a difficult decision. Joliot-Curie's daughter Hélène later suggested that he stayed to help keep French science alive during what he thought would be a long occupation. If such was his main motivation, it had some resemblance to Heisenberg's reasons for remaining in Germany. His decision would certainly confront him with similar moral dilemmas. How far, for instance, should he collaborate with the Nazis to preserve his beloved research facilities for his nation's science?

The collier was the last cargo ship to sail from Bordeaux. According to Kowarski, Suffolk, who had not shaved for days and whose bare arms were covered with tattoos – symbols of his exuberant eccentricity – 'had got the crew too drunk to sail until the machinery and ourselves were aboard'. The bemused Russian thought 'Suffolk was straight out of Wodehouse . . . There was sea-sickness: there were 25 women aboard. Suffolk was pouring them champagne. "This is the perfect remedy," he said.'* When the ship docked at Falmouth, the containers of heavy water, which had been strapped to a raft in the hope that they could be salvaged if the *Broompark* was torpedoed, were transferred first to Wormwood Scrubs prison and then, perhaps most incongruously of all, into the custody of the royal librarian at Windsor Castle.

As Otto Frisch had feared, Denmark too had quickly fallen. On the evening of 8 April 1940, while Niels Bohr was being entertained by King Haakon of Norway at the Royal Palace in Oslo, Nazi forces were preparing to invade his homeland as well as Norway. Unaware of what was about to happen, Bohr boarded the night train for Copenhagen. As the train was shunted off the ferry which had carried it across the Kattegat, Bohr was awakened by Nazi warplanes streaking overhead and shouts that the Germans were coming. At 4.20 a.m. that morning Hitler had presented the Danish government with an ultimatum: accept the protection of his Third Reich without resistance or face all-out attack. While the Danish king and his government agonized, Nazi aircraft flew very low over Copenhagen, their roaring engines emphasizing the Danes' lack of choice. By noon on 9 April, Denmark was an occupied country.

Bohr hurried to the chancellor of the University of Copenhagen and to members of the Danish government to seek protection for the Jewish scientists at his institute, some of them refugees from Nazi racial persecution elsewhere in Europe, and to urge them to resist the imposition of race laws. In turn, officials from the American Embassy sought out Bohr to offer him and his family sanctuary in

* Suffolk would die the following year, 1941, while defusing a bomb.

the United States. Bohr knew that with a Jewish mother he was in personal danger but insisted he must remain to look after his staff. Somehow he found time to send an urgent telegram to Otto Frisch warning him to remain in England.

Another telegram also reached England which caused some puzzlement. It was from Lise Meitner, who had arrived in Copenhagen just twelve hours before the German occupation began and who had also been woken up by the noise of aeroplanes. Since the Germans initially allowed the Danes to retain a degree of self-rule in return for their bloodless surrender, Meitner was able to remain in Copenhagen unmolested for three weeks and to meet Niels Bohr. On her return to neutral Stockholm, she despatched, at Bohr's request, a telegram to his friend the British physicist Owen Richardson reassuring him that the family was all right. The text read 'Met Niels and Margrethe recently both well but unhappy about events please inform Cockcroft and Maud Ray Kent'. John Cockcroft jumped to the conclusion that Meitner's words contained a hidden warning. He wrote anxiously to James Chadwick suggesting that the final three words, 'Maud Ray Kent', were code for 'uranium taken'; others speculated that they were an anagram for 'Make Ur Day Nt' – 'Make Uranium Day and Night'. Only later did they learn the simple truth that Maud Ray had been the Bohr children's governess. She lived in Kent, and her address had mistakenly been omitted from the telegram.

However, the telegram solved one problem, that of choosing a suitably coded name for the group set up in April 1940 by the British government in response to Otto Frisch's and Rudolf Peierls' memorandum to consider the possibility of constructing a uranium bomb. The group decided to call themselves the 'Maud Committee' – formally the 'M.A.U.D. Committee'. Many who became associated with it were convinced that the letters stood for 'Military Applications of Uranium Disintegration'.

The Maud Committee was chaired by George Thomson of Imperial College, London, and members included Mark Oliphant of Birmingham University and James Chadwick of Liverpool University, who was to co-ordinate the laboratory research across the various universities. The committee did not, however, include Otto Frisch or Rudolf Peierls, who had been anxiously awaiting a reaction to their memorandum. While they waited, Frisch was summoned by the police as an enemy alien and interrogated. Security concerns meant that by this time many aliens were being interned, some in camps on the Isle of Man and others overseas in countries such as Canada. Even though Frisch was sent home, he felt that 'all those questions really added up to the simple question, "Is there any reason not to intern that chap?" '. Genia Peierls, who, according to Frisch, 'ran her house with cheerful intelligence, a ringing Manchester voice and a Russian sovereign's disregard of the definite article', was so convinced that the impractical Frisch was about to be locked up that she bought him 'some shirts of sea-island cotton which could be washed by a bachelor' such as himself.

Frisch was spared internment, but at first it did seem that he and Peierls would be barred from working on the project they had initiated. Mark Oliphant told the incredulous pair that the government was grateful to them for their analysis but that, since enemy aliens and recently naturalized British citizens could not be employed on sensitive war work, they would not be consulted further. The normally quiet, equable Peierls was angered by such idiocy, certain that he and Frisch had 'the answers to important questions' likely to perplex and delay the committee. Peierls wrote politely but firmly to Thomson, who acknowledged the logic of his argument and won agreement for Peierls and Frisch to be consulted on the Maud Committee's progress, and later to become members of a technical sub-committee.

Frederick Lindemann, Professor of Experimental Philosophy (Physics) at Oxford, later Lord Cherwell, also attended the technical sub-committee. He was both friend and adviser to Winston Churchill, who had replaced Neville Chamberlain as Prime Minister during the crisis of May 1940. Churchill, who always referred to him as 'the Prof', appreciated Lindemann's ability 'to decipher the signals from the experts on the far horizons and explain to me in lucid, homely terms what the issues were'. The British son of a naturalized Franco-Alsatian father and an American mother, Lindemann had studied physics in England and Berlin. He was also an ace tennis player who competed at Wimbledon. During the First World War, disturbed that pilots had no guidance on what to do if their planes went into a spin, he had studied the mathematics of spin until he believed he had the solution. Determined to test his conclusions without hazarding the lives of others, he learned how to fly, put his plane through a systematic series of spins and, applying his theory, succeeded in straightening it out again. His work saved many lives. Like Churchill, he doubted whether Germany was working on atomic weapons but thought it vital that Britain was not outflanked.

The Maud Committee worked quickly, aware that with

Britain battered, devoid of European allies and facing invasion, time was not on their side. They were also aware of the desirability of greater contact with the United States. Since the start of the war there had been few scientific exchanges between Britain and the neutral United States. However, in the late summer of 1940 Churchill decided to send a scientific delegation under Sir Henry Tizard to woo America by revealing Britain's technical secrets. His team, which included John Cockcroft, sailed with a black-metal steamer-trunk packed with tempting models and blueprints.

Once in the United States, they briefed American scientists on subjects from the design of the Rolls-Royce Merlin engine – powering the Spitfires currently confronting the Luftwaffe in the skies over southern England in the Battle of Britain, and later used in the American P-50 Mustang and the British Lancaster bomber and Mosquito intruder – and the cavity magnetron vital for enhancing radar performance to the emerging evidence of the feasibility of an atomic bomb. The Tizard mission also attended a meeting of the Uranium Committee – the body set up by President Roosevelt in the aftermath of Albert Einstein's warning. It was chaired by Lynam J. Briggs, originally a government soil scientist who had become director of the National Bureau of Standards. The other members were experts in military ordnance with little expertise in nuclear physics.

The British mission returned home and reported to the Maud Committee that America was not pursuing nuclear research with any great urgency. It was, however, impressed by the evidence it had seen of the United States' great productive capacity for costly experimental work. This reinforced the view expressed by Mark Oliphant and shared by many British scientists that 'if things go really badly with this country there is a great deal to be said for investigating any possibility which offers a chance of hitting back from the New World'.

Nuclear research was being pushed forward in Britain, but under increasingly difficult conditions. In July 1940, Otto Frisch joined Chadwick's team in Liverpool to work on isotopic separation. Soon after his arrival he heard 'the wailing of air-raid sirens' for the first time in his life. Within weeks the city began to suffer heavy air raids, and night-time was dominated by the 'popping of anti-aircraft guns' and the 'clatter of falling shrapnel'.

The bombing intensified in November when Hitler ordered a series of bombing raids on British cities. Liverpool was badly hit, but a worse sufferer was Coventry where many of the city's buildings, including the cathedral, were destroyed or badly damaged and 568 people were killed. The Germans invented a new word, 'Koventrieren' – 'to Coventrate' or raze to the ground. Some of the fires joined together to produce greater intensity of heat, a fact not lost on the future Air Marshal 'Bomber' Harris, then working in the Air Ministry. He would later recall that Coventry taught British planners the 'principle' of the fire storm, igniting 'so many fires at the same time'. It was, nevertheless, the Japanese who, a year earlier in 1939, could be said to have begun strategic bombing of undefended civilian cities and the creation of fire storms by dropping numerous incendiaries on the Chinese provisional capital, Chungking. A *Times* reporter described how the timber houses 'burned like tinder . . . the phosphorus kept the fires raging and a breeze extended them, three quarters of a square mile of houses were in flames'.

In the early months of 1941, the German Luftwaffe attacked Liverpool with high-explosive bombs, parachute landmines, oil bombs and incendiary bombs. In March, a parachute landmine hit the courtyard of Chadwick's physics department and blew out all the windows. Scientists hurried to the engineering department to find hammers and nails for makeshift repairs to their labs. Luckily, Chadwick's cyclotron, deep in the basement, was unharmed. Frisch and the fellow occupants of his boarding house spent many nights huddling under the staircase. After one particularly frightening raid they emerged to find that their landlady had fled. Frisch packed a case and scrambled through inner-city streets littered with debris to seek sanctuary with friends in the suburbs. The Chadwicks, who had sent their daughters to Canada for safety, were sleeping on the ground floor of their house for greater protection. Chadwick was discreetly going out with a Geiger counter and checking bomb craters to reassure himself that the Germans were not mixing radioactive material with the explosive in a kind of 'dirty bomb'.

Despite the dangers and difficulties of living in a city under attack, Frisch settled down in Liverpool. As aliens, he and Joseph Rotblat were formally subject to restrictions on their movements, but Chadwick persuaded his friend the Chief Constable of Merseyside to exempt them from what Rotblat called the more 'ridiculous' strictures. Frisch was thus allowed to own a bicycle, and found himself being fined ten shillings for riding without turning on his lamps. He enjoyed working for Chadwick, who encouraged members of the team to discuss their work, 'putting no great trust in the bogus security which relies on compartmentalising knowledge, on letting every scientist know only what he needs to know'. Rotblat was lecturing openly on chain reactions.

Frisch's task was to test the thermal diffusion method for separating isotopes, pioneered by German scientist Klaus Clusius, which he and Peierls had recommended in their memorandum. Frisch told Chadwick that to do this he needed uranium hexafluoride, the only gaseous compound of uranium stable enough to put into a tube. According to Frisch, Chadwick sat for about thirty seconds 'turning his head side to side like a bird', then said simply, 'How much hex do you want?' Frisch set to work with a student assistant, John Holt – the pair were soon nicknamed 'Frisch and Chips' – but they discovered that the process would not work with uranium hexafluoride. As Peierls put it, 'the effect happens to be practically zero'.

Working with fellow refugee the German-born Franz Simon, Peierls thought up another diffusion method for separating isotopes. This involved forcing atoms of uranium hexafluoride gas through fine holes in a porous barrier or membrane made from nickel. Peierls hoped that the lighter U-235 would pass through more quickly than the heavier U-238, and that by repeating the exercise again and again a U-235-rich gas would result. The process was difficult because the gas was highly corrosive and broke down on contact with moist air, but it seemed to work. Their research suggested that an industrial separation plant covering forty acres could yield one kilogram of 99 per cent pure U-235 a day. The huge complex would take eighteen months to construct.

Chadwick was feeling the pressure. With his overview of all the experimental work, it was becoming ever clearer to him that 'a nuclear bomb was not only possible – it was inevitable'. Yet he felt that he had 'nobody to talk to'. Although he had a high regard for his chief helpers, Frisch and Rotblat, he was conscious that they 'were not citizens of this country' and that the other scientists were 'quite young boys'. Isolated and anxious, Chadwick found 'the only remedy' was to take sleeping pills – a habit that remained with him for life.

Chadwick also bore the burden of deciding how to prioritize the research. Back in June 1940, on the day after the Germans marched into Paris, a letter, published in the US journal *Physical Review* by American scientists Edwin McMillan and Philip Abelson, had reported results from working with the largest cyclotron yet built by Ernest Lawrence at Berkeley. It was a giant device with a sixty-inch vacuum chamber, compared with the four-inch chamber of Lawrence's first model. This machine provided a source of high-energy particles which, when they hit beryllium or a similar target, produced a copious stream of neutrons. Using these neutrons, McMillan and Abelson had bombarded uranium and created a hitherto unknown radioactive element. This element, with atomic number 93 – named neptunium after the planet next in line to Uranus – decayed into another unnamed element occupying slot 94 in the Periodic Table. Joseph Rotblat recognized at once that, since the mysterious element shared characteristics with uranium, it would be likely to fission under neutron bombardment. If so, it could be an alternative to U-235 as atomic bomb fuel. He asked Chadwick to allow him to use the Liverpool cyclotron to produce and explore the new element.

With the British effort focused on research on separating U-235, Chadwick decided resources could not be spared. However, worries that his decision was mistaken gnawed at him. In December 1940 he learned that there might be an alternative way of producing element 94. Franz von Halban and Lew Kowarski, now working for the Maud Committee at Cambridge University, were continuing their investigations, initiated in Paris with Frédéric Joliot-Curie, into producing chain reactions by bombarding natural uranium with slow neutrons using heavy water as a moderator. They concluded that, given enough uranium and heavy water, a chain reaction would indeed be possible. Although their primary interest was harnessing the chain reaction to produce nuclear power, they saw the potential military applications of their process: that neutrons could convert the heavy and easily obtainable isotope U-238 into the new element 94. Like Rotblat, they believed that it was fissionable and could be used to fuel a bomb.

A few months later, in March 1941, research in America brought further confirmation. At Berkeley, also using Lawrence's new sixty-inch cyclotron, young chemist Glenn Seaborg and Italian physicist Emilio Segrè, who had emigrated from Italy in 1938, isolated and analysed a tiny amount of the new element for the first time and confirmed that, like U-235, it would fission. Seaborg named it after the planet Pluto, itself only discovered in 1930 – plutonium.

In July 1941, the Maud Committee submitted its final report to the British government, concluding that 'an atomic bomb was feasible'. Written largely by Chadwick, it was in two parts. The first explained with compelling clarity how initial scepticism had turned into conviction that a 'very powerful weapon of war' could definitely be made using U-235. Given 'the destructive effect, both material and moral', 'every effort should be made to produce bombs of this kind'. Some twenty-five pounds of U-235 would be needed and the project would take two years. The second part discussed the possible peaceful applications of nuclear energy: the generation of power by 'uranium boilers' as envisaged by von Halban and Kowarski, the use of nuclear energy for ship propulsion and the production of radioisotopes for medical purposes. The report made no reference to plutonium.

On 30 August 1941, Winston Churchill assented to the proposal to build the atomic bomb with typically mordant wit: 'Although personally I am quite content with the existing explosives, I feel we must not stand in the way of improvement.'

CHAPTER ELEVEN

'HITLER'S SUCCESS COULD DEPEND ON IT'

GENERAL ERICH SCHUMANN, HEAD OF GERMAN WEAPONS research and a descendant of the composer Robert Schumann, was also sceptical about the prospect of a revolutionary new weapon. Although a professor of physics, he knew little of atomic science. The letter sent in April 1939 to the army by Professor Paul Harteck had left him unmoved, despite its tempting suggestion that nuclear explosives would confer an 'unsurpassable advantage' on the country which possessed them.

While Harteck waited impatiently for a reply, he succeeded in coaxing a private company to give him $5,000 to initiate some research into fission since, as he later recalled, 'in those days in Germany we got no support for pure science. We were very, very poor.' Harteck's motive for alerting the German Army to the potential of fission was, he claimed, financial. He was not a Nazi and his sister, who had married into a prominent Jewish family in Vienna, had fled to the United States with her husband and son. What mattered most to Harteck was that 'The War Office had the money and so we went to them. If we had gone somewhere else, we would have got nothing.'

By August 1939, having still received no reply, Harteck wrote again. Unknown to him, Schumann had referred the problem to Kurt Diebner, one of his juniors in Army Ordnance. Diebner was an expert in both atomic physics and explosives and he took Harteck's letter, with all its implied threat and promise, seriously. His first move was to summon to Berlin an able young physicist, Erich Bagge, then working as Werner Heisenberg's assistant at the University of Leipzig and whose work on heavy water had come to the army's attention. A nervous Bagge arrived, expecting to be despatched to the front. Instead, Diebner instructed him to draw up an agenda for a meeting at the War Office to discuss how best to exploit nuclear fission before the war ended. Bagge noticed that the list of invitees consisted almost entirely of experimentalists. He urged that they must have 'a theoretical physicist with a big name', and that it 'should be Heisenberg'. Diebner refused. The German programme would, he insisted, be experimental only. He seems to have been partly motivated by pique that in former years Heisenberg had faulted his scientific work. Heisenberg certainly had little time for Diebner, later describing him as a 'decent physicist' but 'not absolutely first rate . . . one of the many people who had come from a low class level into rather high responsibility through the [Nazi] Party'.

On 16 September an initial meeting took place at the War Office. Those present included Carl Friedrich von Weizsäcker, Otto Hahn, Hans Geiger, Walther Bothe and Paul Harteck. War Office officials instructed them that their task was to determine whether it was feasible that Germany, or her enemies, could harness fission to produce power or bombs. It was not an easy question and the group debated for several hours. At the end of this time, Geiger, who had until then remained silent, rose to his feet. Once a pupil of Rutherford and now seemingly a convinced Nazi, he stated that if there was 'the slightest chance' of releasing nuclear energy through fission then 'it must be done'. Bothe echoed this zeal, declaring, 'Gentlemen, it *must* be done.'

Otto Hahn was much less certain. According to von Weizsäcker, Hahn took much convincing to have anything to do with the project. Von Weizsäcker pleaded, 'Please join . . . not to help us, but to help yourself, because you will protect your Institute by doing so. You will be doing something which is officially judged to be important for the war effort, and therefore your Institute will continue. Your people will not be dispersed to other projects or to the front.' Hahn replied, 'Well, I think you are right, I shall,' but then became 'quite emotional', privately saying, 'But if my work leads to a nuclear bomb for Hitler, I will commit suicide.'

Having agreed with mixed feelings and motivations to study the potential applications of nuclear fission, Kurt Diebner's scientists turned to practicalities – what studies should be undertaken and by whom. Bagge returned to his argument that his mentor Werner Heisenberg must be involved. Not only did the project need his intellect, but there was a serious risk that he might otherwise be called up and perhaps killed in the fighting. This time Diebner assented, and on 20 September Heisenberg was finally ordered to Berlin.

In the fortnight since the war began, Heisenberg had been waiting anxiously with his wife and family at Urfeld in the Bavarian Alps. He had learned of Germany's invasion of Poland from the proprietor of the local hotel who assured him cheerily that it would 'all be over and done with in three weeks' time'. Heisenberg had expected immediate orders to join the Mountain Rifle Brigade, with which he had been training, but days passed and he heard nothing. He wrote to his former professor Arnold Sommerfeld that his call-up 'strangely enough has not yet come through . . . I have no idea what will happen to me.' The summons to Berlin must have been both a relief and a puzzle.

Heisenberg reported to the War Office where, as he later wrote, he was told that he had been conscripted into the new nuclear physics research group 'to work on the technical exploitation of atomic energy'. The group became known, with surprising casualness about security, as the Uranverein – the Uranium Club. According to von Weizsäcker, Heisenberg joined the club without hesitation in order to protect German science. His argument was, 'Well, we must do it. Hitler will lose this war. It is like the end game in chess, with one castle less than the others . . . Consequently, much of Germany will be destroyed, or its value will have disappeared. The value of science will still be there and it is necessary that science should live through the war, and we must do something for that.'

Heisenberg's own subsequent recollections described sessions of deep soul-searching with von Weizsäcker during which both men agreed that the prospect of successfully building an atomic bomb was very remote. The technical problems were formidable, probably insuperable, at least over the likely lifespan of the war. The greatest challenge was obtaining enough fissionable material to create an explosive device. Niels Bohr and John Wheeler had shown that it could not be done with natural uranium; only a sufficient quantity of the rare and highly fissionable isotope U-235 would do. But to separate this from U-238, the less fissionable isotope of which natural uranium was chiefly composed, would, in Heisenberg's words, require 'a gigantic technical feat' that would take until 'the distant future'. However, according to Heisenberg's post-war account, the two men agreed that it might well be possible to use natural uranium to trigger a chain reaction capable of yielding controllable amounts of energy which could be used for 'power stations, ships and the like'. They also agreed that when the war ended such technology would be important for the rebuilding of Germany. They could, they convinced themselves, work on it 'with a clear conscience'.

The two broad thrusts of the Uranium Club's research were how to separate enough U-235 and how to build a chain-reacting nuclear pile – a 'reactor'. Meanwhile, Army Ordnance swiftly requisitioned the Kaiser Wilhelm Institute for Physics, which became the heart of the army project. They gave the institute's Dutch director Peter Debye an ultimatum: renounce his Dutch nationality and take German citizenship or resign his directorship. Debye departed for the United States to teach at Cornell University.

Heisenberg's role was to drive the theoretical side of the project. Still only thirty-seven years old and brimming with drive and energy, the man whom James Chadwick would identify later in the war as 'the most dangerous possible German in the field because of his brain power' got quickly to work. His first priority was to develop a theoretical basis for a workable reactor. By December 1939, just weeks after his appointment, he submitted a secret twenty-four-page report to the army suggesting that the production of power through nuclear fission in a reactor was technically possible using natural uranium. But 'enriched' uranium, where the percentage of the rare isotope U-235 had been increased by means of isotopic separation, would be better. He offered an alluring scenario: enriched uranium could be used to run a smaller reactor at a higher temperature than achievable with natural uranium and to generate enough power to drive German warships and submarines. Heisenberg also suggested, in a statement in his report somewhat at odds with his later justification of his motives, that enriching natural uranium could create a new explosive surpassing 'the explosive power of the strongest existing explosive materials by several orders of magnitude'. Isotopic separation was, he said, the 'surest method' for achieving a nuclear reactor but the 'only method for producing explosives'.

Heisenberg's report and a follow-up paper in February 1940 would provide the template for the Nazi fission research programme until the end of the war. However, he made a critical misjudgement over the choice of a suitable material to use as a moderator to slow neutrons down and thus to enhance their chances of hitting their target uranium nuclei, causing fission and thereby triggering the release of more neutrons to sustain a chain reaction. Heisenberg had initially focused on two substances as a moderator, heavy water or carbon, which, as he later wrote, 'I had suspected, for theoretical reasons . . . could be used as a moderator in place of heavy water'. However, in his second report to the army he declared it doubtful whether the uranium machine – that is, a reactor – could be built with carbon.

Heisenberg had been misled by imprecise data from experiments he had had conducted. The error was compounded by von Weizsäcker, whose calculations in Berlin supported Heisenberg's views. So did measurements made in early 1941 by Walther Bothe, by then Germany's leading experimental physicist despite some difficult times. In 1933 he had been ejected from his professorship at Heidelberg University for failing to show due enthusiasm for the Nazi Party. However, he had managed to obtain a post at the Kaiser Wilhelm Institute for Medical Research in Heidelberg. At first, Bothe believed that carbon was a promising material for a moderator: it did not absorb neutrons and it was freely available. However, just as von Weizsäcker had done, Bothe chose, as his form of carbon, industrial graphite. Both men failed to realize that even the best industrial graphite contains too many impurities to function well as a moderator. In particular it contains boron, which absorbs or mops up neutrons. Had they experimented with completely pure graphite they would have discovered, as had Enrico Fermi in his experiments at Columbia University, that it was an excellent moderator. Thanks, however, to Leo Szilard's persistence, Fermi's results had not been published so Bothe remained unaware of his mistake.

In the spring of 1940 at Hamburg University, Paul Harteck came close to devising a carbon-based moderator. He conceived the brilliant notion of using carbon dioxide and persuaded industrial giant IG Farben to loan him a chunk of frozen carbon dioxide – dry ice. However, the dry ice, whose excellent credentials as a moderator would have been revealed in experiments, arrived before Harteck could obtain sufficient uranium. Consequently, the limited tests he was able to perform were inconclusive.*

---

* Ironically, Heisenberg refused to lend him any of his own uranium stockpile.

The net result, as Heisenberg wrote, was that German scientists 'abandoned the whole idea' of carbon 'prematurely' and turned, instead, to heavy water. Had they pursued carbon, the first self-sustaining chain reaction using a carbon-based moderator might have been achieved not in the United States but in Nazi Germany.

The German scientists' immediate problem in the early stages of the war was how to obtain sufficient stocks of heavy water. In April 1940, after invading Norway, the Germans had seized the Norsk-Hydro plant at Vemork, where they quickly increased production from twenty litres a year to one ton. However, the amount the Germans estimated they needed for one reactor per year was closer to four or five tons. Paul Harteck designed a catalytic exchange process to increase the plant's production to those levels, but it would still take time for significant quantities of heavy water to be produced and shipped.

In July 1940, Walther Bothe arrived at the Collège de France in occupied Paris, followed soon afterwards by Kurt Diebner and Erich Schumann. The three men were keenly interested in the fate of the heavy water shipped out of Bordeaux on the *Broompark* a month earlier by von Halban and Kowarski. A nervous Frédéric Joliot-Curie, who had recently returned to Paris leaving a frail Irène to continue her recuperation in the country, convinced them that the heavy water had been loaded onto another ship known to have been sunk by the Germans. He also persuaded them that a substantial quantity of uranium ore purchased by the French from the Belgians before the war had been taken south by the fleeing French government. He assured his visitors that its whereabouts were unknown although, as he knew, it was in fact in Algeria, where it would remain throughout the war.

Bothe, Diebner and Schumann also wanted to know about Joliot-Curie's cyclotron. Though unfinished, it was one of only two in occupied Europe; the other was in Niels Bohr's laboratory in Copenhagen. The visitors realized they could not reveal the military-related motives behind their interest in Joliot-Curie's facilities, but they also knew they needed his co-operation. They therefore blandly proposed some joint nuclear studies and offered Joliot-Curie a compromise: they would leave him in virtual control of his laboratory and help him complete his cyclotron; in return, he had to agree to accept a German research team under the direction of Wolfgang Gentner.

Gentner had worked at Berkeley with Ernest Lawrence and his motivation for returning to Germany had been, like that of Heisenberg, to protect German science rather than any enthusiasm for the regime. He had worked with Joliot-Curie in the mid-1930s and regarded him as a friend. At a private meeting he sought and received Joliot-Curie's blessing to come to his laboratory. Despite their friendship, both their situations were fraught with ambiguity. Gentner might not always be able to protect Joliot-Curie, and the results of Joliot-Curie's work would inevitably be known to the Germans.

Although in 1940 German scientists were short of heavy water, they had excellent sources of raw uranium. The Nazi occupation of the formerly Czechoslovak Sudetenland on the borders of Bohemia had delivered them the world's richest uranium mines at Joachimsthal. From 1940 onwards, slave workers mined the uranium ore for the Nazis. The Auer company, which had produced the radioactive toothpaste used by James Chadwick in his experiments during his internment in the First World War, organized the processing of the uranium into a usable form at their works at Oranienburg near Berlin. Their labourers included two thousand female inmates from the Sachsenhausen concentration camp. In April 1940, when Heisenberg complained about the time it was taking to obtain processed uranium, Auer requisitioned more slave workers and stepped up production.

While awaiting sufficient quantities of suitable materials, German scientists addressed two main tasks: assessing techniques for separating U-235 from natural uranium, and working out the optimum size and configuration for a reactor. Of nine research teams controlled by Kurt Diebner, two were detailed to work on reactor construction: the experimental physics section of Heisenberg's Physics Institute at Leipzig University, and the Kaiser Wilhelm Institute for Physics in Berlin. Heisenberg, who was also exploring the properties of heavy water, commuted between the two.

By the summer of 1940, a new laboratory for reactor experiments was under construction among a pleasant grove of cherry trees in the grounds of the Kaiser Wilhelm Institute for Biology and Virus Research in Berlin, located next to the Institute for Physics. To deter unwanted visitors, the wood-framed building was named the 'Virus House'. Rumours spread that scientists there were conducting deadly experiments with bacteria. In fact, Heisenberg was directing some early reactor experiments using whatever was available – paraffin as a moderator and limited amounts of uranium. The results suggested that, with heavy water and enough uranium, a self-sustaining chain reaction might indeed be achievable.

German scientists were also exploring the potential applications of elements heavier than uranium – the 'transuranics' that had so fascinated and perplexed Lise Meitner, Otto Hahn and others. Some had spotted the article by the Berkeley cyclotroneers Edwin McMillan and Philip Abelson published in the American journal *Physical Review* in June 1940 reporting the discovery of the transuranic element 93, neptunium, created when U-238, the most common isotope in natural uranium, captures a neutron and transmutes into U-239, which in turn decays into element 93. However, at the Kaiser Wilhelm Institute for Chemistry, others were already and independently on the trail. Young radiochemist Kurt Starke had stumbled on element 93, and Otto Hahn and Fritz Strassmann immediately began dissecting the new element's chemical characteristics.

Strassmann had remained true to the principles which had first endeared him to Lise Meitner. He not only despised the Nazi regime but was prepared to risk his life and that of his wife Maria and baby son Martin to protect others. At the very time he was working on one of the most sensitive and secret projects of the German war effort, he was secretly sheltering Jewish pianist Andrea Wolffenstein in his Berlin apartment. She later wrote that he and his wife helped her 'in full knowledge of all the dangers' they were running, sharing their meagre food with her and taking her to safety during air raids, and that Otto Hahn knew that the Strassmanns were hiding her. The risks to them all were heightened by the presence of a staunch and watchful Nazi living in the flat directly beneath, so that when all the Strassmanns were known to be out Wolffenstein had to be careful not to make a sound. After she managed to escape from Berlin undetected, the Strassmanns also helped her sister Valerie.*

If Hahn's and Strassmann's preoccupation was, as Hahn claimed after the war, simply with the chemistry of element 93, von Weizsäcker, at least, was working on a broader canvas and he quickly grasped the element's bomb-making potential. He deduced that it was highly fissionable and could be manufactured in a reactor and used to fuel an atom bomb. Because element 93 could be separated by conventional chemical rather than isotopic processes it would be easily retrievable from other fission products. This would overcome the greatest technical obstacle in the path of a German atom bomb – developing isotopic separation techniques to squeeze enough of the rare isotope U-235 out of natural uranium to fuel an explosive device. On 17 July 1940, von Weizsäcker wrote a five-page paper to the army authorities which he also copied to Werner Heisenberg. In it, he suggested that element 93 could be as useful as U-235 in making 'Sprengstoff' – 'explosive'.

Like their counterparts in America and Britain, German scientists also began seeking element 93's fissionable, longer-lived, more stable daughter, element 94 – plutonium. In early 1941, theoretical physicist Fritz Houtermans concluded that a reactor fuelled by natural uranium could manufacture plutonium which could be chemically extracted and used to make a bomb – a discovery that both excited and disturbed him.

---

* Fritz Strassmann is commemorated at the holocaust centre in Jerusalem, Yad Vashem, as one of the 'righteous gentiles' who came to the aid of persecuted Jews.

Houtermans – in Otto Frisch's words an 'impressive eagle of a man' but 'not quite adult' and with 'an over-developed sense of humour which he often exercised at the expense of his colleagues . . . and no discipline' – was fortunate to be alive. He had been born in Danzig* to a wealthy Dutch banker and his half-Jewish Viennese wife. He had rejected his father's bourgeois values but was proud of his Jewish ancestry. He had grown up in Vienna and, as a young man, had been psychoanalysed by Freud until he admitted he had been making up the dreams he so vividly related.

While visiting his father in Germany, Houtermans, who had become a communist, had come to the attention of the Gestapo, who arrested and interrogated him. After his release, he fled first to Britain and then to the Soviet Union. However, in 1937 he fell victim to Stalin's purges, was arrested by the Soviet secret police and spent the next two and a half years in prison where he was tortured and questioned relentlessly. Made to stand for days on end and revived with buckets of icy water when he fainted, his feet became so swollen his shoes had to be cut off. Sometimes he was stretched against a wall and his feet kicked back until his whole weight rested on his fingertips – an agonizing position for any length of time. Ernest Rutherford's favourite protégé Peter Kapitza helped Houtermans' wife and their two children to get out of Russia, but he could do nothing for Houtermans himself. Thin and broken – a 'former human being', as he introduced himself to another prisoner – Houtermans kept himself sane in the appalling conditions by performing complex mental mathematics and scratching

* Danzig is now Gdansk in Poland.

equations with a matchstick on scraps of soap.

In 1940, as a result of Stalin's pact with Hitler, Houtermans had been taken to the border town of Brest-Litovsk, handed back to the Nazis as a 'German' and immediately arrested by the Gestapo as a suspected Soviet agent. He managed to send a brief message, 'Fizzl [Houtermans' nickname] is in Berlin', to a friend, who guessed he must be in prison. This man hurriedly enlisted the help of Max von Laue, who was also Houtermans' friend and who used his influence to secure his release. Houtermans found himself a job with inventor and scientist Manfred von Ardenne, who had a private laboratory in a suburb of Berlin. Von Ardenne was interested in fission studies and, perhaps surprisingly, had persuaded the German Post Office to divert some of its large but mostly unallocated research budget to him. It was in von Ardenne's laboratory that Houtermans made his perturbing discovery.

Houtermans decided that he must get a warning out of Germany and chose as his messenger Jewish scientist Fritz Reiche. Reiche was still living in Berlin with his family, but in circumstances of such stress and isolation that his daughter had had a breakdown. He had finally, after repeated desperate efforts, secured visas for the family to emigrate to America. They departed just six weeks before the implementation of new laws forbidding any further Jewish emigration. Reiche reached the United States safely in April 1941 and passed on Houtermans' message to physicist Rudolf Ladenburg at Princeton. Because of the risks of carrying anything on paper, Reiche had committed Houtermans' words to memory. As he later recalled, Houtermans had asked him to say, 'We are trying here hard, including Heisenberg, to hinder the idea of making the bomb. But the pressure from above . . . Please say all this; that Heisenberg will not be able to withstand longer the pressure from the government to go very earnestly and seriously into the making of the bomb. And say to them, say they should accelerate, if they have already begun the thing . . .'

Ladenburg handwrote a note to Lynam Briggs, head of the US Uranium Committee, reporting Houtermans' warning. He also organized a dinner in New York for Reiche to meet his fellow refugees, including Eugene Wigner, Wolfgang Pauli, Hans Bethe and John von Neumann. Reiche told them what Houtermans had said. As he recalled, 'They listened attentively and took it [in]. They didn't say anything but were grateful.' No doubt for all those at the dinner events in Germany were gathering an ominous momentum. The nightmare of an atomic bomb in Nazi hands might indeed become a reality, justifying Leo Szilard's bleak conviction that 'Hitler's success could depend on it'.

Two other countries, both shortly to become combatants, had also been assessing the potential of nuclear fission during the previous two years.

In the Soviet Union, official interest in fission was slow to ignite. Under Stalin's pact with Hitler, the Soviets had occupied part of Poland in September 1939 and had fought a brief war with Finland, which had ended in March 1940. However, at that time the Soviet Union remained on the sidelines of the European war – a position Stalin intended she should occupy as long as possible – with no particular impetus to explore the threats or possibilities of atomic weapons.

Most Soviet scientists were anyway sceptical about the

immediate applications of nuclear fission. They were also still reeling from Stalin's purges and reluctant to draw attention to themselves by promoting initiatives that might not succeed. At a conference held in Kharkov in November 1939, scientists had concluded that although 'the possibility of using nuclear energy' had been discovered, the chances of achieving it were 'fairly fantastic'. Peter Kapitza agreed, believing that separating isotopes of uranium would require 'more energy than one could count on obtaining from nuclear reactions'.

Nevertheless, Russian scientists continued to conduct some experiments to test atomic theories, and, despite Szilard's attempts at censorship, enough articles and papers appeared in the American press to rouse their interest. These included a report by William Laurence in the *New York Times* of 5 May 1940 describing experiments with U-235 and suggesting that the implications of nuclear fission could be enormous, which was posted to a prominent Russian scientist by his historian son who was working at Yale. This new information, coupled with their own new experimental findings, convinced some Soviet scientists that their earlier reactions to fission had been too casual. In the summer of 1940 they began to lobby a more receptive government which set up a Uranium Commission, including Kapitza, and instructed it to draw up a research programme. Yet less than a year later, when on 22 June 1941 Hitler invaded the Soviet Union, reneging on their neutrality pact, Soviet scientists were immediately diverted from atomic research to other work perceived as more pressing.

In 1940 in Japan, Lieutenant-General Takeo Yasuda, a research engineer and director of the Aviation Technology Research Institute of the Imperial Japanese Army, had also noted reports on fission appearing in the foreign press. At his request, Yoshio Nishina – Niels Bohr's former pupil who had become Japan's leading physicist – began to look into the potential applications of fission. Nishina was currently building a large 250-ton cyclotron at Tokyo University – a successor to a smaller twenty-eight-ton device – using plans provided in a spirit of comradely co-operation by Ernest Lawrence's team at Berkeley. On the basis of advice from Nishina and others, in April 1941 the Imperial Army Air Force authorized the establishment of an atomic bomb project. The result was that with the incipient Russian programme wavering and Britain and America still pondering the way ahead, Germany and Japan were the only countries with military research projects specifically dedicated to establishing the feasi-bility of an atom bomb.

On 13 April 1941, Japan and Russia signed a five-year neutrality pact. On 2 July, ten days after the German attack on Russia, an imperial conference was held in Tokyo to discuss Japan's territorial aspirations. Among the decisions taken by the conference and approved by Emperor Hirohito were measures to hasten the end of the protracted war in China and for an advance south 'in order to establish a solid basis for the nation's preser-vation and security'. The goal was the establishment of a Greater Asia Co-Prosperity Sphere under which Japan would satisfy her aspirations for more land and be guaranteed access to natural resources such as oil and iron lacking in her home islands by imposing a hegemony over much of her region. The initial step would be an early advance into French Indo-China. The secret documents outlining the conference's decisions for the first time also explicitly referred to 'war with Britain and the United

States' if they continued to block Japanese ambitions, and to the desirability of an attack on Russia if Germany destroyed her armies in the West.

Britain and America quickly protested at the Japanese advance into Indo-China, which the Vichy French authorities did not resist. America imposed economic sanctions on Japan, including a freeze on Japanese assets in the US and an oil embargo. Britain had already placed sanctions on Japan because of her earlier alliance under the Anti-Comintern Pact with Britain's enemies Germany and Italy, but now she also froze Japanese assets. As a consequence of American actions, the Emperor ordered the abandonment of any attack on Russia and the creation thereby of a multi-front campaign. However, in September 1941 he approved the stepping up of plans for an attack on America if diplomatic negotiations failed to secure a sufficiently free hand for Japan in the Far East as well as the removal of the oil embargo. The latter was biting hard: if not lifted, it would cause Japan's armies to run out of fuel oil within two years.

Negotiations with the United States remained dead-locked, the Americans insisting on full Japanese withdrawal from China as the price for the removal of sanctions. America also declined a Japanese suggestion of a summit meeting between President Franklin Roosevelt, increasingly preoccupied with support of Britain in the West, and Prime Minister Fumimaro Konoe. War became even more likely when, in mid-October, Emperor Hirohito replaced Konoe with General Hideki Tojo, one of the strongest proponents of war and expansion. On 8 November the Emperor received plans for an attack on Pearl Harbor.

CHAPTER TWELVE

'HE SAID "BOMB" IN NO UNCERTAIN TERMS'

WINSTON CHURCHILL HAD GIVEN THE GREEN LIGHT TO THE British atom bomb project at the end of August 1941, but the scale and ambition of the Maud Committee's recommendations worried many of those involved. Building an atomic bomb could cost millions of pounds and it seemed doubtful whether Britain, suffering sustained and heavy bombing and short of manpower, could construct the necessarily enormous plants in time to affect the outcome of the war. She needed help, and the obvious place to seek it was across the Atlantic.

News of the Maud Committee's work had reached the United States quickly, even before Churchill had given the go-ahead. A copy of the draft report was sent to Lynam Briggs as chairman of the Uranium Committee, but he proved unresponsive. In the summer of 1941, Mark Oliphant, a passionate advocate of complete Anglo-American co-operation, was sent to America to discover why nothing had been heard from Briggs. He was 'amazed and distressed' to discover that Briggs had simply tossed the report into his safe without showing it to the other members of the committee. Perhaps Briggs thought he was being discreet, or perhaps his action – or lack of it – reflected the reality that not a single member of his committee was truly convinced that uranium fission had military potential.

Oliphant attended a meeting of the Uranium Committee to convince them otherwise. Extremely short-sighted and totally deaf in one ear, the Australian was outwardly an unlikely emissary. However, as Samuel Allison of Chicago University recalled, 'he said "bomb" in no uncertain terms. He told us we must concentrate every effort on the bomb and said we had no right to work on power plants or anything but the bomb. The bomb would cost twenty-five million dollars, he said, and Britain didn't have the money or the manpower, so it was up to us.' Leo Szilard was so impressed by Oliphant's passion as the latter toured key laboratories around the States cajoling his fellow physicists into action that he later joked that Congress should create a special medal to recognize 'distinguished services' by 'meddling foreigners'.

A few weeks earlier, in July 1941, a draft copy of the Maud Report had, however, also reached Vannevar Bush, former Dean of Engineering at the Massachusetts Institute of Technology and then president of the Carnegie Institution. A man of vigour and vision, in June 1940 he had talked President Roosevelt into appointing him head of a new National Defense Research Committee (NDRC) and had swiftly assumed oversight of the Uranium Committee, whose torpor annoyed him. Roosevelt had subsequently appointed him director of the new Office of Scientific Research and Development (OSRD), reporting directly to the President and responsible for the NDRC.

Roosevelt had thus made Bush, in effect, the United States' science chief. Even so, Bush found it hard to galvanize the Uranium Committee.

Bush's successor at the NDRC and his overall deputy was organic chemist James B. Conant, president of Harvard. A self-confessed Anglophile, Conant was a modest man with an excellent analytical brain who during the First World War had worked on the army's gas warfare programme. Like Bush, he was critical of the piecemeal way in which fission research was being conducted by universities and private and public institutions across the United States.

As a result of his own concerns, Bush had, in April 1941, requested the National Academy of Sciences – America's scientific elite – to appoint a committee of physicists to review uranium research. However, their two reports, focusing principally on the prospects for generating power from uranium fission, had disappointed him. The creation of violently explosive devices was, they said, a possibility, but too many uncertainties remained for them to make firm recommendations. Some National Academy physicists even thought the whole idea should be 'put in wraps' until the war was over.

Ernest Lawrence disagreed. In May 1941 he pointed out that bombs could be made without the complex processes necessary to separate U-235. Unseparated uranium could, he insisted, provide excellent bomb fuel. He reminded the committee of the recent success of two members of his team, Glenn Seaborg and Emilio Segrè, in creating, isolating and analysing a new element, which they had named plutonium and which fissioned almost twice as easily as U-235.

Against this background of cautious inertia in some

quarters and passionate advocacy in others, Conant and Bush privately put out feelers to Charles Darwin, director of the British Central Scientific Office in Washington, a former researcher of Rutherford's and grandson of the famous natural scientist. In a letter of 2 August 1941 to the British government, written by hand because of what he called the 'extreme secrecy' of its contents, Darwin reported that Conant and Bush had not only raised the issue of atomic bombs but had proposed a joint US/UK atomic bomb programme. Darwin also revealed that he himself had raised a wider issue – would any government ever deploy such a weapon in reality? 'Are', he wrote, 'our Prime Minister and the American President and the respective general staffs willing to sanction the total destruction of Berlin and the country round when, if ever, they are told it could be accomplished at a single blow?' His words echoed both those of Rudolf Peierls and Otto Frisch when they'd suggested to the British government that the civilian casualties resulting from an atom bomb 'may make it unsuitable as a weapon for use by this country', and those of President Roosevelt who on the outbreak of war in 1939 had urged belligerents to refrain from 'bombardment from the air of civilians or unfortified cities'.

However, for many on both sides of the Atlantic, a more pressing issue in 1941 was to determine whether an atomic bomb could actually be built.

When it landed on his desk in July 1941, Vannevar Bush found the draft Maud Report compelling reading. Not only did it give a cogent summary of the underlying science, it defined a concrete programme for taking the project forward that chimed with his own desire for action. By the time that, on 3 October, Conant and Bush received an official copy of the Maud Report, they had already decided to show it to the President and urge close collaboration between the United States and Britain. Mark Oliphant had travelled to Berkeley specifically to brief Ernest Lawrence and had spoken to him frankly about both the British work and the German threat. Further energized by these discussions, Lawrence had then gone out of his way to assure Bush and Conant that American experimental results confirmed the British conclusions.

To give themselves even more ammunition, Bush and Conant asked the National Academy of Sciences to carry out a fresh review to validate the British claims that an atomic bomb was feasible. On 9 October, while still awaiting the results of this review, Bush took the Maud Report to Roosevelt. He underlined the British conviction that a bomb with a destructive power equivalent to 1,800 tons of TNT could be made with just twenty-five pounds of active material and that the first bombs could be available by the end of 1943, but that achieving this feat, in particular building a plant to separate sufficient quantities of U-235, would require a huge and expensive industrial effort.

Impressed by Bush's advocacy, the President endorsed 'complete interchange with Britain on technical matters'. He also agreed that Bush could expand current fission research and assured him of sufficient funding from a special source without the need for explicit Congressional approval. Anxious to ensure political control of the programme and to restrict the scientists to their own sphere, Roosevelt decided to limit consideration of policy to a tiny inner circle. In addition to himself, Bush and Conant, the members were Vice President Henry Wallace, Secretary of War, the white-haired, seventy-seven-year-old Henry Stimson, and Army Chief of Staff General George C. Marshall. It became known as the Top Policy Group.

Bush was well pleased with the outcome. Roosevelt had recognized the imperative of determining the feasibility of an atomic bomb, had agreed to make the necessary resources available and had sanctioned collaboration with the British. On 11 October, the President offered Britain a partnership deal, writing to Winston Churchill, 'It appears desirable that we should soon correspond or converse concerning the subject which is under study by your Maud Committee and by Dr. Bush's organisation in this country in order that any extended efforts may be co-ordinated or even jointly conducted.'

In Germany that autumn the initiative rested with Germany's scientists rather than with their political masters. As Werner Heisenberg later told a British historian, 'It was from September 1941 that we saw an open road ahead of us, leading to the atomic bomb.' A month earlier, in August, Manfred von Ardenne had suddenly decided to circulate Fritz Houtermans' report 'On Triggering a Nuclear Chain Reaction', revealing that atom bomb fuel in the form of plutonium could be made in a reactor. Heisenberg claimed that this had induced a 'panic reaction' in him. Not only had this sensitive information been widely revealed among the German scientific community, but it made Heisenberg worry whether scientists abroad had also discovered this and were even then planning massive plants to manufacture plutonium. A letter to a friend revealed Heisenberg's nervous frame of mind. He wrote, 'perhaps we humans will recognize one day that we actually possess the power to destroy the earth completely, that we could very well bring upon ourselves a "last day" or something closely related to it'.

Houtermans' findings actually caused few ripples in Germany where, in the late summer of 1941, as their troops advanced ever deeper into Russia, most people were convinced of an early and victorious end to the war. However, Heisenberg later claimed, the progress of nuclear research forced him to confront certain moral issues at that time. Should he and others disengage from fission research? Alternatively, should they try to ensure their efforts focused on nuclear power not nuclear weapons? In his memoirs, Heisenberg wrote, 'We all sensed that we had ventured onto highly dangerous ground.' He also recalled a conversation with Carl-Friedrich von Weizsäcker during which the two men discussed their worries. According to Heisenberg, 'Von Weizsäcker said something like, "At present, we don't have to worry about atom bombs, simply because the technical effort seems quite beyond our resources. But this could easily change. That being so, are we right to continue working here? And what may our friends in America be doing? Can they be heading full steam toward the atom bomb?" '

Heisenberg remembered that he tried to put himself into their position, acknowledging that refugee scientists must be firmly convinced that they were 'fighting for a just cause' and that 'even the good fight invariably involves some bad means'. However, he suggested, 'is there not a point beyond which [the scientist] cannot go under any circumstances?' 'All in all,' he concluded, 'I think we may take it that even American physicists are not too keen on building atom bombs.' But, he added, 'they could, of course, be spurred on by the fear that we may be doing so'.

It was then, according to Heisenberg, that von Weizsäcker suggested a solution to their dilemma. ' "It might be a good thing," Carl-Friedrich told me, "if you could discuss the whole subject with Niels in Copenhagen. It would mean a great deal to me if Niels were, for instance, to express the view that we are wrong and that we ought to stop working with uranium." '

There had been no direct contact between Heisenberg, von Weizsäcker and Bohr for nearly a year after the occupation of Denmark. Then, in March 1941, von Weizsäcker had been invited to Copenhagen to lecture at the newly opened German Cultural Institute, a propaganda organization to promote Germanic 'values' among the conquered Danes. There he had met Cecil von Renthe-Fink, the German plenipotentiary in Denmark, a friend of von Weizsäcker's high-ranking father in the German Foreign Office, who left the visitor in no doubt of Bohr's uncompromising attitude towards his country's occupiers. Bohr, he said, would have absolutely nothing to do with the Germans.

Nevertheless, Heisenberg sought a way to engineer a private meeting with Bohr. He later claimed that, without revealing his true purpose, he asked the German Embassy in Copenhagen to organize a visit for him and officials arranged for him to speak at a series of lectures on astrophysics at the German Cultural Institute. The series, a propaganda exercise intended to follow up von Weizsäcker's March visit, had in fact been in gestation for some months, possibly before Houtermans' discoveries about plutonium, and Heisenberg's participation had already been discussed. It is therefore unclear which came first, Heisenberg's decision that he had to speak to Bohr or the invitation to visit occupied Denmark. What is clear from the records is that the authorities were initially wary of allowing Heisenberg out of the country. Only when the German Foreign Office, perhaps at von Weizsäcker senior's prompting, suggested that allowing him to go to Copenhagen would be a good test of his suitability to lecture at other propaganda events in occupied Europe was permission finally given.

On 14 September 1941, accompanied by von Weizsäcker, Heisenberg caught the night train to Copenhagen for a meeting that would damage a twenty-year-long friendship and spawn enduring controversy over what he actually said and why.

Denmark's eight thousand Jews were, for the most part, still living unmolested. At this stage in the war the Germans were treating the Danes with care, anxious not to provoke resistance on their doorstep while their forces were busily, and successfully, engaged elsewhere. Denmark was allowed a degree of automony, and Niels Bohr and his Institute for Theoretical Physics were permitted to function relatively normally. On the surface, life went on. However, as one Nazi propagandist observed, 'A feeling of quiet rage prevails here, which only comes to the fore when the Danes believe themselves alone and unobserved.' If the Germans did observe overt dissidence they dealt with it mercilessly. Just before Heisenberg arrived in the city, a group of Danish communists and other known opponents of the Nazis had been deported to Germany.

Despite repeated encouragement from Allied agents to escape, Bohr stayed, determined to protect his institute and the people in it. However, as he must have known, his position was precarious. His name had long been on Gestapo files, not just as a prominent scientist who was half-Jewish but because he had spoken out against Nazi ideology. At an international congress on anthropology on the eve of the war he had denounced prejudice and argued that 'different human cultures are complementary to each other'. He had spoken feelingly of the 'unlimited richness and variety' of human life, at which point the German delegates had walked out.

Bohr refused to attend Heisenberg's lecture. Heisenberg, however, lunched several times at Bohr's institute with Bohr's staff and, according to a post-war account by Bohr's assistant, Polish émigré Stefan Rozental, appeared to feel little awkwardness about being in occupied Copenhagen. He 'spoke with great confidence about the progress of the German offensive in Russia' and stressed 'how important it was that Germany should win the war'. While regretting the occupation of western countries such as Denmark and Holland, he had no doubt that German rule was 'a good development' in eastern Europe 'because these countries were not able to govern themselves'.

Bohr was, however, clearly prepared to welcome Heisenberg privately as an old friend, and as the man his sons thought of as their German uncle. But their reunion inevitably took place against 'a background of extreme sorrow and tension for us here in Denmark', as Bohr later wrote. The two met and talked at the institute and may have dined together at Bohr's house, although accounts conflict. They certainly went for a walk, probably to frustrate Gestapo surveillance, and it was then that their critical discussion seems to have taken place. The risks of discovery were too great for either man to commit anything to paper at the time. Any reconstruction of the meeting therefore relies on explanations and interpretations made many years later in a world where the atom bomb had been dropped and the full scope of Nazi atrocities revealed; moreover, in a world where Heisenberg was anxious to distance himself from the Nazis while Bohr's anxiety about the Cold War arms race may have influenced his memories.

Heisenberg's first written account of what happened was in a letter of 1948 to a Dutch friend. He recalled that he had asked Bohr whether a physicist had a moral right to work on problems in atomic physics relevant to the war. Bohr, in turn, had asked Heisenberg whether military applications of atomic power were feasible. When Heisenberg replied that they were, Bohr's apparent response was that a mobilization of physicists on both sides was unavoidable and therefore justified. Heisenberg believed that Bohr thereby dismissed his implicit suggestion that physicists of the world should band together against their governments.

Heisenberg provided a more detailed account of the meeting in a letter to writer Robert Jungk in the 1950s. Before launching into his version of events in Copenhagen, he summarized the state of German fission research in the autumn of 1941. Firstly, German scientists believed it was possible to build a reactor that could generate energy. Secondly, although they had not yet solved the problem of separating U-235, they knew that they could produce plutonium, which could, in turn, be used to fuel a bomb. However, this could only be accomplished in 'huge reactors' which would have to operate for years. The production of bombs was therefore only possible with 'enormous technical resources'. Heisenberg believed that this put scientists in a 'favourable' position. Had bombs been easy to make, then physicists 'would have been unable to prevent their manufacture'. Instead, they could play a decisive role by deciding what advice to give their governments. They had two choices: to say 'that atomic bombs would probably not be available during the course of the war', or to say that 'there might be a possibility of carrying out this project if enormous efforts were made'. It was against this background that he had gone to Copenhagen to seek out Bohr.

Heisenberg claimed, in his letter to Jungk, that he opened the discussion by asking Bohr whether he believed it was right in time of war for physicists to devote themselves 'to the uranium problem – as there was the possibility that progress in this sphere could lead to grave consequences in the technique of war'. Heisenberg realized from Bohr's 'slightly frightened reaction' that he had immediately understood what Heisenberg meant. As Heisenberg recalled, the Dane responded by asking, 'Do you really think that uranium fission could be utilized for the construction of weapons?' According to Heisenberg, he 'may have replied, "I know that this is in principle possible, but it would require a terrific technical effort, which, one can only hope, cannot be realized in this war." ' Bohr was 'shocked', assuming that Heisenberg had 'intended to convey to him that Germany had made great progress' towards atomic weapons. Heisenberg tried to correct this false impression but could not.

In his later memoirs Heisenberg dealt more briefly with

the Copenhagen meeting, but the same leitmotifs appear – his attempt to raise the moral dimension with Bohr, his admission that atomic weapons were possible but only with 'a tremendous technological effort', and his insistence that scientists were in a pivotal position, able to 'advise their governments that atom bombs would come too late for use in the present war, and that work on them therefore detracted from the war effort, or else contend that, with the utmost exertions, it might just be possible to bring them into the conflict'. Once again, he claimed Bohr was too shocked to take in what he was saying. Heisenberg left Copenhagen knowing that his mission had failed. He blamed himself for having spoken too guardedly because he was afraid for his life. Had he been more explicit, Bohr would have understood him.

Bohr's picture of Heisenberg's visit, which his wife Margrethe later unequivocally described as 'hostile', was quite different. He was immediately struck by something peculiar in Heisenberg's demeanour and their meeting deteriorated rapidly. Bohr's earliest account is reported in a letter written in 1946 by American physicist Rudolf Ladenburg, who repeated Bohr's comments that Heisenberg had expressed the 'hope and belief' that if the war lasted long enough atomic weapons would give Germany victory.

Nearly a decade later, when Bohr read Heisenberg's account of events reproduced in Robert Jungk's book, he was shocked and angered. In a recently released letter to Heisenberg, drafted but never sent and discovered tucked into Bohr's copy of Jungk's book after Bohr's death in 1962, Bohr wrote, 'I am greatly amazed to see how much your memory has deceived you'. He taxed Heisenberg with expressing the 'definite conviction that Germany would win', and that it was 'therefore quite foolish for us to maintain the hope of a different outcome of the war and to be reticent as regards all German offers of co-operation'. He also remembered 'quite clearly' that Heisenberg spoke in a manner that could only give Bohr 'the firm impression' that under his leadership 'everything was being done in Germany to develop atomic weapons'. Furthermore, Heisenberg had said that 'there was no need to talk about details [of atomic weapons]' since Heisenberg was 'completely familiar with them and had spent the past two years working more or less exclusively' on them.

Bohr also refuted Heisenberg's suggestion that the news that atom bombs were possible had stunned him into silence, insisting that the possibility of nuclear weapons had been 'obvious' to him for a while. The reason he had not spoken was twofold. Firstly, a 'great matter for mankind was at issue in which, despite our personal friendship, we had to be regarded as representative of two sides engaged in mortal combat'. Secondly, he was dismayed to learn 'that Germany was participating vigorously in a race to be the first with atomic weapons'.

Further unsent letters written in subsequent years reveal a softening of Bohr's attitude to Heisenberg – some affection still remained. A draft letter to Heisenberg congratulating him on his sixtieth birthday ended with 'fondest greetings and warmest wishes for many happy years'. However, Bohr continued to agonize over Heisenberg's motives. He asked Heisenberg to clarify on whose authority he had come to Copenhagen, writing, 'I have often wondered from which official police agency permission was given to talk to me about a question which was surrounded by such great secrecy and held such great dangers'. He wanted Heisenberg to tell him 'what purpose lay behind' his visit. It was 'quite incomprehensible' to him how Heisenberg could claim to have suggested to Bohr 'that German physicists would do all they could' to prevent an atomic bomb. On the contrary, 'you [Heisenberg] informed me that it was your conviction that the war, if it lasted sufficiently long, would be decided with atomic weapons and I did not sense even the slightest hint that you and your friends were making efforts in another direction'.

Over the years many suggestions have been made about Heisenberg's true purpose. Some have claimed that Heisenberg was trying to discover what Bohr knew about the Allied programme; Bohr later told Oppenheimer that Heisenberg and von Weizsäcker had come 'less to tell what they knew than to see if Bohr knew anything that they did not'. It certainly appears that during his visit to Copenhagen in March 1941 von Weizsäcker had been fishing for information. He reported to the Nazi authorities on his return that 'concerning the more technical questions' Bohr 'knew a great deal less than we'.

Others have suggested that Heisenberg was trying to pass messages about the German programme to the Allies. According to von Weizsäcker, this was true. In a letter in 2002, he wrote that 'we hoped that Bohr could tell colleagues in England and the USA that we were no longer working on a bomb'. Elizabeth Heisenberg, in her book *Inner Exile* about life with her husband, similarly suggested that Heisenberg 'saw himself confronted with the spectre of the atomic bomb, and he wanted to signal to Bohr that Germany neither would nor could build a bomb. That was his central motive. He hoped that the

Americans, if Bohr could tell them this, would perhaps abandon their own incredibly expensive development. Yes, secretly he even hoped his message could prevent the use of an atomic bomb on Germany one day. He was constantly tortured by this idea.'

A particularly intriguing dimension is a story that during their meeting Heisenberg gave Bohr a drawing which, later in the war after his escape from Denmark, Bohr sent to the United States. Robert Oppenheimer, Edward Teller and Hans Bethe puzzled over the sketch, which showed a box-like structure with stick-like objects projecting from the top. Bethe recalled, 'as far as we could see, the drawing represented a nuclear power reactor with control rods. But we had the preconceived notion that it was supposed to represent an atom bomb. So we wondered, "Are the Germans crazy? Do they want to drop a nuclear reactor on London?" ' After further study, he and Teller concluded that it 'was clearly a drawing of a reactor'. Some have claimed that Heisenberg gave the sketch to Bohr as proof that the Germans were working on peaceful not military applications of nuclear power. However, Aage Bohr, in whom his father confided, always maintained that Heisenberg gave Bohr no such thing. If the drawing was indeed of German origin, it must have been provided by someone else, although the risks would have been enormous. Otherwise, it may have represented Bohr's own interpretation of German thinking based on his discussion with Heisenberg and others. To heighten the mystery, despite extensive searches in archives in Denmark, the United States and the United Kingdom, no trace of the sketch can now be found.

According to Hans Bethe, the Copenhagen meeting was doomed to fail. 'It was impossible that the two of them could understand each other. Heisenberg knew about plutonium and was convinced it was the key to the whole business. Bohr didn't know about plutonium. Therefore there was a technical misunderstanding.' In Bethe's view, practical considerations also intervened: 'Bohr was much better at speaking than listening and he mumbled.'

No-one will ever know the full truth of what happened and why. Perhaps Heisenberg himself did not know exactly what he was trying to achieve. Perhaps he felt intuitively that to talk to his father-figure, Bohr, would clear his mind, as it had previously on scientific questions. An English officer after the war recorded private conversations between Heisenberg and von Weizsäcker and reported that 'They seem to consider international physics as being almost synonymous with work under the leadership of Niels Bohr.' Aage Bohr suggested that, as well as respect for his father, affection also played a role: 'Undoubtedly one of the reasons why Heisenberg went to Copenhagen was to see if there was anything he could do for his Danish physicist friends living as they were in an occupied country. Heisenberg had a strong sense of loyalty towards them.'

However, Heisenberg certainly did not appreciate the sensitivities of the situation. Although eager in his memoirs to portray his readiness to put himself in others' shoes, he was not good at it in reality. He could not place himself in Bohr's position as a half-Jewish citizen of a country occupied by a brutal regime standing for everything Bohr despised. He also had an ability to ignore unpleasant truths and unconsciously to twist them into a more palatable form.

According to Hans Bethe's account of a post-war

discussion with Heisenberg, the latter believed passionately that Germany should win the war. 'He said he knew
that the Germans had committed terrible atrocities
against the populations on the Eastern Front – in Poland
and Russia – and to some extent in the west as well. He
concluded that the Allies would never forgive this and
would destroy Germany as a nation – that they would
treat Germany about the way the Romans had treated
Carthage. This, he said to himself, should not happen;
therefore, Germany should win the war, and then the
good Germans would take care of the Nazis.' Bethe found
it 'unbelievable that a man who has made some of the
greatest contributions to modern physics should have
been that naive'.

Heisenberg's reasons for visiting Copenhagen were
undoubtedly complex, quite possibly confused and quite
likely a combination of the various motives alleged. He
may have convinced himself that building an atom bomb
before the war was likely to end was currently beyond
the capability of any country. However, to protect the
Germany he still loved he had to be sure that Britain and
the United States were not making faster progress. Until
he knew whether Germany was at risk he could not decide
how he, and his fellow scientists, should act.

Whatever his motives, Heisenberg clearly never got the
chance to say all he wanted because Bohr became so
agitated. All he achieved was to convince Bohr that
Germany was actively pursuing the atomic bomb.

Suspicion about Heisenberg's visit spread quickly. Lise
Meitner was alarmed to learn of it from the young Danish
physicist Christian Møller when he visited Stockholm
several months later. He reported Heisenberg to be

'entirely filled with the wish-dream of a German victory'. Meitner wrote hastily to Max von Laue in Berlin, who, unknown to his colleagues, had been corresponding regularly with her. They took the precaution of numbering their letters so that they would know whether any had been intercepted. Like Fritz Strassmann, risking his life and that of his family to save the Wolffenstein sisters, von Laue stayed true to his principles, visiting the elderly and lonely Jewish former editor of the scientific journal *Naturwissenschaften*, Arnold Berliner, marooned in his apartment by fears of anti-Semitic violence until his imminent deportation induced him to commit suicide. Meitner warned von Laue, in guarded, elliptical language, to be wary of Heisenberg and von Weizsäcker. She conveyed that she had once thought very highly of the two, but added grimly, 'it was a mistake'.

Von Laue was not particularly surprised by Meitner's warning, replying to her with remarkable perception, 'I have often wondered about the inner attitude of Werner and Carl-Friedrich, but I believe I understand their psychology. Many people, especially young ones, cannot reconcile themselves with the great irrationality of the present, and so in their imagination they construct castles in the air. It is an enormous task they have undertaken, to find a good side in things they can do nothing about.' Heisenberg and von Weizsäcker were not, he added, alone in this.

CHAPTER THIRTEEN

'WE'LL WIPE THE JAPS OUT OF THE MAPS'

FRANKLIN ROOSEVELT'S OFFER TO BRITAIN IN OCTOBER 1941 of a partnership received a lukewarm reception. Churchill's preference, like that of his chiefs of staff and some scientists including James Chadwick, was that any nuclear bomb should be developed in Britain. Not only did Churchill wish to keep control, but he mistrusted American security. In fact, the desire for secrecy was guiding much British thinking. The research director of Imperial Chemical Industries, Wallace Akers, had recently been appointed to head work on both civil and military uses of nuclear energy. He and officials racked their brains for a name for his new organization that had 'a specious air of probability about it' while concealing its true purpose. They came up with the plausible-sounding 'Directorate of Tube Alloys'.

Churchill allowed two months to elapse before replying to Roosevelt's letter. He wrote blandly, 'I need not assure you of our readiness to collaborate with the United States Administration in this matter'. He told the President that he had arranged for the US scientific liaison officer in

London, Mr Hovde, to have full discussions with Sir John Anderson, the government minister responsible for 'Tube Alloys', and Churchill's adviser, the recently ennobled Frederick Lindemann, now Lord Cherwell. When they met in November 1941, Anderson, an able man but so haughty and inflexible in manner that he was nicknamed 'God's butler', coolly informed Hovde that, while the British were anxious to collaborate, they were 'disturbed about the possibility of leakage of information to the enemy' through the officially neutral United States. Britain wanted strict assurances that the American project would be run in such a way as to preserve maximum secrecy. Unsurprisingly, given Anderson's approach, instead of agreeing a joint project the meeting ended with a simple agreement to share some information and with condescending British offers to help 'improve' the American organization.

As Britain would soon discover, collaboration restricted to an exchange of information would not be enough to keep her in the race. Ironically, the brilliant Maud Report had provided the impetus for an American programme which would soon surge ahead with increasingly little need of, or desire for, British assistance. French scientist Bertrand Goldschmidt, observing from the sidelines, later wrote that 'in this ballet of Anglo-American nuclear relations, the Americans [Vannevar] Bush and [James] Conant revealed themselves to be by far the best and most perspicacious advisers . . .' By contrast, British officials, in his view 'imbued with the antiquated dogma of imperial superiority', missed the boat.

By late October 1941, the National Academy of Sciences submitted its new report, which, Conant wrote approvingly, 'radiated a more martial spirit than the first two'. It concluded that a fission bomb 'of superlative destructive power' was possible. Based on calculations by Enrico Fermi at Columbia, it estimated that the amount of U-235 required to achieve this could 'hardly be less than 2 kg nor greater than 100 kg'. It predicted a somewhat lower destructive force than the Maud Committee, but that the loss of life from the effects of radioactivity might 'be as important as those of the explosion itself'. Furthermore, fission bombs might be achievable 'in significant quantity within three or four years' at a cost of between $80 million and $130 million. It defined the priority tasks, in particular assessing the different techniques for separating isotopes of uranium and understanding the engineering requirements of separation plants. Like the Maud Report, it did not mention plutonium.

Exactly as German scientists had told the German Army, the report warned that in years to come, military superiority would depend on who had nuclear bombs and that 'adequate care for our national defense seems to demand urgent development of this program'. On 27 November, Bush handed the report to President Roosevelt.

In Russia, Hitler's exhausted troops were making one last attempt to capture Moscow before winter closed in, inevitably prolonging the war in the East. On Friday, 5 December 1941, three feet of snow fell. Nevertheless, eighty-eight new Russian divisions attacked on a five-hundred-mile front and broke through for eleven miles in places before the Germans could stabilize their lines. The next day, 6 December, Hitler was forced to accept that Moscow would not fall quickly.

On that same Saturday, Vannevar Bush told members of the Uranium Committee, gathered in Washington, how he had allocated tasks within the expanded programme sanctioned by Roosevelt – known as the 'S-1 Project'. He appointed three 'programme chiefs', all Nobel Prize winners: Ernest Lawrence was to explore electromagnetic techniques for separating U-235 at Berkeley, and to supply the first samples of enriched uranium for experiments; Harold Urey, discoverer of deuterium, was to develop gaseous diffusion separation methods, as advocated by the Maud Committee, at Columbia University; and Arthur Compton, pioneer of gamma ray research, was to be responsible for theoretical studies and bomb design at the University of Chicago, where he himself was based, and elsewhere. In addition, industrialist Eger V. Murphree, Standard Oil of New Jersey's research director, was to oversee the development of high-speed centrifuge technology for separating U-235 and to take responsibility for broad engineering issues such as procuring materials and constructing pilot and production plants.

Lawrence had become an increasingly strong proponent of electromagnetic separation as the simplest route for U-235 production. The process used magnetic force in a device called a mass spectrograph to bend beams of charged uranium particles into circular paths. The precise path of each particle was determined by its mass. Therefore, the heavier isotope U-238, with its greater mass, took a different path to the lighter U-235, making it possible to collect the two in separate containers. Lawrence ordered his team to convert Berkeley's thirty-seven-inch cyclotron into a mass spectrograph. On the very morning that Bush was assigning work to his top scientists, news came that the Berkeley spectrograph had started work. It was producing a microgram of U-235 an hour – not much, but a start.

Also in Washington that busy Saturday, in the US Navy Cryptographers' Department, a young woman translated a decoded secret Japanese telegram sent four days previously from Tokyo, asking the Japanese consul in Hawaii to report on US berthing positions, ship movements and torpedo netting at Pearl Harbor. Concerned, she took the message and some related ones to her head of department, who told her that he would get back to it on Monday.

Just over twenty-four hours later, in Hiroshima Bay, the Commander-in-Chief of the Japanese Fleet was piped aboard his flagship, moored among other battleships, to hear the first messages coming in about the success of the surprise attack on the anchored US fleet in Pearl Harbor by planes from his carrier fleet. That afternoon, Congress declared war on Japan. On Thursday, 11 December, Hitler declared war on the United States. Winston Churchill's reaction was that Britain had won the war after all. Only a few weeks later, Enrico and Laura Fermi heard their five-year-old son Giulio cheerily singing a verse he had picked up from other little boys:

> We'll wipe the Japs
> Out of the maps.

The people of Hiroshima welcomed the massive extension of the war, which for them had been under way since the invasion of Manchuria more than ten years previously. They rejoiced at the continuing success of their forces as they conquered Hong Kong, Malaya, the Philippines,

Singapore, Borneo, the Dutch East Indies (today's Indonesia) and many Pacific islands such as Guam. They shared their Emperor's view that 'the fruits of war are tumbling into our mouth almost too quickly'. They hummed patriotic songs such as 'Divine Soldiers of the Sky', about Japanese paratroopers descending on the foe like 'pure white roses' from heaven. They read cleverly drawn and widely circulated comic strips showing the victorious Japanese forces 'saving the country from foreigners' by cutting 'the iron chain with which the Anglo-Americans had surrounded Japan'. The comic strips also showed how Japan was heading the Greater Asia Co-Prosperity Sphere, eliminating pernicious Western influence throughout the region and replacing it with Japan's own 'imperial way'. People flocked to Hiroshima's cinemas where newsreels showed advancing Japanese troops welcomed by smiling local people. Propaganda features followed, such as *Suicide Troops of the Watchtower*, about a Korean guerrilla who came to appreciate the justice of the Japanese cause and slew his own comrades.

At the same time, Hiroshima geared up for the expanding war. A new industrial port and an army airport were swiftly constructed. Extra workers were recruited for the naval dockyard established on reclaimed land in the south-west of the city in 1939 by Mitsubishi Heavy Industries. Small workshops sprang up in houses all over the city to produce parts or simple military equipment. Among the new workers were many young women. They dressed not in kimonos but in more practical tunics and work pants and cheerfully attended the mandatory military drills. Other women, members of the Defence Women's Association, donned their purple and white sashes to stand outside Hiroshima's department stores soliciting help with their *sennimbari* – strips of white cloth decorated with a thousand red stitches, each sewn by a different woman. The *sennimbari* were presented to soldiers off to the war to wish them good luck and long life.

Many of the city's male inhabitants had been conscripted and boarded transports in Hiroshima Bay for duty overseas. Their wives and families could not help but worry. The situation of the family of tailor Isawa Nakamura was not unusual. His wife Hatsuyo heard no news from him for a long time. Then, in March 1942, came a brief telegram: 'Isawa died an honourable death at Singapore'. Promoted to corporal, he had been killed on 15 February, the day Singapore fell. Army payments to Mrs Nakamura ceased on his death and she had no alternative but to use her husband's sole legacy – his sewing machine – to get work to sustain herself and her three children, Toshio, Yaeko and Myeko.

Government information was cascaded down to the people through a series of organizations stretching from the prefecture to town associations and then to neighbourhood associations. The latter might consist of only ten or twenty households, while a town association might have seventy neighbourhood associations within it. The neighbourhood associations were already implementing food rationing, which had begun in the largest cities in 1940 and a little later in Hiroshima. Now they redoubled their efforts to eliminate the black market and, aided by dramatic posters captioned 'Donate to Win', encouraged the collection of metals for the war effort. Families piled their cooking pots up along with the iron railings from their verandahs to be melted down for weapons.

One Hiroshima resident recalled how much she 'missed the sound of the temple bells' since even they were demounted and fed to the furnaces.

The expansion of the war also gave a fillip to Japan's nuclear programme. A week after the attack on Pearl Harbor, the Imperial Navy – traditionally a fierce rival of the Imperial Army and resentful of the atomic bomb project that Lieutenant-General Takeo Yasuda had asked Yoshio Nishina to initiate – convened a meeting of scientists and technicians. Captain Yoji Ito of the Navy Technology Research Institute announced to the gathering that the navy too wished to develop a nuclear weapon. He asked Nishina, who was present, to get involved in this project also.

Nishina agreed. Although he personally believed the attack on Pearl Harbor to have been insane, he was a patriot. Furthermore, the army's interest in atomic research was waning as a result of his predictions of the long timescales involved. The navy had more money and, he hoped, perhaps more patience. Although Nishina was not enthusiastic about building a bomb, he recognized that naval support would enable him to continue his own nuclear research. He was joined on the naval project by the elderly Hantaro Nagaoka, who in gentler days had known Ernest Rutherford.

In Germany, the atomic bomb programme was about to fall victim to a military and political situation which by the end of 1941 had changed dramatically, with German troops bogged down outside Moscow and with Hitler's declaration of war against the United States in the aftermath of Pearl Harbor. Hitler ordered the total mobilization of the German economy to focus on an all-out pursuit of 'total war' to secure the victory that at the time of Heisenberg's visit to Copenhagen had seemed so nearly his. One consequence was that Army Ordnance ordered a review of all its research programmes, and Erich Schumann told the Uranium Club that fission research could continue 'only if a certainty exists of attaining an application in the foreseeable future'.

In February 1942, Army Ordnance called a conference on fission to which the scientists submitted a cautiously optimistic 144-page report. Provided that the army supplied them with appropriate materials, a successful working reactor to generate atomic energy could be expected shortly and would have the potential to power submarines and other warships. The prospects for nuclear weapons depended on developing techniques for separating uranium, but, the scientists reminded the army, there was another option for producing weapons material – generating plutonium in a reactor.

This conditional optimism was not enough. To many in the army, the fission project had only ever been 'Atomkakarei' – 'atomic babble'. The army cut its funding, abandoned research at the Kaiser Wilhelm Institute for Physics and restricted its programme henceforth to a modest reactor project in a Berlin suburb under Kurt Diebner. This left a vacuum which Minister for Education Bernhard Rust, the virulent anti-Semite who had taken great pride in expelling Jewish academics from their posts in 1933, was eager to fill. He wished the Reich Research Council, pushed out of fission research by the army at the start of the war, to take control. On the day that the Army Ordnance conference started, Rust organized a series of non-technical lectures, and several senior scientists, including Heisenberg, attended.

Heisenberg gave a lecture on the potential of nuclear power, into which he interpolated occasional allusions to possible weapons uses. 'Pure uranium-235', he announced, was 'an explosive of quite unimaginable force', although separating it was a problem. He then spoke of plutonium, explaining that the construction of a viable reactor was key: 'through the transmutation of uranium inside the pile [reactor], a new element is created which is in all probability as explosive as pure uranium-235, with the same colossal force'. Rust was impressed, but his attempt to put the Reich Research Council in charge of fission research sparked a battle with the Kaiser Wilhelm Society, sponsoring organization of the Kaiser Wilhelm Institute for Physics. Delighted to have got rid of the army, the society was hoping to regain control of its own programmes.

This power struggle would continue for many months, but in April 1942 the German government sanctioned Heisenberg's appointment as director of the institute and professor of theoretical physics at Berlin University. No longer regarded with suspicion by the Nazis as a 'white Jew', he was being welcomed into the fold, despite his continuing refusal to join the Nazi Party. According to his wife, he regarded it as a victory for 'modern physics' – what his critics had tried to dismiss as 'Jewish physics' – over 'German physics'.

In early June, Heisenberg was summoned to brief Albert Speer, the young and newly appointed Reichsminister for Armaments and Munitions, and military heads of weapon production on the seemingly encouraging prospects for the bomb. The subject had also caught the attention of Joseph Goebbels, who a few weeks earlier had noted in his diary, 'I received a report about the latest developments in German science. Research in the realm of atomic destruction has now proceeded to a point where . . . tremendous destruction, it is claimed, can be wrought with a minimum of effort . . . It is essential that we should keep ahead of everybody . . .' No formal record of the meeting survives, but according to Heisenberg he reported definite proof that atomic energy could be created in a reactor. He also said that it was theoretically possible to produce a powerful explosive in a reactor, and that, though many uncertainties remained, a single bomb could blow up a city the size of London. Afterwards, Speer, although by profession an architect not a scientist, perhaps scented some contradictions, for he questioned Heisenberg carefully about the real prospects for nuclear weapons. According to Speer, Heisenberg's answer was 'by no means encouraging'. He replied that 'the technical prerequisites for production would take years to develop, two years at the earliest, even provided that the pro- gramme was given maximum support'.

Speer was sufficiently intrigued to invite Heisenberg to go for a walk with him after the meeting so that he could quiz him over what level of resources constituted such support. Heisenberg asked for only a surprisingly modest increase in funding to cover the construction of a radiation-proof bunker and a cyclotron. He also asked for high-priority ratings for acquiring materials. A few days later he submitted an estimate of 350,000 marks for the coming year, a sum Speer found 'ridiculously tiny' com- pared to Heisenberg's aspirations.* The meeting, his private walk with Heisenberg and the subsequent funding request apparently convinced Speer that German scientists

*350,000 marks at that time roughly equated to £17,000 or $82,000.

were not serious about building a bomb. At a meeting with Hitler a few days later he reported only briefly on splitting the atom. Projects such as Wernher von Braun's V-2 rocket programme, now awarded top-priority status, seemed more deserving to Speer of the Führer's attention and the Reich's diminishing resources.

This may have been what Heisenberg intended. If he genuinely believed the technical problems implicit in building a bomb to be insurmountable, at least while the war was in progress, then he must have believed Germany, whose fission programme might even be ahead, to be safe from Allied atom bombs. If that was the case, he presumably did not want the Nazi high command pressuring him to deliver the impossible. It was far safer and more comfortable for Heisenberg if the Nazi bomb project was officially on the back burner and he was absolved from pursuing it and confronting in acute form the moral dilemmas involved. He was now free to concentrate on the construction of a working reactor using uranium and heavy water, the two key constituents of the German atomic programme.

Heisenberg's memoirs, written nearly thirty years later, convey his relief that 'no orders were given to build atom bombs, and none of us had cause to call for a different decision'. 'As a result,' Heisenberg wrote comfortably, 'our work helped to pave the way for a peaceful atomic technology in the postwar period.' With this neat explanation Heisenberg brought down the curtain on a period of his life riddled with ambiguity and which, in later years, he would repeatedly be asked to justify.

But this was not, as he implied, the end of the German atomic bomb programme.

CHAPTER FOURTEEN

'V.B. OK'

IN THE UNITED STATES, THE BOMB PROJECT WAS GATHERING momentum. On 19 January 1942, six weeks after the attack on Pearl Harbor, President Roosevelt returned the 'martial' National Academy of Sciences report submitted to him by Vannevar Bush the previous November under a short cover note on which he simply scrawled 'V.B. OK', adding, 'I think you had best keep this in your own safe'. As Bush knew, these brief words were the sanction to build the bomb.

The S-1 Project under its programme chiefs made swift progress. In May, James Conant, having reviewed the entire nuclear programme, reported to Bush the scientists' latest views that there were five basic ways to produce bomb fuel. U-235 could be separated by the centrifuge, diffusion and electromagnetic processes; plutonium could be manufactured from uranium in reactors moderated by either graphite or heavy water. All five methods were sufficiently advanced for the building of pilot plants and possibly for the preliminary design of production plants.

A key question was how to control this work and, given

its sensitivity, how to camouflage the expenditure. It was essential to prevent sharp-eyed Congressmen spotting something unusual about government budgets and asking awkward questions. As early as 6 December, the day before the Pearl Harbor attack, Vannevar Bush and James Conant had discussed the desirability of placing the project under the army, whose huge budget could easily conceal the S-1 Project. Now, Bush told President Roosevelt that the project might well determine the course of the war and recommended that 'the whole matter should be turned over to the War Department'. The President had no objection, providing Bush was certain that the War Department could guarantee absolute secrecy. Three months later he gave his formal approval to the army taking over the project. As a result, the majority of the project's funding would be hidden for the duration of the war in the army's huge Corps of Engineers' budget under such bland entries as 'procurement of new materials' and 'expediting production'.

In June 1942, the army appointed Colonel James Marshall of the Corps of Engineers to head the project and instructed him to form a new 'district' – the unit of organization used by the corps. Marshall discussed his new role with fellow officers, including Colonel Leslie Groves, the corps' deputy chief of construction, who had helped supervise the building of the Pentagon. Marshall set up his headquarters in Lower Manhattan, sparking a debate on what the project should be called. One suggestion was the 'Laboratory for the Development of Substitute Materials'. Groves wisely objected that this would only attract attention. He suggested, instead, that the project be named 'Manhattan', after the convention of naming new engineer districts after the city in which they were based. Groves also advised Marshall on how to seek approval for the purchase of a large site in Tennessee.

Groves was about to become far more closely involved with the Manhattan Project than he either guessed or desired. Marshall was competent, but Bush and Conant worried that he lacked sufficient driving force. They suggested to Secretary of War Henry Stimson the appointment of a more energetic officer, and Stimson agreed. In September 1942 the forty-six-year-old Leslie Groves was confidently expecting a transfer to what he anticipated would be 'an extremely attractive assignment overseas' in command of combat troops. All he needed was the approval of his commanding officer, General Brehon Somervell. Instead, Somervell told Groves that he could not leave Washington; 'The Secretary of War has selected you for a very important assignment, and the President has approved the selection.' When Groves objected that he did not want to stay in Washington, Somervell replied, 'If you do the job right it will win the war.' Groves' spirits did not improve when he learned the nature of his assignment. 'Oh, that thing,' he said.

Groves was by his own admission extremely disappointed, writing, 'I did not know the details of America's atomic development program . . . what little I knew of the project had not particularly impressed me, and if I had known the complete picture I would have been still less impressed.' Groves' pill was sweetened by the promise of promotion to brigadier-general, although, as Groves later observed, 'it often seemed to me that the prerogatives of rank were more important in the academic world than they are among soldiers'.

If Groves doubted whether he was the right man for the job, so did Bush. Bush had wanted Somervell but, as a

three-star general, he was too senior. Somervell, meanwhile, had decided that Groves possessed the ideal
qualities for the task and had not waited for Bush's formal
agreement before telling the disgruntled Groves of his
selection. Unaware that Groves had already been
appointed and anxious to appraise him as a candidate,
Bush summoned him. The details of their uneasy
encounter have become part of the folklore of the bomb.
When Bush realized what had happened, he reacted
angrily. His interview with Groves convinced him that
Groves was tactless, aggressive and unlikely to be able to
deal with scientists. Furthermore, he felt he had been
bounced by the army into accepting him. In a terse note to
one of Stimson's aides, he wrote, 'I fear we are in the
soup.'

Time would, however, show that the stout, ebullient,
rough-edged Leslie Groves was an excellent choice. He
might not be particularly good with people, but he had an
impressive facility for grasping essentials, identifying
problems and securing solutions.

Born in Albany, New York, in 1896, Groves – known
to his family as Dick – was the descendant, on his father's
side, of Protestant Huguenots who had fled religious
persecution in France and emigrated to America in the
seventeenth century. Their family name, originally Le
Gros, became first La Groves and then simply Groves. On
his mother's side, Groves came from Welsh farming stock.
Groves' father, also named Leslie, was a conscientious but
indecisive man who attempted various professions with
limited success, moving from teacher to lawyer to
Presbyterian minister to army chaplain. His limited means
made him frugal, a trait he passed to his son. He also
bequeathed his son a distrust of the British. As army

chaplain to the US Fourteenth Infantry, he had marched with the international force sent to Beijing in 1900 to relieve the foreign legations under siege there during the Boxer Rebellion. The British contingent reached Beijing first, convincing Groves senior that they had selfishly kept intelligence about the best route into the city to themselves.

Early exposure to army life confirmed to the young Groves that it was the career he wanted. He worked single-mindedly to secure a cadetship at the US Military Academy at West Point, succeeding on his second attempt. He gained the nickname 'Greasy Groves', perhaps because of the sweet tooth he would retain all his life and his fleshy physique. However, he did well at the academy, graduating fourth in his class ten days before the end of the First World War. He chose a commission in the Corps of Engineers and moved on to train at the Engineer School, where the assistant commander noted his 'very keen mind'. In 1922 Groves wed his childhood sweetheart Grace Wilson, who as a teenager had shrewdly captured two of his overriding characteristics, stubbornness and love of sweet things, in a little rhyme:

> This is Dick and this is fudge.
> From it little Dick won't budge.

In September 1942, Groves told Grace and their children that he had a new job, 'that it involved secret matters and for that reason was never to be mentioned. The answer to be given if they were asked what I was doing was, "I don't know, I never know what he's doing." ' Groves' passion for secrecy would be a hallmark of his running of the Manhattan Project.

Groves embraced his new role with energy and resolve. The very day after his appointment he took steps to secure stocks of uranium ore, despatching a subordinate to New York to see Edgar Sengier, managing director of Union Minière du Haut-Katanga which owned the Shinkolobwe uranium mine in the Belgian Congo. Sengier had left Brussels for the United States in October 1939. Towards the end of 1940, fearing that the Germans might invade the Belgian Congo, he had ordered his staff to ship to New York all the uranium ore in storage there. As a result, more than 1,250 tons of ore were sitting snugly in two thousand steel drums in a warehouse on Staten Island. Groves immediately arranged to buy the entire stock, as well as a further three thousand tons to be shipped from Africa, thereby securing two thirds of the US bomb project's required stock of uranium ore.

Two days after his appointment, Groves authorized the purchase of the site in Tennessee on which a few weeks earlier he had been advising Marshall. On 23 September, the day his promotion came through and he formally took charge of the Manhattan Project, Groves caught the train to Tennessee to see the 56,000-acre site for himself. It lay along the Clinch River in rolling hills near the small rural town of Clinton and would soon be named Oak Ridge after the ridges overlooking the site. It would eventually house the project's massive U-235 production plants.

Groves returned to Washington and moved into an austere suite of rooms in the War Department building. As he later wrote, 'It was undoubtedly one of the smallest headquarters seen in modern Washington. Nevertheless, I fell far short of my goal of emulating General Sherman who, in his march from Atlanta to the sea, had limited his headquarters baggage to less than what could be placed in a single escort wagon.' He organized his small team on 'simple and direct' lines so he could take 'fast, positive decisions'. Realizing that delays in obtaining resources could prove fatal, he insisted on being given a top priority – 'AAA' – rating. When officials objected, he threatened to advise the President that the project would have to be abandoned. The threat worked.

Routinely working fourteen hours a day, Groves reviewed all the scientific work currently under way and visited all the key laboratories. He was particularly concerned about the work on plutonium development – in his view 'an even greater venture into the unknown than the first voyage of Columbus'. In October 1942 he arrived on the neo-Gothic campus of the University of Chicago, where in early 1942 Arthur Compton had relocated research directly related to making plutonium for the bomb. The focus of this work was to achieve the chain reaction needed to create plutonium. The project, led by Enrico Fermi, who had moved from Columbia, was being undertaken in the so-called Metallurgical Laboratory, or 'Met Lab' – a name chosen, like the British 'Tube Alloys' and the German 'Virus House', to deter the curious.

Despite such measures, word inevitably spread among the relatively small scientific community. In the autumn of 1942, Philip Morrison, a young physics instructor at the University of Illinois, was visiting Chicago at Thanksgiving and called in on fellow physicist Robert Christy, who had studied with him at Berkeley. As Morrison later wrote, Christy asked, 'Do you know what we're doing here?' Morrison 'admitted that it was easy to guess: this must be the hidden uranium project to which so many others had gone. "Yes," he said, in his familiar style of calm speech, "we are making bombs." ' Within weeks, Morrison too had moved to Chicago to join him.

Groves, who would have been horrified at such casual conversations about classified topics, joined the Met Lab scientists at one of their meetings to quiz them 'about the plutonium process, and the anticipated explosive power of an atomic bomb, as well as of the amount of fissionable material that a single bomb would require'. He was shocked by their answer to the latter. He had expected their estimate to be accurate 'within twenty-five or fifty per cent'. Instead, 'they quite blandly replied that they thought it was correct within a factor of ten . . . while I had known that we were proceeding in the dark, this conversation brought it home to me with the impact of a pile driver'. Groves accepted reluctantly that they could not yet be more precise, but, as he later wrote, uncertainty about this critical aspect would plague the project until, in 1945, they could at last test a bomb.

Groves nevertheless left Chicago somewhat reassured about plutonium. What he had heard from the scientists convinced him that it offered 'the greatest chances for success in producing bomb material'. Although the process was 'extremely difficult and completely unprecedented', it seemed to Groves more feasible than trying to extract U-235 from U-238 where success depended on the scientists' skill in separating materials with what he thought were 'almost infinitesimal differences in their physical properties'. At the same time, with characteristic common sense, he decided to try all routes.

The meeting had been important in another respect. It drew battle lines between the army and the scientists that would endure until the project's end. Groves was well aware 'that scientists didn't like me'. His characteristic response was, 'Who cares?' He was determined to assert his authority and show that, despite his lack of scientific background, he would not allow science or scientists to overawe him. He was fond of saying that 'atomic physics is not an occult science'. As a result, he made the scientists a speech which did him few favours: 'There is one thing I want to emphasize. You may know that I don't have a Ph.D. . . . But let me tell you that I had ten years of formal education after I entered college. *Ten* years in which I just studied. I didn't have to make a living or give time to teaching. I just studied. That would be the equivalent of about two Ph.D.s, wouldn't it?' This crass self-advertisement left many of the scientists dismayed that such a man was in charge of their work. As they came to know him better, some would revise that opinion, but a common view of Groves as a bully and a boor had been born.

Groves, in turn, had formed a fairly poor view of the Met Lab scientists, later writing that 'the unique array of scientific talent that had been collected there was imbued with an active dislike for any supervision imposed upon them and a genuine disbelief in the need for any outside assistance'. To an extent he was right. Leo Szilard, who had moved from Columbia to Chicago with Fermi, was among those who felt it would be impossible to work with such a man, and said so. Groves, in turn, had no intention of working with Szilard if he could help it. Their brief meeting had convinced him that the Hungarian was an interfering know-all of possibly dubious loyalty. He tried to persuade Henry Stimson to lock Szilard up for the duration of the war as an enemy alien. Stimson refused, on the grounds that such an act would infringe the Constitution. Groves later wrote that this was the reply he had expected but it had been worth a try. In fact, Szilard would remain a thorn in Groves' side throughout the project.

According to some accounts, within days Groves had also alienated Ernest Lawrence. At their first meeting at Berkeley he warned the Nobel Prize winner that he had better do a good job since his reputation depended on it. Lawrence replied, 'My reputation is already made. It is *yours* that depends on the outcome of the Manhattan project.'

Everything Groves saw and heard in these early months convinced him that the project was a very much bigger undertaking than he had previously thought. This, in a sense, cheered him up. One of his objections to taking responsibility for the bomb project had been its apparently small scale compared to engineering projects he had directed. Groves decided, in the interests of efficiency and simplicity, to appoint the Du Pont company to take charge of the engineering, construction and operation of the industrial-scale plutonium plants that would have to be built. Although Bush and Conant were content, Groves knew that he would face a battle with the scientists about industrial involvement.

The Du Pont company was not initially enthusiastic either. Their senior executives pointed out that their expertise was chemistry, not physics, and that 'they were incompetent to render any opinion except that the entire project seemed beyond human capability'. Groves required all his guile and energy to win their agreement. He clinched it by appealing to their patriotism and their purse. The President, he assured them, considered the project to be of the utmost national urgency and provision would be made to protect the company against financial

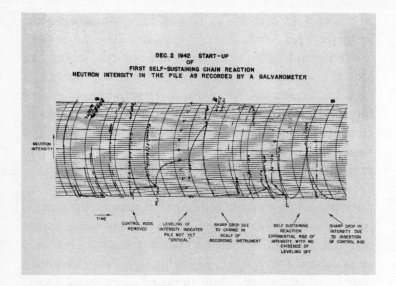

loss. There would also be a government fund to compensate any employee injured because of 'the entirely unpredictable and unprecedented hazards involved'. As Groves later wrote, the fate of the luminous dial painters of the 1920s who had licked radium-tainted brushes had not been forgotten.

While final negotiations with Du Pont were still under way, news came of a great scientific breakthrough at the Met Lab in Chicago. On 2 December 1942, Arthur Compton telephoned James Conant with the news, 'The Italian Navigator [Fermi] has reached the New World.' When Conant asked, 'How did he find the natives?', Compton replied, 'Very friendly.' What this actually meant was that Enrico Fermi had achieved the world's first self-sustaining chain reaction. The nuclear pile he had built at a cost of a million dollars in a squash court under the west stands of the disused university football stadium had gone critical.

It was a formidable achievement. Over four weeks, Fermi and his students had positioned fifty-six tons of uranium and uranium oxide between black graphite blocks, creating a great layer cake measuring twenty feet high and twenty-five feet wide. As they sought to achieve the self-sustaining reaction, Fermi and his team progressively removed from the reactor a series of control rods, allowing more and more neutrons to be released from the uranium. By 11.35 a.m. on that December day the counters were clicking rapidly, but then, with a loud clap, the mechanisms designed to ensure safety slammed the control rods back home. As Fermi later recalled, 'the safety point had been set too low'. It seemed a good time to go to lunch.

That afternoon the team changed the settings. Fermi was highly nervous as the last control rod, nicknamed 'Zip', was slowly withdrawn and the moment of criticality approached. He had constructed his machine in the heart of the city and was attempting something never done before. Three scientists, known as 'the suicide squad', stood by with buckets of a cadmium salt solution, a substance that sucked up neutrons, which they were ordered to throw on the pile if the nuclear reaction showed signs of getting out of control. But all went well, and an audience of forty gasped as the Geiger counters clicked and the line on the graph paper shot up as the chain reaction began. 'The event was not spectacular,' Fermi wrote a decade later, 'no fuses burned, no lights flashed. But to us it meant that release of atomic energy on a large scale would be only a matter of time.' After a run of twenty-eight minutes the pile was safely closed down again by reinserting the control rods. Eugene Wigner produced a bottle of Chianti in Fermi's honour. After drinking the contents out of paper cups, those present signed their names on the bottle's straw covering.

With hindsight, this event would be seen as the great turning point, the release of a transcendental source of energy. Laura Fermi would call it the atomic age's only true birthday. Many would dismiss what followed as mere engineering. However, that was not how things seemed to Groves in late 1942 with his great task still ahead of him. Despite Fermi's success, there was still no firm proof that controlled chain reactions in a nuclear pile could be used to produce plutonium on a large scale, or that a bomb using plutonium or U-235 would explode. Fermi's experiments were based on using 'slow neutrons' – that is, neutrons that had been slowed down using a moderator, in this case graphite. In a bomb the neutrons would be 'fast', because it would be technically infeasible to include a moderator.

Nevertheless, the Chicago experiment was highly encouraging. On 28 December 1942, President Roosevelt formally approved funding for industrial-scale production plants for plutonium and U-235. Groves was content for the uranium separation plants and a small pilot plutonium production plant to be built at Oak Ridge, where construction would start in February 1943. However, he decided not to locate the huge plutonium-producing reactor plants there. There would be practical problems over acquiring more land and ensuring sufficient power and water supplies, but more importantly he worried 'about the possible danger to the surrounding population . . . If because of some unknown and unanticipated factor a reactor were to explode and throw great quantities of highly radioactive materials into the atmosphere . . . the loss of life and the damage to health in the area might be catastrophic.'

Groves drew up a list of criteria for the plutonium site, one of which stipulated that no town of a thousand or more people should be within twenty miles. Everything suggested that a location should be found in the west. For two weeks a reconnaissance party searched from Washington State to the Mexican border. The site Groves finally selected was near the small town of Hanford in Washington State, a scrubby, infertile area of sagebrush along the Columbia River mostly used for grazing sheep. The population was small and land values were low. In January 1943, Groves initiated what would be one of the largest land purchases of the war: more than four hundred thousand acres and space enough for the gargantuan physical and intellectual challenge ahead. Hanford would one day have 540 buildings, 600 miles of roads, 158 miles of railway track and 132,000 people working there.

In Russia, the Nazi invasion had interrupted the already very limited fission research programme. Facilities and personnel were transferred to Kazan and other industrial cities beyond the Urals, and scientists diverted to more urgent defence projects such as devising ways to protect ships from magnetic mines. However, in early 1942 a sharp-eyed physicist, twenty-eight-year-old Georgii Flerov, had noticed that the names of all the well-known scientists understood to have been working on atomic fission had disappeared from international academic journals. Personally and passionately convinced of the feasibility of constructing a nuclear weapon, and suspicious about 'dogs that did not bark', Flerov wrote to Stalin urging that the Soviet Union should build a uranium bomb without delay.

Flerov's messianic enthusiasm chimed with reports

which had begun arriving in Moscow a few weeks earlier from a Soviet agent in London – Anatolii Gorskii, code-named Vadim. These reports detailed top-level British discussions of the Maud Report. They revealed to the Russians that Britain had decided to build an atomic bomb, that it would be likely to take between two to five years and that some of the necessary plant would be built in the United States. The source of this information was probably John Cairncross, the so-called 'Fifth Man' of the group of spies recruited by the Soviets at Cambridge University in the 1930s and at that time private secretary to Lord Hankey, then a minister in the War Cabinet.

At first the Soviet Union was too preoccupied with holding back the invading Germans to react to the startling intelligence, but in March 1942 government interior minister Lavrentii Beria, Stalin's notorious police chief, ordered a thorough review of the Maud inform-ation. Eminent Soviet scientists, including Abram Joffé and Peter Kapitza, were consulted in strictest secrecy. Although Kapitza believed a nuclear bomb to be theoretically possible, he said that the Soviet Union was not ready for such a step; an atom bomb was not a weapon for the war with Germany but a matter for the future. However, others, including Igor Kurchatov, who at the outbreak of war had adopted the Roman custom of refusing to shave until the enemy was defeated and was growing ever more hirsute, were more enthusiastic. Their comments convinced Stalin to revive the Soviet Union's nuclear research programme and, in the words of Stalin's devoted aide Vyacheslav Molotov, 'to realise the creation of an atomic bomb'. By a strange irony the Maud Report was thus the catalyst not only for the US bomb pro-gramme but for the Soviet one as well.

Stalin authorized the building of a new laboratory to take charge of all nuclear research and placed it under the scientific direction of Igor Kurchatov. Work began in March 1943, just over a month after German troops surrendered at Stalingrad after a Soviet counteroffensive codenamed 'Uran', meaning 'Uranus' or 'uranium' – which was, perhaps, not a coincidence.

CHAPTER FIFTEEN

'THE BEST COUP'

IN BRITAIN AND THE UNITED STATES THE OVERRIDING WORRY was how far the Germans had progressed. For some, every scientific advance by Allied scientists was double-edged. As Leo Szilard pointed out tirelessly, anything the Allies did the Nazis could do too. On 22 June 1942, a worried Arthur Compton wrote from Chicago to Vannevar Bush that 'we have just recognised how . . . a small heavy-water plant can quickly supply material for a high power plant for producing [fissionable material]. If the Germans know what we know – and we dare not discount their knowledge – they should be dropping fission bombs on us in 1943, a year before our bombs are planned to be ready.'

The British were increasingly anxious about intelligence reports that the Nazis were stepping up production of heavy water at the Norsk-Hydro plant at Vemork, near Rjukan, in occupied Norway. They knew that Jomar Brun, a member of the Norwegian resistance and an engineer at the plant, was feeding castor oil into the production process to ensure frequent breakdowns. However, realizing that local sabotage could not succeed for ever,

Churchill's war cabinet ordered the plant to be destroyed in an operation codenamed 'Freshman'.

On the bitterly cold night of 19 November 1942, two four-engine RAF Halifax bombers, each towing a glider holding seventeen British commandos, took off for Norway's remote Hardanger Plateau. On the plateau a team of four British-trained Norwegian commandos, codenamed 'Grouse', listened carefully for the sound of approaching aircraft engines. The team, led by Jens Anton Poulsson, accompanied by radio-man Knut Haugland, Claus Helberg and Arne Kjelstrup, had parachuted in a month earlier. Several times Haugland thought he heard through the headphones of his radio direction-finding equipment the buzzing that would announce Freshman's arrival. His comrades flashed signal lights into the sky but no gliders floated silently in to land. Shortly before midnight, the Grouse team returned frustrated to their base-hut.

Radio messages from London soon told them that both gliders and one of the Halifaxes had crashed in sudden bad weather. The fate of the survivors would only emerge after the war. The glider that had been released by the surviving plane crashed on a mountain-top near Stavanger, killing eight outright. The Germans quickly captured the nine survivors. They took four severely injured commandos first to hospital and then to Gestapo headquarters for interrogation. Afterwards, a German medical officer gave them a series of lethal injections. When they failed to die quickly enough, Gestapo men stamped on their throats. They then flung the four bodies into the sea. The five uninjured men were sent to Grini concentration camp north of Oslo. Two months later, in January 1943, the Germans tied the men's hands behind their backs with barbed wire and shot them. The other glider and its mother Halifax crashed soon after crossing the Norwegian coastline. All aboard the plane died instantly, but on the glider only three were dead. Of the remaining fourteen, three were badly injured but the rest were in reasonable shape. Two commandos struggled through the deep snow to a farmhouse to beg for help. The frightened farmer, knowing the Germans would shortly arrive, refused, sending them instead to the local sheriff, who at once phoned the German authorities. The Germans quickly captured the men and executed them all a few hours later.

The total failure of Operation Freshman posed a stark dilemma to the British. Dare they hazard more men, especially now that the Germans had been alerted to British interest in the Rjukan area? Yet, how could they allow German heavy water production to continue? They decided to try again, using Norwegian commandos, familiar with the terrain, who would parachute in.

The man selected to lead the new expedition, code-named 'Gunnerside', was Joachim Ronneberg, who had fled Norway after the German occupation. In early December 1942 he was training Norwegian resistance fighters at a Special Operations Executive (SOE) camp in the west of Scotland. He was ordered to pick five men to accompany him and to be ready in two weeks' time. The twenty-three-year-old Ronneberg appointed as his second-in-command Knut Haukelid, who had plotted un-successfully to kidnap the Norwegian puppet prime minister Vidkun Quisling, before himself escaping to Britain.*

---

* Vidkun Quisling was the origin of the word 'quisling' to describe a person collaborating with an enemy occupier.

The team trained at a secret SOE school at Farm Hall, a country house near Cambridge, where, ironically, captured German nuclear scientists would one day be interned. Using microphotographs of blueprints of the Norsk-Hydro plant smuggled out of Norway in fake toothpaste tubes, the British had reconstructed key parts of the plant, including wooden replicas of the eighteen cells that produced the heavy water. Unknown to the team, their training was being guided by Jomar Brun, the ingenious castor oil saboteur, also recently smuggled out of Norway on Churchill's express orders.

Meanwhile, still in Norway, the Grouse team was surviving high in the mountains while awaiting fresh orders from London. The failure of Operation Freshman had been, as Poulsson wrote in his diary, 'a hard blow'. Since then they had been pushed to their limits physically and emotionally, dodging German patrols, bivouacking in remote huts and eating anything they could find – sometimes just 'reindeer moss', the soft, green moss beneath the snow on which reindeers grazed, so acid as to be barely digestible by humans even when boiled into a soup. Some of the men became almost too weak to stand, and their skin turned yellow. Poulsson's timely shooting of a reindeer on Christmas Eve probably saved them, providing protein and vitamins. They ate every part of the animal including eyes, brains and stomach. Even the reindeer moss predigested in the animal's stomach proved more palatable than the fresh version. A message from London of a new operation heartened them, only for them to be disappointed again when in January 1943 the pilot of the plane bringing the Gunnerside men aborted the mission after failing to locate the drop site in the shadowy, moonlit maze of snowy mountains.

On 16 February, a fresh message announced that the Gunnerside commandos were coming. The plane again missed the drop point, but the men parachuted anyway, landing on the Hardanger Plateau with containers of arms and explosives and packs containing skis and sledges. They buried their equipment in the snow and found a hut to shelter them while they worked out what to do. A map in the hut showed they were some miles from the rendezvous point, but three days of vicious snow storms kept them pinned down.

Finally, on 23 February, skiing over the frozen terrain in their white camouflage suits, they spied the tiny dot of a distant figure. Ronneberg ordered Haukelid to ski ahead and investigate. Drawing nearer, pistol at the ready, he saw not one man but two, both heavily bearded. He was within fifteen yards before he recognized the ragged, wan-faced men with drooping shoulders as Claus Helberg and Arne Kjelstrup of Grouse, and rushed forward to embrace them.

That night at Grouse's headquarters, a remote hut at Svensbu near Lake Saure, the commandos celebrated with a dinner of reindeer meat supplemented by chocolate and dried fruit brought by the new arrivals. The next morning, 24 February, they began to plan the attack. The location of the heavy water plant, on a lip of rock jutting out from a three-thousand-foot-high mountain and five hundred feet above a river gorge, could hardly have been more impregnable. The only direct route across the gorge was a heavily guarded suspension bridge. The strategy agreed at Farm Hall was that the commandos should cross the gorge somewhere between Rjukan and Vemork and then follow the railway line that ran around the side of the mountain into the plant. First, though, they needed more

detailed information. Ronneberg despatched Claus
Helberg to seek details of the latest German deployments
from a contact in the Norwegian resistance, and then to
rendezvous with the main group later that day at another
hut nearer the plant.

Helberg returned with important news: amazing
though it seemed, the railway line into the plant was
unguarded. However, the critical question remained:
where could the men climb up and out of the river gorge
onto the railway? Scrutinizing aerial photographs, they
noticed bushes and trees growing up the side of the gorge
at a single point. Reckoning that where plants could grow
men could climb, Ronneberg again sent Helberg to
reconnoitre. Slipping and sliding down into the ice-bound
gorge at a safe distance from the plant, he crept along the
frozen river at its base until he reached the bushes and
identified 'a somewhat passable way' up to the factory.

Ronneberg made his final preparations. He would lead
a four-man assault team to break into the plant and
destroy the heavy water cells. Haukelid would command
a five-man support party. The tenth man, radio operator
Knut Haugland, would remain concealed in a nearby hut
to inform London of developments.

In the early evening of 27 February, the commandos put
on their British Army uniforms. All were carrying 'the
death pill' – cyanide encapsulated in rubber and
guaranteed to kill them in three seconds if captured. At
eight p.m. the nine men of the attack and support parties,
weighed down by sixty-five-pound packs, skied out into
the darkness. They slithered down onto the road leading
from Rjukan to the plant, nearly colliding with buses
carrying shift workers; only by thrusting their ski sticks
hard into the snow did they brake in time. The buses

rumbled past and off into the night, leaving the intruders undetected.

Descending to the bottom of the gorge was dangerous and difficult in the dark for men thrown off balance by the movement of their heavy, unwieldy packs. Because there had been a slight thaw and the river ice had thinned, their next worry was whether it would bear their weight as they crossed. Treading cautiously, Helberg found a strong enough area of ice for them to cross, then guided his comrades to the route he had discovered up out of the gorge. Grasping at snow-covered rocks, shrubs and branches of birch with frozen fingers, by eleven p.m. the men were on the railway line, following it silently and in single file towards the factory.

At 11.30 they halted five hundred yards from the gate leading into the plant. From there they had a clear view of the suspension bridge where they knew the guard would change at midnight. They waited for the change to take place and for the new guard to settle down, huddled in their guardhouse. At 12.30 the Norwegians advanced, moving cautiously at first for fear of land mines. Haukelid and the support party cut the padlocked iron chain on the gate, ran inside and, well armed with tommy-guns and sniper rifles as well as pistols, knives and hand grenades, took up position to provide covering fire if necessary.

Ronneberg and his demolition team slipped past them towards the building housing the heavy water cells. They found the steel door locked, but after a frantic search Ronneberg and one of his men discovered a cable duct through which they could climb. It led them thirty yards over rusty pipes and tangled cables down to a semi-basement room. Carefully opening a door in the room marked 'no admittance except on business', they peered into the heavy water production chamber itself. An elderly watchman was sitting with his back to them.

The two commandos rushed inside. While Ronneberg secured the door from the inside, his comrade pointed his pistol at the watchman's head. He told him in Norwegian that they were British commandos on a mission to destroy the plant; so long as the watchman co-operated he would not be harmed. Pulling on rubber gloves in case of electric shocks, Ronneberg began swiftly fastening explosive charges to each of the eighteen steel-clad, four-foot-high heavy water cells. Then the sudden sound of smashing glass in a skylight made him freeze. The face of another member of the demolition team appeared in the jagged opening. Unaware of the duct, he had broken the glass in a desperate bid to get inside. The commandos waited for alarms to sound, but, to their surprise, none did. Ronneberg hurriedly completed placing the charges. Just as he was about to ignite the fuses the old watchman implored him to help him find his spectacles, pleading that he was almost blind without them. Touched by the man's desperation, Ronneberg sacrificed precious moments to find them. He was again just about to light the fuses when the sound of boots descending the steps from the floor above made him pause anew. It was the Norwegian night foreman, who gazed at the intruders in astonishment.

Hesitating no longer, Ronneberg lit the fuses. There were two sets – thirty-second fuses with, as back-up in case they should fail, two-minute fuses. Shouting to the two employees to take cover higher in the building, the commandos rushed upstairs and out through the steel door, which they had unlocked with a key taken from the old watchman, pulling it shut behind them. They were just twenty yards from the building when the explosion came.

To Haukelid, waiting in the shadows with the support team, it seemed 'astonishingly small'. Indeed, the noise was so muted that several minutes passed before an unarmed German soldier came outside to take a look. After a cursory glance around the compound he returned, satisfied, to the warmth of the guardhouse.

Meanwhile, the nine commandos regrouped outside the plant and embraced. They retraced their path back along the railway track and reached the bottom of the gorge before the wailing of air-raid sirens – the signal for general mobilization in the Rjukan area – announced that the Germans had finally realized what had happened. Once again, luck was with them. The Germans had installed powerful floodlights to illuminate the gorge but in the confusion no-one could find the switch. The commandos slipped away into the darkness, some to ski to safety in neutral Sweden, others to hide out in the mountains, all to survive to fight another day.

Despite the muted sound of the explosion, their attack had been entirely successful. The heavy water cells were wrecked and nearly half a ton of the precious liquid had leaked away. The commander of the German occupying forces in Norway, General Nikolaus von Falkenhorst, summoned to inspect the damage, conceded that 'the English bandits performed the finest coup I have seen in this war'.

CHAPTER SIXTEEN

BEAUTIFUL AND SAVAGE COUNTRY

LESLIE GROVES MET ROBERT OPPENHEIMER FOR THE FIRST time in October 1942 at Berkeley during his initial inspection tour of the key laboratories. They could not have been more different – Groves the supremely practical human bulldozer, Oppenheimer the intellectual sophisticate. Nevertheless, Groves took to the thirty-eight-year-old scientist, recognizing a man who could penetrate a problem swiftly and would give of himself unstintingly. Soon after, Groves invited Oppenheimer, then leading a small team of theoretical physicists set up by Arthur Compton at Berkeley to look at bomb design, to Washington. Groves asked his views on the type of laboratory needed to design and build the bomb. Oppenheimer suggested that, rather than choose an existing location like Chicago University, the laboratory should be built somewhere remote where scientists could work freely but securely.

Groves agreed and began doggedly seeking a suitable site. His criteria were not easy to satisfy: the location had to be isolated but still accessible by car, train and plane, with a good, year-round climate and enough power and water. New Mexico seemed promising, and initial surveys suggested potentially suitable locations near Albuquerque. Groves arrived on an inspection tour, accompanied by Oppenheimer. The first site they visited was hemmed in on three sides by high cliffs, which Oppenheimer argued would depress the work force. Knowing the region well from vacations at Perro Caliente, his ranch near the Sangre de Cristo mountains – named by the Spanish conquistadors after the blood-red glow that stained them at sunset – he recommended a site facing them and about thirty-five miles north-west of Santa Fe belonging to the Los Alamos Ranch School.

The school lay on a seven-thousand-foot-high mesa, a tableland formed by the flattened cone of a long-extinct volcano, whose red and gold striped walls plunged to the Rio Grande valley below. The valley was pure desert except for a fecund strip along the water's edge, dotted with Indian villages. Across the valley Oppenheimer's snow-tipped Sangre de Cristo mountains swept skywards. To the west lay the slopes of the green-domed Jemez hills. The mesa itself was covered with sweet-scented, long-needled pine trees.

Groves and Oppenheimer arrived in November 1942 as light snow was falling, dusting the trees. Despite this, the schoolboys and their masters were out on the playing fields in shorts. Looking around, Groves noted the school's neat buildings of wood and stone that could be used to house people until more accommodation could be built. The mesa itself was riven with deep canyons, suitable for containing special laboratories. A narrow, rutted mountain road connected the site with the highway to Santa Fe. According to one of Groves' party, he announced, 'This is the place.' Barely a week later, the War Department ordered the purchase of nearly fifty thousand acres there. It was, Oppenheimer wrote to a colleague, 'a lovely spot'.

Robert Oppenheimer had not been Groves' immediate choice as director of the new laboratory. He had first reviewed the more obvious candidates. Despite their initial prickly encounter, he already regarded Ernest Lawrence as an outstanding experimental physicist but did not believe Lawrence could be spared from the work on electromagnetic separation of U-235 at Berkeley. Arthur Compton was also highly competent, but he was at full stretch running the Met Lab at Chicago. Chemist Harold Urey at Columbia University, the discoverer of deuterium, was a possibility but, in Groves' view, too weak a personality. Oppenheimer was, Groves concluded, the best man available.

Groves knew he was taking a considerable risk. As Hans Bethe recalled, 'Oppenheimer had never directed anything – he was a pure theoretical physicist interested in the most advanced ideas – nobody trusted him except Groves.' Still, Groves brushed aside the reservations and alternative suggestions of Compton and Lawrence, who argued that Oppenheimer lacked the experimental and administrative experience to run a laboratory, was not a Nobel Prize winner and would find it hard to impose his authority. Oppenheimer's sheer intellectual ability would, Groves believed, drive the project on. However, he promised Lawrence and Compton that, should Oppenheimer prove inadequate, they could take over.

Far more surprisingly, Groves ignored evidence of Oppenheimer's left-wing sympathies and communist connections. His wife Kitty, whom he had married in

November 1940, was a former communist who had been married twice before. Her first husband was an American Communist Party member, killed while fighting for the Republicans in the Spanish Civil War. She had then married an English doctor who had moved to the United States with his new wife. Less than a year later she and Oppenheimer had fallen in love, and within another year she had obtained a Nevada divorce. Oppenheimer himself had earlier been engaged to a Berkeley professor's daughter, Jean Tatlock, a committed member of the Communist Party with whom he was still in touch.

Groves was an arch conservative with an inherent distaste for liberal thinkers and an obsessive attitude towards security. He infiltrated counter-intelligence officers among the work force of the Manhattan Project. He even had himself tailed to see whether he was under enemy surveillance, and carried a small automatic pistol in his trouser pocket when travelling. He ordered any failure of plant or machinery to be rigorously examined in case it was the result of sabotage. Yet he dismissed the assertions of US Military Intelligence that Oppenheimer was 'playing a key part in the attempts of the Soviet Union to secure, by espionage, highly secret information which is vital to the security of the United States'. After personally reviewing the evidence against Oppenheimer, he concluded that 'his potential value outweighed any security risk'. He demanded that Oppenheimer be given security clearance, insisting, 'He is absolutely essential to the project.' Nevertheless, as Groves well knew, Oppenheimer would remain under surveillance by military intelligence throughout. They tapped his phones and tailed his movements.

Groves also believed it essential to find a prime

contracting agent for Los Alamos to function formally as employer of the staff and procurer of whatever was needed, and he appointed the University of California. Obtaining the right experimental equipment was one of the first challenges, but the project progressively begged, borrowed and leased from universities across the United States, acquiring a cyclotron and several linear accelerators from Harvard and the universities of Illinois and Wisconsin.

Oppenheimer, meanwhile, was identifying the team he wished to bring to Los Alamos. A potential difficulty was that Groves wanted to draft the laboratory's scientists into the army, believing it would contribute to discipline and security. Oppenheimer initially supported him. According to Hans Bethe, 'Oppenheimer was eager to do this – he would have been a lieutenant-colonel'. However, others were much less certain, and they found a champion in the highly respected American physicist Isidor Rabi. Although Rabi did not plan to work at Los Alamos himself, believing that his present work on radar had more short-term importance for the war effort, Bethe recalled that he 'came to us and said, "Don't do that. If you make this a military laboratory nothing will ever, ever happen. You will need hundreds of permissions just to buy a screw of one diameter rather than another and you'll be commanded by Groves, who'll boss you around. You won't be able to refuse, you'll have to do it because he is the general, even if you know the experiments are pointless." '

Oppenheimer and Groves agreed on a compromise. During the experimental stage of the project the laboratory would remain under civil administration, but when large-scale testing began – and, whatever happened, not before 1 January 1944 – scientists and engineers would become commissioned officers. In fact, this never occurred. Groves wisely did not raise the militarization question again. He did, however, establish two lines of command at Los Alamos. Oppenheimer would be scientific director, but there would also be a military commander – Lieutenant-Colonel John M. Harmon.* The site itself would be a military reservation, fenced and guarded. The technical facilities and laboratories would be housed within an inner protected zone – the 'Technical Area'.

By April 1943, the new laboratory was beginning to function, although it was still, essentially, a building site. Some three thousand construction workers, billeted there since the previous December in cramped trailers, had made remarkable progress on a main building, five laboratories, a machine shop, a warehouse and the first accommodation blocks. However, most of the buildings were not ready. The roads oozed with mud when it rained. When it was fine, building dust blew everywhere. To one new arrival the site looked 'as raw as a new scar'. A few scientists moved into the old school buildings but the rest lodged in dude ranches and were bussed daily to Los Alamos along bumpy dirt roads where surprised chickens ran for cover.

Oppenheimer organized a series of lectures to review the latest state of knowledge in atomic physics and to thrash out a detailed experimental programme. As well as those already on the site, he invited others such as Enrico Fermi and Hans Bethe whom he hoped to attract to Los

* Harmon was replaced after four months because of a weakness for alcohol and difficulties in his relations with non-military personnel. His successor was Lieutenant-Colonel Whitney Ashbridge.

Alamos or whose work elsewhere would contribute to the design and building of the bomb. Those in academic posts were paid the equivalent of their university salaries. Others were remunerated according to their qualifications. Oppenheimer's own pay was $10,000 a year – a sum he considered excessive. His sustained attempts to have it lowered failed. The scientists discussed everything, from how best to determine the detailed characteristics of chain reactions – including how rapidly new neutrons would be released in each fission – to how much fissionable material it would take to make a bomb. A fundamental question was whether to use U-235 or plutonium, or indeed both, as bomb fuel. As Groves and Oppenheimer knew, it would be some time before sizeable amounts of either material became available from Oak Ridge and Hanford. In the meantime, all investigations of the chemical properties of U-235 and plutonium would have to be carried out on microscopic samples, and it would be necessary to pursue both types of bomb in tandem.

Another key question was how to ensure that a nuclear reaction culminated in the desired huge explosion. Groves later wrote that 'two opposing considerations came into play. The violence of the explosion was dependent upon the number of neutrons released by the chain reaction. This number increased geometrically with each generation of the chain. Yet to allow the reaction to progress through a number of generations took a certain amount of time during which the energy already released by previous generations could blow the bomb apart and terminate the chain reaction before any major detonation was achieved.' The crux of the problem was how to bring the critical mass together quickly enough. At this early stage,

Hans Bethe sketched by Otto Frisch

most believed the fastest method was the gun-assembly
technique whereby one subcritical mass of fissionable
material – U-235 or plutonium – was fired into another to
produce one critical mass.

The task ahead of the scientists was clearly so huge and
complex that Oppenheimer decided to establish specialized
divisions for theoretical physics, experimental physics,
chemistry and metallurgy, and ordnance. He asked Robert
Bacher, working on radar at MIT, to head up experimental
physics. Bacher agreed but stipulated that the minute the
army took over the work he would resign. Oppenheimer
wanted to lead the theoretical physics work himself but
accepted he could never combine this with his responsibilities as director. Instead, he appointed Hans Bethe.

Sketch of John van Neumann

Bethe had a powerfully logical mind. After fleeing Nazi Germany, first for England and then for Cornell University, he had made his mark in 1936 and 1937 with the publication of three encyclopaedic reviews of nuclear physics which together had come to be known as 'the Bethe Bible'. However, Oppenheimer's first approach caused Bethe some soul-searching. Bethe's wife Rose, guessing that the project was connected with some new form of weapon, asked him during a long walk in the Yosemite National Park whether he really wanted to become involved. Bethe reflected carefully but concluded that 'the fission bomb had to be done, because the Germans were presumably doing it'.

Much of the project's success would be due to Bethe.

There were no blueprints on how to build an atom bomb. The selection of materials, design, size and properties of the bomb would all have to be based on theoretical judgements derived from whatever experimental results were available. Bethe divided his team, which included Edward Teller, Victor Weisskopf, Robert Serber and a precocious young scientist by the name of Richard Feynman – whom Bethe recalled as 'more eager than almost anybody, and extremely ingenious' – into smaller groups. Anticipating the amount of calculation to be done, Bethe also set up a unit composed primarily of scientists' wives who punched the numbers into hand-held computing machines. These manual machines were later replaced by faster IBM machines, the sight of which would, as Bethe recalled, inspire John von Neumann with the ambition 'to change these machines and make them much faster and electronic'.

Edward Teller had wanted to head theoretical physics himself and did not relish the prospect of Bethe as his boss. Teller wrote in his memoirs, 'I was a little hurt. I had worked on the atom bomb project longer than Bethe.' He also thought Bethe plodding and overly focused on ' "little bricks", work that is methodical, meticulous, thorough and detailed'. His own approach was, he believed, more visionary. Teller was by then a commanding physical presence, with, as Laura Fermi described, thick and bushy eyebrows that 'jutted out so much above his green eyes that they looked like gables over the stained windows of some old church. When he was absorbed in thought, he thrust them up, and his face acquired a strange intensity.' Although Teller agreed to work for Bethe, he later wrote that disagreements over his tasks 'marked the beginning of the end of our friendship'.

Some of the friction centred on the low priority given to Teller's special interest – the development of an explosive weapon based on fusing light hydrogen atoms rather than fissioning heavy elements. Enrico Fermi had made the original suggestion for such a weapon – nicknamed the 'Super' – somewhat in passing, but it had caught Teller's imagination. He had hoped and assumed that he would be able to work almost exclusively on the Super – an early version of the hydrogen bomb. However, Bethe, backed up by Oppenheimer, made it clear that, for now, a bomb based on fission was the priority.

Forty-one-year-old Captain William S. 'Deak' Parsons of the US Navy was appointed head of ordnance – a vital role. As the project moved to fruition, he would be responsible for ballistic testing and the planning for, and perhaps the actual use of, the bomb. Groves selected him, on Vannevar Bush's recommendation, after the briefest of interviews, recognizing his grasp of both theoretical and practical ordnance including high explosives, guns and fusing. As the work progressed he would also show skill in melding together a mixed team of scientists, engineers and explosives experts. Parsons arrived at Los Alamos soon after his appointment, the first naval officer to be assigned there. His naval summer uniform caused consternation then suspicion at the entry gate. An army guard telephoned his sergeant to report, 'Sergeant, we've really caught a spy! A guy is down here trying to get in, and his uniform is as phoney as a three-dollar bill. He's wearing the eagles of a colonel, and claims that he's a captain.'

Despite their respective talents, at the outset neither Groves nor Oppenheimer predicted the scale of operations and the population that would be required to sustain it. Los Alamos grew at phenomenal speed from just a few hundred people in the spring of 1943 to well over three thousand by January 1944. Before long, extraordinary stories began circulating in Santa Fe about what was really going on at 'the hill', as Los Alamos became known. Townspeople could see smoke curling up from the site in daytime and lights at night. Some believed that the army was operating a home for pregnant WACs. When naval officers were spotted, a rumour spread that a new type of submarine was being perfected there. Local people took a particular interest in the enclosure fences, wondering whether they were designed to keep people in or out. The advice given to Los Alamos staff was not to confirm or deny anything. The wilder the rumours, the easier it would be to obscure the truth.

The trip up to Los Alamos from Santa Fe enthralled newcomers. Ruth Marshak, accompanying her physicist husband, wrote, 'As we neared the top of the mesa, the view was breathtaking. Behind us lay the Sangre de Cristo mountains, at sunset bathed in chan-ging waves of color – scarlets and lavenders. Below was the desert with its flatness broken by majestic palisades that seemed like ruined cathedrals and palaces of some old, great, vanished race. Ahead was Los Alamos . . .'

There, the majesty ended abruptly. Seven-foot-high fences topped with barbed wire surrounded the site itself. Notices read:

U. S. Government Property
DANGER! *PELIGRO!*
Keep Out

Military policemen in battle helmets – a 'formidable-looking bunch of young men', as another woman recalled – inspected passes. To some of the refugee scientists at Los Alamos, the stark fences, strict security, dog patrols and heavily guarded Technical Area held disturbing echoes.

Life was certainly strange for the new arrivals as they adjusted to existence behind the wire. They could tell no-one where they were; the only address they were allowed to quote was 'Box 1663, Santa Fe'. The site itself was confusing. Barracks-like buildings stood at odd angles on streets without names, all alike and all painted green, camouflaged among the green pines. They were so uniform that it was easy to get lost. People used the cylindrical wooden water tower on the site's highest point to orientate themselves.

Rose Bethe was appointed head of the housing office and was therefore responsible for allocating accommodation – a task, as another Los Alamos wife put it, requiring every ounce of her 'self-reliance, efficiency and stubbornness'. There were a few ground rules to help her. Childless couples were only entitled to a one-bedroom apartment; couples with one child were allocated two bedrooms; families with two were given three-bedroomed dwellings. There were, nevertheless, perpetual problems to solve and people to soothe. Edward Teller and his beloved, monumental piano were placed immediately below a quiet, contemplative bookworm who relished silence rather than Teller's nocturnal sonatas. An enthusiastic chemist with a passion for conducting explosive experiments in his apartment lived adjacent to a large brood of children. One childless couple asked Rose for a two-bedroom apartment. When she enquired whether they were expecting a baby, the blushing pair replied, 'No, but we let nature take its course.'

Babies were, in fact, a prominent feature of life at Los Alamos, which was, above all, a young site – the average age was twenty-seven. Many couples decided to start their families there. Medical care in the one-storey hospital was free and it was especially strong in paediatrics and gynaecology. Groves later wrote wryly, 'Apparently we provided adequate service, for one of the doctors told me later that the number and spacing of babies born to the scientific personnel surpassed all existing medical records.' Some of the scientists blamed Groves for a perennial shortage of nappies which they believed he had arranged on purpose. They also believed he had ordered Oppenheimer to discourage people from reproducing, but Oppenheimer's own daughter Toni was born at Los Alamos. Like the others, her place of birth was simply listed as Box 1663, Sandoval County Rural.

There were tensions between parents and dog-owners. Many had brought their pets with them and the animals roamed the mesa at will. One dog started biting people and was found to have rabies. Rules were hastily introduced to keep dogs under control, but, as a Los Alamos mother recalled, 'when the dog owners got tired of keeping their pets inside or on a leash, they suggested putting the children on leashes and letting the dogs go free'.

The most desirable residences were in 'Bathtub Row'. These were the attractive, sturdy stone and log cottages which had belonged to the school. Their great attraction – hence the nickname – was that they possessed baths, whereas the new army-built accommodation had only showers. At first, only the most senior people like Oppenheimer and his wife, who settled into the erstwhile headmaster's house, lived there. But later, as others moved in, 'it became uncertain in envious minds whether Bathtub Row derived its lustre from its residents or whether the residents acquired distinction from living in it', according to one wife. Apartments in nearby 'Snob Hollow' were also highly prized.

Snobbery was a genuine issue, as American physicist Luis Alvarez discovered shortly after arriving at Los Alamos. As news spread that a family called Alvarez was moving in, other wives in the apartment building hurried to the housing office to complain about living next to Spanish-Americans. They were reassured to learn that the tall, blond Alvarez was only partly Spanish. The shortage of domestic help was another potential source of discord. Indian girls from the nearby villages were assigned by need rather than by the ability to pay. Kitty Oppenheimer, who had a full-time maid, took a role in the allocation. As an incentive to wives to work, those who did volunteer were given priority with household assistance.

Wives were not the only women working at Los Alamos. Promising young female scientists were recruited, such as Joan Hinton, a graduate physics student from Wisconsin University, who worked on the design and construction of research facilities. By October 1944 there would be twenty women scientists and about fifty women technicians in total working on the site, in addition to nurses, teachers, secretaries and clerks.

Frenetic partying became an established feature. As Emilio Segrè recalled, 'The isolation of Los Alamos pushed families to an active social life: there were many dinner parties; many people for the first time took up poker and square dancing.' Amateur dramatics flourished. Edward Teller played a corpse in a production of *Arsenic and Old Lace*. Oppenheimer also negotiated with a local woman, Edith Warner, who agreed to provide dinner three nights a week for small groups of scientists and their wives at her little house on the banks of the Rio Grande. As Oppenheimer had hoped, it gave them a brief respite from the stressful claustrophobia of the site.

The extraordinary surroundings, the ever-changing colours of mountains, sky and desert, the clear, crisp air and the vivid flow-ers that bloomed from early spring to late autumn also helped invigorate people. Scientist Philip Morrison, summoned to Los Alamos, was seduced by 'the utterly enchanting landscape'. To Robert Christy, it was 'a wonderful environment for anyone who liked the out-doors. The only ones who didn't like it were the complete New Yorkers.' There were trails to ride and hike, streams

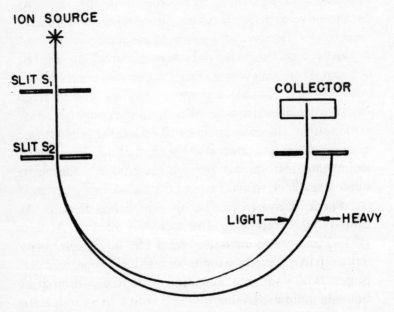

The electromagnetic separation process used in a calutron

to fish, Indian ruins to visit and, in the winter, snowy slopes to ski. Leo Szilard, still in Chicago, had warned departing Met Lab colleagues that 'Nobody could think straight in a place like that. Everybody who goes there will go crazy.' But he would, for once, be proved wrong. Despite the pressures, many would remember their time at Los Alamos as the most stimulating and enjoyable of their lives.

Meanwhile, rapid progress was being made at the two giant industrial sites of Oak Ridge and Hanford. At Oak Ridge, where construction was shared between several contractors, work began in early 1943 on Lawrence's electromagnetic uranium separation plant, codenamed 'Y-12'. It was based on cyclotrons modified into mass spectrographs which were known as 'calutrons', after the University of California. They were put together in great 'racetracks', each containing ninety-six calutrons. Eventually, fifteen such racetracks would be built. From the outside, the complex of concrete and brick buildings connected by a maze of streets with gantries of pipework and electrical wiring passing overhead resembled a conventional chemical plant.

In June, the first ground was broken at Oak Ridge for the gaseous diffusion plant, codenamed 'K-25' and so immense that it would consume more electricity than a small city. At half a mile long it was probably the largest chemical engineering plant ever built. In the interests of safety, Y-12 and K-25 were located in valleys seventeen miles apart. With no time to build pilot plants, the respective designs were based on Lawrence's research at Berkeley and Harold Urey's at Columbia. As Groves later wrote, 'research, development, construction and operation all had to be started and carried on simultaneously and without appreciable prior knowledge'.

At Hanford, the uncertainties were the same. According to Teller, the plutonium-producing reactors were built in under eighteen months 'on the basis of a theory proposed by physicists that no engineer had thoroughly checked'. In April 1943, Du Pont began work on three industrial-scale reactor piles based on a design developed by Eugene Wigner using graphite as a moderator. For safety's sake, they were constructed six miles apart. Each reactor was a giant block, forty-six feet wide, forty-six feet high and forty feet deep. Inside was a thirty-six-foot-high stack of a hundred thousand graphite blocks encased in six thousand tons of cast-iron and steel. The uranium fuel, sealed in eight-inch-long aluminium cylinders and then assembled into batches, was pushed through tubes running from the front to the back of the pile, irradiated, and then discharged ready for reprocessing to retrieve the plutonium produced by the controlled fission the fuel had undergone. This reprocessing, using chemicals to dissolve the uranium and to extract, concentrate and purify the plutonium, was the most hazardous part of the operation and was carried out in windowless separation plants built in isolation more than ten miles away.

Hanford itself, with some twenty thousand construction workers, had swiftly developed the feel of a Wild West frontier town. 'There was nothing to do after work except fight,' exaggerated one physicist, if only slightly, 'with the result that occasionally bodies were found in garbage cans the next morning ... It was a tough town.' There was also racial segregation: the site administrators bowed to local sentiment and provided separate accommodation and amenities for black workers. At Oak Ridge, local Tennessee law decreed that the twenty-five thousand construction workers had to be segregated.

From the very start, Oppenheimer justified Groves' selection of him as director of Los Alamos. He showed himself to be a disciplined, inspirational leader with breadth of vision and a facility to appreciate, assimilate and analyse issues and then to take the right decision. Hans Bethe praised him in the following terms: 'A physicist like Fermi would delight in solution of a single problem, I admired him to idolatry, but there is another type of mind which is equally needed. Oppenheimer . . . worked at physics mainly because he found physics the best way to do philosophy. This undoubtedly had something to do with the magnificent way he led Los Alamos.' Above all, Oppenheimer inspired trust in his team.

Groves was also impressed with Oppenheimer's unemotional objectivity. In 1943, James Conant was asked to lead a study into the prospects for developing radiological weapons. Learning of the project, Oppenheimer told Groves that he and Fermi had been discussing a scheme of their own to pollute German food supplies with a lethally radioactive fission by-product, beta-strontium. Groves' response was apparently enthusiastic, but Oppenheimer wrote coolly to Fermi that the idea was probably not worth pursuing 'unless we can poison food sufficient to kill a half a million men'.

However, there was one area where he and Groves did not agree. To protect security, Groves wanted a system of 'compartmentalisation'. The aim, as he later wrote, was that 'each man should know everything he needed to know to do his job and nothing else'. The system worked in the industrial environment of Hanford and Oak Ridge;

it was, however, anathema to scientists used to a free exchange of ideas. Groves particularly disapproved of the colloquia which Oppenheimer asked Teller to organize and at which scientists discussed their respective progress and problems. From a scientific perspective they were creative and valuable exercises, producing cross-specialization synergy. From a security perspective they were, Groves believed, highly dangerous. However, as Teller recalled, Oppenheimer 'fought hard for an open exchange so that everyone could contribute, and he won'.

Most of the Manhattan Project scientists considered Groves' attitude to security obsessive, even childish. The Fermis were amused by the personal protection rules he established for Enrico at the Met Lab in Chicago. Laura Fermi considered that they would have done credit to the nervous mother of a teenage girl: 'Enrico was not to walk by himself in the evening, nor was he to drive without escort . . .' The pile which had gone critical in December 1942 to cheers and sips of Chianti had been moved to the newly built Argonne Laboratory some twenty miles away. By mid-1943 Fermi was driving there almost daily, but, at Groves' insistence, he was always accompanied by his powerfully built bodyguard who looked 'as if he had sufficient strength to wring the neck of any evil-minded spy or saboteur'.

Groves' fears about spies and espionage would, however, be vindicated after the war, when the extent of spying at Los Alamos was revealed. Groves would rightly claim that Soviet spy David Greenglass, recruited to Los Alamos as a machinist, passed information to the Russians to which he should never have had access. Greenglass was, in fact, the brother of Ethel Rosenberg, whom he would later denounce, with her husband Julius, as responsible for his acts. They were executed for spying in 1953.

Groves would also bitterly recall the spying activities of Klaus Fuchs at Los Alamos, but for those he would blame the British.

CHAPTER SEVENTEEN

'MR BAKER'

THE ARRIVAL OF A TEAM OF BRITISH SCIENTISTS IN THE UNITED States in the autumn of 1943 to work on the bomb project was the result of the three-page Quebec Agreement, signed on 19 August 1943. After snubbing the American offer of partnership in 1941, the British had found themselves increasingly marginalized, in part because the United States believed she no longer needed Britain. By the end of October 1942 Henry Stimson had felt confident enough to advise the President that the US should proceed 'for the present without sharing anything more than we could help'. However, it was also a security issue. In May 1942 Britain and the Soviet Union had signed a twenty-year mutual assistance treaty, the Cripps–Molotov Agreement, and several weeks later a specific scientific exchange agreement. Groves was convinced that information about the bomb project would inevitably reach the Soviet Union and had been doing what he could to restrict the flow of information to the British.

Churchill pressed Roosevelt, first at the Casablanca conference in January 1943 and then in Quebec, for a greater role for Britain. To Groves' dismay, the President yielded. The treaty provided for the two countries to pool their nuclear research but crucially stipulated that neither country would pass information to a third party without the other's consent. It also provided that neither country would deploy the atomic bomb without the other's agreement.

The British were now so eager to collaborate that even before the treaty was formally signed James Chadwick, Rudolf Peierls, Franz Simon and Mark Oliphant were on their way to the United States aboard the Pan American flying-boat service from Ireland. It was a far more luxurious trip than the usual form of wartime trans-atlantic transport experienced by British scientists – flying in a bomber.* The British team spent a few days in New York, amazed by the abundant food in the shops. Chadwick, with his legacy from the First World War of an impaired digestion, unwisely visited Grand Central Station's oyster bar and suffered agonizing consequences. A few days later, according to Oliphant, he still looked 'like death'. However, news that the Quebec Agreement had indeed been signed revived him. On Monday, 13 September 1943, at a meeting at the Pentagon, Chadwick and the British team learned for the first time of the existence of Los Alamos. General Groves suggested that

---

* Passengers on the bomber spent the sixteen-hour flight in the bomb bay, lying in the freezing cold on a rough mattress resting on the bomb doors. They had to wear full flying gear, including oxygen mask, helmet and parachute. It was impossible to read because the plane was blacked out. Sometimes people lost their head in the cold, noisy dark-ness. Oliphant recalled a man who, threshing about in a panic, inflated his life jacket and passed out. As Oliphant struggled to help him, the rip cord of the man's parachute caught on something, filling the bay with rippling silk.

Chadwick and Oliphant should go there at once because of the lack of experienced experimental physicists.

Chadwick's initial impression of Groves, who forcefully outlined his views on secrecy and compartmentalization at the meeting, was that he was 'the dominant personality' of the American project – in fact, 'a dictator'. Chadwick and Groves would later come to a mutual respect, even admiration, but for the time being Chadwick found his discussions with Robert Oppenheimer more productive. The British team also toured several US laboratories where, according to Peierls, the American scientists revealed Groves' instructions that the British 'could be told everything, but must not be shown anything'. However, as nobody could understand this perplexing order, 'it caused no problem'. By the time Chadwick returned to Liverpool in late September, arrangements for the revived Anglo-American collaboration were largely in place.

Chadwick began assembling a team to go to America. The Quebec Agreement required all its members to be British citizens. For many this was no problem. Among Chadwick's own group at Liverpool, Otto Frisch happily agreed to take British nationality. His aunt Lise Meitner could have accompanied him. She was invited to leave Stockholm and join the British team, but her response was that 'I will have nothing to do with a bomb'. Her views on the moral duty of scientists had altered since the early days of the First World War when she reassured Hahn about his chemical warfare work with the words 'any means which might help shorten this horrible war are justified'. She later explained, 'I hoped that the newly discovered source of energy would be used only for peaceful purposes. During the war, I used to say . . . "I hope they will not succeed in making an atomic bomb, but I fear they will." ' It also seemed that Chadwick's favourite protégé, Joseph Rotblat, would not be coming to the United States. Deeply attached to his Polish nationality, he refused to renounce it. However, Chadwick was so eager to bring Rotblat with him that he obtained a special dispensation from Groves, assuring him of Rotblat's complete loyalty.

The next sticking point was over security. Groves demanded US security checks on every scientist the British proposed sending. The British were affronted and offered, instead, to guarantee that every member of their team had been thoroughly vetted by British intelligence. Groves attributed their reaction to 'the attitude then prevalent in all British officialdom that for an Englishman treason was impossible, and that when a foreigner was granted citizenship he automatically became endowed with the qualities of a native-born Englishman'. He was forced to accept the British position but tried to ensure that, as far as possible, British scientists did not gain access to the most sensitive areas of the project. Even Chadwick was not allowed to visit Hanford.

The British scientists were allocated to various teams and locations. Chadwick, as leader of the British team, would base himself at Los Alamos. Mark Oliphant was to work with Ernest Lawrence at Berkeley on electro-magnetic isotope separation. Rudolf Peierls was to work on gaseous diffusion theory in New York. Among those also cleared to go to the United States by British intelligence, which failed to spot his communist allegiances, was Klaus Fuchs,* a British citizen since 1942.

---

* The pronunciation and spelling of Fuchs' name was a source of difficulty and embarrassment to English-speakers. Even his fellow German-speaker Rudolf Peierls signed one letter addressed to 'Dear Fucks'.

Brought up in a left-wing German family with deeply rooted socialist and Christian beliefs – his father was a pastor – Fuchs had become a communist in 1932, believing that only a united working class could stop the rise of the Nazis. He later wrote, 'I was ready to accept the philosophy that the [Communist] Party is right and that in the coming struggle you could not permit yourself any doubts . . .' After Hitler came to power and the Nazis began arresting known communists, Fuchs fled Germany, reaching England in the autumn of 1933. His orders from the Communist Party were to complete his education to prepare himself for the struggles ahead.

Fuchs duly found a position as a research assistant in Bristol University's Physics Department and, after completing his doctorate, he worked with Max Born at Edinburgh University. In 1940, the British briefly interned him as an enemy alien but soon released him. Realizing that Fuchs had the kind of ability he was looking for, Rudolf Peierls offered him a post at Birmingham University as assistant in theoretical physics, writing, 'I cannot now describe the nature or purpose of the work, but it is theoretical work involving mathematical problems of considerable difficulty, and I have enjoyed doing it, quite apart from its extreme importance.' He obtained official clearance for Fuchs to join the nuclear project and put him to work on gaseous diffusion techniques for isotopic separation. His contributions were so significant that when Peierls was invited to go to America, his gifted young colleague naturally went too. Before leaving for the United States, Fuchs contacted his Soviet 'handler', a woman codenamed 'Sonia', who promised that a new agent would contact him there, a man he would know only as 'Raymond'. They would make contact in February 1944.

Under the terms of the Quebec Agreement, the United States also promised to underwrite the Anglo-Canadian nuclear project. In late 1942 the British had established an Anglo-Canadian laboratory in Montreal, which later moved to Chalk River, and despatched a team to Canada, including Hans von Halban, Lew Kowarski and Bertrand Goldschmidt. Its primary purpose was to study the effectiveness of heavy water at slowing down neutrons. Hans von Halban was director until, in April 1944, John Cockcroft took over. Groves allowed some low-level exchanges with the Montreal team but, deeply distrustful in particular of the French contingent, forbade direct contact with the US scientists. His reservations would ultimately be proved correct. The team seconded to Montreal by the British included two men – one a Briton, Alan Nunn May, and the other a refugee, Bruno Pontecorvo – who would later be unmasked as ideologically motivated Soviet agents.

The Chadwicks arrived at Los Alamos in early 1944 and moved into a two-bedroom log cabin on 'Bathtub Row'. When Rotblat arrived a few weeks later he moved in with them. Also in early 1944, Mr Nicholas Baker and his son arrived to join the British contingent. Laura Fermi, when she arrived at Los Alamos later that year, would be struck by how

In the Los Alamos array of faces wearing an expression of deep thought at all hours and under all circumstances, whether the men they belonged to were eating dinner or playing charades, Mr Baker's face stood out as the most thoughtful, the one expressing the gravest meditations. He appeared to be dedicated to a life of the intellect alone, which allowed no time for earthly concerns ... Mr

Baker's eyes were restless and vague. When he talked, only
a whisper came out of his mouth, as if vocal contacts with
his fellow-men were of little consequence. He was a few
years older than the other scientists – close to sixty in
1944 – and all looked at him with reverence . . .

'Mr Baker' was, of course, Niels Bohr. He had been given
a pseudonym, just as Oppenheimer was 'Mr Smith',
Lawrence was 'Mr Jones' and Fermi was 'Mr Farmer'.

Bohr was fortunate to be alive. In early 1943 British
intelligence had received a warning from Denmark that
Bohr was likely to be deported to Germany. A message
from James Chadwick inviting Bohr to England was
smuggled into Denmark on microdots concealed in two
ordinary-looking doorkeys. Chadwick promised 'a very
warm welcome and an opportunity of service in the
common cause'. Bohr, however, felt unable to leave
Denmark, knowing his flight would expose family and
colleagues to Nazi reprisals. His reluctant refusal, reduced
to a microdot measuring two by three millimetres, was
smuggled out of Denmark in the hollow tooth of a
resistance worker.

The rapidly worsening position in Denmark altered
Bohr's view. News of the Russians' victory at Stalingrad at
the very end of January 1943 had encouraged the Danish
resistance to launch a series of sabotage attacks. The
Germans responded by shooting hostages, prompting a
series of strikes which the Germans again savagely
suppressed. On 28 August the Danish government resigned,
and the following day the Germans declared martial law.
The British sent a further, urgent message to Bohr, passed to
him by word of mouth: 'We are still waiting for you.'

A few days later Bohr learned from informants that the

Germans planned to deport 'undesirable aliens'. Realizing that this meant Jewish refugees in Copenhagen, he warned those of his staff who were at risk, helped them contact the Danish underground who would assist them to flee to Sweden, and gave them money. He expected at any moment to be arrested himself. In fact, as it emerged at the Nuremberg war crimes trials, the Germans had intended to seize him the day they declared martial law. They had, however, changed their mind, fearing that it would attract too much attention. They decided instead to arrest Bohr during a general round-up of Denmark's Jews.

Bohr hurriedly destroyed his papers. He also dissolved in acid the gold Nobel medals that James Franck and Max von Laue had left with him for safekeeping. They lived out the war in an innocent-looking bottle on a cluttered shelf, and the gold was later retrieved and re-cast. On 29 September, Margrethe Bohr's brother-in-law brought the news Bohr had been expecting. According to a contact in the German diplomatic corps, Berlin had ordered the deportation of Niels and his brother Harald to Germany. Bohr knew that he and Margrethe had to leave at once. Friends arranged for a boat to take them to Sweden and promised to send their sons after them.

Copenhagen was under strict night-time curfew. Anyone out on the streets after the deadline was shot on sight. The Bohrs therefore had to try to reach a beach undetected while it was still daylight. In the late afternoon they walked down a still-crowded street, carrying only a small bag. A scientist friend, standing on the corner, gave Bohr a surreptitious nod – the signal that everything was in place for the escape. The Bohrs made their way to fields beyond the city and hid in a shack until dark. They were supposed to make their escape at nine p.m., but as

Margrethe Bohr recalled, when the time came the Nazis 'had come out so that we had to wait until late in the night'. When at last the coast was literally clear, the Bohrs hurried down to the beach. It was, Margrethe remembered, 'very dramatic' – 'you had to throw yourself down to the ground not to be seen'. They clambered gratefully aboard a small motor boat waiting to take them out to the fishing boat that would carry them to Sweden.

Safely arrived near Malmö, Margrethe waited for their sons while Bohr hurried to Stockholm. His mission was to plead for Denmark's Jews, who he knew were about to be rounded up and shipped to concentration camps. The neutral Swedish government, which had tried unsuccessfully to intercede on behalf of Norway's Jews, agreed to help and broadcast a formal announcement offering sanctuary to Danish Jews. This offer prompted one of the most honourable and courageous acts of the war. The Danish underground assembled a fleet of small boats and ferried their Jewish countrymen to safety in Sweden. The dangerous shuttle operation saved nearly six thousand lives. The Nazis were able to deport only 472 Jews, many of them elderly, helpless and living in old people's homes. One of Niels Bohr's aunts was among them. She did not survive.

The rest of Bohr's family reached Sweden safely. Margrethe and their younger sons would remain there for the rest of the war, but within weeks Bohr received a telegram from Lord Cherwell inviting him to England. This time he accepted. On 6 October 1943, a British Mosquito fighter bomber – painted in civilian livery, unarmed and flown by two civilian crew to avoid violating Swedish neutrality – landed in Stockholm. The only available space for the large-framed Bohr was in the empty bomb bay, which had been specially padded to take a passenger. He was equipped with flying suit, parachute and a set of distress flares and told that if the Luftwaffe attacked the plane the pilot would open the bomb-bay doors, jettisoning Bohr, who was to parachute into the sea and send up the flares. He was also given a helmet fitted with headphones which was the only means the crew had of communicating with him.

To avoid attack, particularly from Luftwaffe bases in Norway, the Mosquito at first flew at very high altitude. The pilot instructed Bohr to turn on his oxygen supply, but unfortunately Bohr's helmet was too small for his gigantic cranium. The headphones did not cover his ears and he never heard the order. He lost consciousness, but as the Mosquito descended he began to revive, and by the time it landed in Scotland he was conscious once more.

Bohr was flown on to London where James Chadwick was waiting to greet him. Since 1940 Bohr had been cut off from information about British and American progress on atomic research. He was amazed by what he soon learned, especially that Enrico Fermi had achieved a self-sustaining chain reaction. Bohr was assigned an office near the London headquarters of the Directorate of Tube Alloys, where he was joined by his son Aage as his assistant. Bohr spent the next few months visiting laboratories across the country and bringing himself up to speed. The reality of a nuclear bomb disturbed as well as fascinated him, and he already foresaw that it could, in the future, prompt an arms race.

The British tried to persuade Bohr to go to the United States as part of their team under the Quebec Agreement. Bohr, who had close personal ties with America as well as Britain, was reluctant to be affiliated to any particular camp. To meet his concerns and to allow him the requisite degree of independence, he was appointed 'Consultant to the British Directorate of Tube Alloys'. His brief was to review the work under way in the United States and decide how he could best assist the common goal.

Niels and Aage Bohr sailed for the New World under their assumed names of Nicholas and James Baker. However, the FBI agents who met them as they disembarked were horrified to see NIELS BOHR written in large black letters on 'Nicholas Baker's' suitcase. General Groves accompanied them on the long train journey from Chicago to Lamy, New Mexico – the nearest station to Los Alamos. To keep the Danes' presence on the train secret, Groves ordered their meals to be served in their compartment. He was chagrined to discover that on both mornings of the journey the Bohrs breakfasted in the dining-car. Groves found the trip stressful in other ways too. During their hours of confinement Bohr was hard to understand. The morning after they reached Los Alamos, Oppenheimer noticed that the general seemed below par and asked him what the matter was. Groves replied, 'I've been listening to Bohr.'

There were, of course, many at Los Alamos only too eager to listen to Bohr. One evening, at Oppenheimer's house, Bohr addressed a small group of European scientists about conditions in Denmark and about his escape. It made a deep impression. As Emilio Segrè recalled, 'For many of us this was the first eyewitness account of what was really happening in a Nazi-occupied country . . . the account left us depressed and worried, and more determined than ever that the bomb should be ready at the earliest date possible.'

Segrè would have been relieved to know of the increasing practical difficulties confronting Germany's scientists.

During the summer of 1943, British night bombing attacks on Germany had achieved a new intensity in an operation codenamed 'Gomorrah'. Aided by the first use of 'window' – strips of aluminium foil designed to confuse German radar when released from bombers – the Royal Air Force had targeted Hamburg. On the night of 27 July, the blast of their high-explosive bombs, combined with incendiaries, created a firestorm. Fires merged, sucking air into the centre where the oxygen was burned out. One pilot simply muttered, 'Those poor bastards.' Another crewman recalled, 'It was as if I was looking into what I imagine an active volcano to be.' Eight square miles of the city were reduced to ashes. Some victims were caught in melting asphalt as they tried to escape. In a raid which lasted only forty-three minutes, 42,000 people were killed, including more civilians than had died as a result of all German raids on London.* Fearing such Allied attacks

*Sir John Colville, Churchill's private secretary, related that Air Marshal Harris had shown Churchill a film of the bombing raids on Hamburg and elsewhere, expecting praise for his efforts. When the lights came up, Colville saw tears running down Churchill's face, and he (Churchill) said, 'Are we beasts that we should be doing these things?' However, Churchill's views on what we would now call 'weapons of mass destruction' varied with his mood and the progress of the war. Later, when German flying bombs were falling on Britain in July 1944, he wrote a memo to his military chiefs of staff: '*I want you to think very seriously over the question of using poison gas. I would not use it unless it could be shown that (a) it was life or death for us, or (b) that it would shorten the war by a year. It is absurd to consider morality on this topic when everybody used it in the last war without a word of complaint from the moralists or the Church. On the other hand, in the last war the bombing of open cities was regarded as forbidden. Now everybody does it as a matter of course. It is simply a question of fashion changing as she does between long and short skirts for women.*' Churchill, of course, on reflection concluded that gas should not be used.

on Berlin, Albert Speer ordered Germany's research institutes to seek new and safer homes outside the capital.

Heisenberg did not share Speer's anxiety. Though forced on one occasion to flee through the burning streets of Berlin, shoes smouldering with phosphorus, he believed that his reactor experiments in a concrete bunker in the 'Virus House' were well protected. Nevertheless, realizing that maintaining essential supplies of electricity and water in Berlin might not be possible for much longer, he decided gradually to relocate the Kaiser Wilhelm Institute for Physics to Hechingen, a small town in south-west Germany. Not only was it quite close to Urfeld in the Bavarian Alps, where he had recently moved his family permanently, but, he reckoned, if the worst should come Hechingen was more likely to fall to invading western Allies than Russians advancing from the east. By the end of 1943 he had sent a third of his institute – those not essential to the fission work – south under Max von Laue as assistant director.

Heisenberg, in the meantime, continued his work in the Virus House, undeterred by the night-time wail of the air-raid sirens and the crump of exploding Allied bombs.

CHAPTER EIGHTEEN

HEAVY WATER

SINCE THE RAID ON THE VEMORK HEAVY WATER PLANT IN February 1943, Knut Haukelid had been living a precarious existence, organizing resistance groups in the mountains. It had, as he later wrote, been very hard surviving in the wilds of the Hardanger Plateau: 'Snow and cold had been our constant companions and we had carried danger with us wherever we went.' Reports that the Germans had immediately started to rebuild the plant perturbed him. Predictably, they had taken precautions against further assault, bricking up doors and windows up to the first-floor level and fortifying entrances with double doors through which only one person at a time was admitted after scrutiny through a peephole. In addition, they had trebled the guard, floodlit the entire area and laid new minefields. A further commando raid seemed out of the question.

In the United States, General Groves also worried about Germany's continued capacity to manufacture heavy water. He was angered too by Britain's attitude towards the problem. Before the Gunnerside attack, the head of the Directorate of Tube Alloys, Wallace Akers, had told him that the British were planning to raid Vemork but had revealed no details of how or when. Groves had only learned the outcome from a translation of an article published in the Swedish *Svenska Dagbladet* on 14 March 1943 reporting that all the apparatus, machines and facilities for the production of heavy water had been blown up.

Groves' annoyance grew as the weeks passed and the British still refused to disclose exactly what had happened. He even suggested to his superiors that the US government should buy the information if that was the only way. He was particularly concerned that the plant had not been knocked out permanently. When the British finally furnished sparse details about the raid, claiming that the plant would not be fully effective for more than twelve months, he was unconvinced. A message to London from the Norwegian resistance on 8 July 1943 that the plant was expected to reach full production again by 15 August, which was passed on to him, proved him right.

Groves convinced Vannevar Bush and Army Chief of Staff General George C. Marshall that the plant had to be bombed from the air. The British at first resisted, arguing that casualties among Norwegian civilians would be heavy, but, seeing no other way, they eventually bowed to American pressure. On 15 November 1943, 388 B-17s and B-24s of the US 8th Air Force took off from English airfields, some to make diversionary raids around Oslo, the remainder to target the heavy water complex. Anticipating just such attacks, the Germans had installed anti-aircraft batteries and stretched cables from mountain to mountain to hinder low-flying aircraft. As a result, the air assault failed. Only two bombs hit the Norsk-Hydro plant. The heavy water cells were undamaged, but twenty-two Norwegians were killed by stray bombs.

Nevertheless, the attack made the Germans re-think the production of heavy water at Vemork. The risks of further air raids and sabotage were, they decided, too great. They considered manufacturing heavy water in an Italian nitrogen plant, then sending it to Germany for purification, but abandoned the idea as too complex. Instead, they decided to ship Vemork's heavy water to Germany and there construct their own heavy water plant. When agents passed rumours of this to London, the Special Operations Executive on 29 December sent a message to the Norwegian resistance: 'We have information that heavy water equipment may be dismantled and sent to Germany. Can you verify this? . . . Can this transport be aborted?'

The first question was easily answered. The resistance checked with their contacts at the plant and two days later confirmed to London that the Germans were indeed planning to remove all the stocks of heavy water and the key equipment. Furthermore, the move was imminent. The second question was more problematic. The Norwegians told London that they could not yet suggest a plan.

Haukelid and his colleagues managed to discover that the Germans intended to transport the heavy water by rail from Vemork to the northern end of a long, narrow inland lake called Tinnsjö. Here, on Sunday, 20 February 1944, they would load the railcars onto the ferry sailing south down the lake to connect with the railhead at Tinnoset, whence the railcars with their cargo of heavy water would continue their journey to the coast for shipment to Germany. Three weeks before the shipment was due the

Norwegians believed they had the answer. Haukelid radioed London that the most reliable solution would be to sink the ferry. Lake Tinnsjö was almost 1,300 feet deep and it would be impossible for the Germans to retrieve the drums from its frigid depths.

They also warned London that 'we must expect reprisals'. A special army detachment together with a company of SS had been drafted in to guard the shipment – a sign both of how seriously the Germans regarded the transport and of their likely response if thwarted. Indeed, the resistance were so worried about what the Germans might do to the civilian population that on 15 February they sent a further message, urgently querying whether the importance of the operation justified the potential consequences. London replied the same day. The answer was perhaps tactlessly breezy, but also unequivocal: 'Matter has been considered. It is thought very important that the heavy water shall be destroyed. Hope it can be done without too disastrous results. Send our best wishes for success in the work. Greetings.'

Haukelid reviewed the options again. Even putting thoughts of reprisals aside, the dangers to innocent passengers aboard the ferry were hard for him to stomach, but there seemed no choice. Disguised as a workman, he made a reconnaissance trip. He carefully timed how long the ferry took to reach the deepest part of the lake. The answer was twenty minutes. He knew that he and his fellow saboteurs would have to be very careful getting to the ferry and boarding it. The Germans were on high alert: 'There were more Germans than Norwegians in the whole valley ... German police stopped everyone that looked suspicious and checked their identity cards and parcels they carried.' On another scouting mission, this time in Rjukan during a local music festival, Haukelid disguised himself as a musician, concealing his machine gun in a violin case like any Chicago gangster.

On the evening of 19 February, twelve hours before the ferry was due to depart, Haukelid and two companions, Rolf Sorlie and Knut Lier-Hansen, dodged through the shadows down to the landing where the ferry was moored for the night. 'The bitterly cold night set everything creaking and crackling; the ice on the road snapped sharply as we went over it. When we came out on the bridge by the ferry station, there was as much noise as if a whole company was on the march.' While his comrades covered him, Haukelid, encumbered by a sack of explosives and detonators as well as his weapons, crept up the ferry gangplank. To his surprise, all seemed quiet apart from the voices of the crew playing poker below decks. The Germans had failed to place guards on the ferry – the most vulnerable link in the whole heavy water transport route.

Haukelid signalled to Sorlie and Lier-Hansen to follow him aboard. The trio crept below to the third-class accommodation and found a hatchway leading down to the bilges. However, before they could raise the steel hatch they heard footsteps and hastily took cover. It was the ferry watchman. According to Haukelid, they told him they were seeking a suitable place to hide. The man replied that he had several times helped conceal 'illicit things' on the ferry and himself showed them the hatch. While Haukelid and Sorlie climbed down and got busily to work fixing the explosives, Lier-Hansen kept the watchman engaged in conversation.

'It was', Haukelid later wrote, 'an anxious job and it took time. The charge and the wire had to be connected; then the detonators had to be connected to the wire and the ignition mechanism. Everything had to be put together and properly laid. It was cramped and uncomfortable down there under the deck, and about a foot of water was standing in the bilge.' It was important that the ferry should sink quickly: Lake Tinnsjö was so narrow that, unless the boat sank within five minutes, the captain might be able to beach her. Haukelid therefore laid the charge, consisting of nineteen pounds of sausage-shaped high explosive, towards the bows. On his reckoning, the blast would punch a hole about eleven feet square in the ship's side and the ferry would sink rapidly by the bows. The railway trucks holding the heavy water would roll off the deck and go to the bottom first. To be absolutely certain that the explosion occurred where the lake was deepest, Haukelid positioned two alarm clocks on a spar of the hull and wired them to the charge. He timed them to go off at 10.45 a.m. the next morning.

The saboteurs withdrew, telling the watchman that they had a few things to fetch and would be back on board in good time before the ferry sailed. Haukelid worried about the man who had been so co-operative, and whom the Germans would be bound to interrogate after the ferry was sunk. The fate of two Norwegian guards at the Vemork heavy water plant, sent to Grini concentration camp after the February 1943 raid, still weighed on his conscience. Yet if they warned the watchman and he was absent when the ferry sailed, this would raise German suspicions. Haukelid contented himself with 'shaking hands with the watchman and thanking him – which obviously puzzled him'.

As the three men ran from the ferry they heard the rumble of the approaching train bringing the heavy water. Haukelid and Sorlie fled at once, Sorlie up into the

mountains and Haukelid to catch a train the next day to
Oslo, and thence to ski to Sweden. Lier-Hansen was determined to remain behind to check that the ferry actually
sailed. If there was any delay he would defuse the bomb
to prevent a premature explosion. The next morning on
the train to Oslo, Haukelid consulted his watch yet again.
It showed 10.45 a.m. If all had gone according to plan the
ferry should now be sinking. A newspaper headline the
next day, RAILWAY FERRY 'HYDRO' SUNK IN THE TINNSJÖ, told
him that the mission had indeed succeeded, though at a
cost. Of the fifty-three people on board, only twenty-
seven survived. However, the canisters containing more
than six hundred kilos of heavy water lay beyond Nazi
reach at the bottom of Lake Tinnsjö.

Despite an initial wave of arrests, the reprisals the
Norwegian resistance had so feared did not materialize.
General von Falkenhorst found it less embarrassing to
maintain the fiction that the ferry's boilers had exploded
than to acknowledge another 'brilliant' act of Allied
sabotage, and another example of German incompetence
and carelessness.

CHAPTER NINETEEN

BOON OR DISASTER?

THE ALLIES HOPED THAT THEY HAD SIGNIFICANTLY DISRUPTED Germany's bomb project. However, in the spring of 1944, their own experienced a crisis. One of the greatest scientific challenges was how to configure the bombs to ensure an explosion of the right force at the right time. Until then, the assumption had been that the two types of atomic weapon on which they were working concurrently – the uranium-fuelled bomb, originally nicknamed 'Thin Man' for Roosevelt but renamed 'Little Boy' when the proposed gun barrel was shortened, and the plutonium bomb, nicknamed once and for all 'Fat Man' for Churchill – would both be detonated by a high-velocity gun. This would fire one subcritical piece of fissile material into another, thereby creating a critical mass, initiating an uncontrolled chain reaction and producing the desired explosion. However, samples of plutonium produced by the Du Pont pilot plant at Oak Ridge, which began reaching Los Alamos at the rate of a gram a day from April 1944, showed an alarming capacity to fission spontaneously. The phenomenon was not entirely unexpected – the possibility of spontaneous fission had been raised in 1939 during discussions about how much material would be required to produce an atomic weapon. However, what worried the scientists was that these samples appeared five times more likely to fission spontaneously than plutonium hitherto produced for experimental purposes in cyclotrons. It was a rate never observed before.

Emilio Segrè and his team, working in an isolated canyon away from the main Los Alamos site to keep their equipment free of radiation from other work, examined the plutonium samples and discovered that they contained a hitherto unnoticed isotope, Pu 240. This 'rogue' isotope, created by the high rate of neutron irradiation of uranium in a reactor pile, spewed forth so many neutrons from spontaneous fission that its presence in the reactor plutonium meant that the gun-assembly technique was useless. It was simply too slow; the spontaneous fission would cause the plutonium to fly apart and vaporize with a modest release of energy before the two parts could combine fully to form a critical mass and produce a full chain reaction. For a while it seemed that the mammoth efforts at Hanford to construct large-scale production plants to ensure a plentiful and timely supply of plutonium had been wasted. The plutonium they would produce would possess the same propensity to fission spontaneously. General Groves seemed to have spent many millions of dollars for nothing.

Several solutions were proposed, including a faster gun assembly or purging the plutonium of Pu 240, but none seemed practicable. In July 1944, an anxious Groves convened an emergency meeting in Chicago attended by, among others, Robert Oppenheimer and Enrico Fermi.

They discussed an ingenious technique proposed by physicist Seth Neddermeyer the previous year but not seriously considered until then. Instead of hurling one chunk of fissionable material into another, Neddermeyer's suggestion was to wrap an outer shell of conventional high explosives around an inner core of plutonium. The explosives would be so positioned that, when they detonated, the shock waves would be channelled inwards, squeezing the plutonium into a small, dense, walnut-sized sphere, forcing it into a critical mass and thus producing a full explosion. The technique was called 'implosion'. The precise details remain classified to this day.

Oppenheimer returned to Los Alamos and ordered work on the plutonium gun assembly to cease. Instead, he gave implosion studies top priority. Explosives expert George Kistiakowsky, whom Oppenheimer placed in charge of the work because he thought Neddermeyer lacked the necessary project management skills, brought together a multidisciplinary team including physicists, machinists and explosives experts to work on what had become a highly resourced priority task instead of an interesting theoretical sideline. By the end of 1944 fourteen different groups would be engaged on implosion studies. Philip Morrison was appointed as one of two 'G-engineers' – the 'G' stood for 'gadget', codename for the implosion bomb. Their work was, as Morrison later recalled, dangerous. It was also arduous. They talked to the heads of all the different groups and studied their reports to see where the gaps and problems were, 'looking for anything that might go wrong or get in the way'. It was also their responsibility 'to certify that all problems and issues had been solved'.

The crux of the implosion problem was how to achieve

a perfectly symmetrical explosion and thereby produce
the perfectly symmetrical pressure waves needed to com-
press the plutonium into a supercritical sphere. Hans
Bethe recalled the first attempts at implosion as 'an utter
failure'. Then James Tuck, a young member of the British
team who was thoroughly in love with Los Alamos,
believing it embodied the spirit of Plato's ideal republic,
suggested using explosive lenses. Just as glass lenses could
be used to focus light waves, high explosives – cast into
special shapes, or 'lenses' – could be used to focus shock
waves, driving them inwards. This was, Bethe recalled, 'a
most important key'.

Meanwhile, Oppenheimer asked another British
physicist, William Penney, to study how waves of highly
compressed air radiated outwards from an explosion.
Penney was one of the few scientists at Los Alamos to
have actually witnessed the effects of blast waves on
human bodies and buildings, having studied the results of
German bombing in Britain. One evening, Penney
addressed one of Oppenheimer's colloquia on the subject.
As Rudolf Peierls recalled, 'His presentation was in a
scientific matter-of-fact style, with his usual brightly
smiling face; many of the Americans had not been
exposed to such a detailed and realistic discussion of
casualties . . . he was nicknamed "the smiling killer".'

Peierls himself, though based initially in New York, had
been advising Edward Teller on the use of Los Alamos's
newly arrived IBM punch-card calculators to compute the
characteristics of implosion. In the summer of 1944, he
and his wife Genia moved to Los Alamos where Hans
Bethe, Peierls' old friend, was anxious for him to replace
Teller and take charge of implosion theory. Bethe's disagreements with Teller had by then come to a head. Teller

was increasingly reluctant to work on implosion calculations, or for Bethe – whom he considered 'over-organised' as well as over-focused on detail – at all. In June 1944 Oppenheimer transferred Teller out of the Theory Division. His replacement, Peierls, brought Klaus Fuchs with him, and the two men, working closely together, shared an office. Ironically, as Bethe later recalled, Fuchs was 'the best of them all in computing just how the implosion wave would proceed'. Enrico Fermi joined Los Alamos just a few weeks later, arriving from Chicago in September 1944. Oppenheimer set up a new division for him. Fermi's task was to investigate problems outside the scope of the other more task-specific divisions. Edward Teller, working on the theory of the hydrogen bomb, became one of Fermi's group leaders. With so many different countries of origin represented on site, Oppenheimer often had to remind his colleagues that the project's official language was English.

Los Alamos was still growing. As Laura Fermi described it, 'The influx of new families on the mesa never ceased, and building went on at a feverish pace, invariably lagging behind the increase in population . . . we found the confusion and disorder that always accompany a fast pace of construction.' Nevertheless, Genia Peierls was delighted to have exchanged hot, humid New York City for the cooler air of the mountains. Earlier that summer she had taken refuge with the children at Cape Cod at a hotel whose brochure stated that it catered to a 'restricted clientele'. As her husband later wrote, 'We were not yet aware that this phrase means "No Negroes, Jews, Italians etc." Had we known, she would not have wished to stay there.' Genia had already been shocked by the more overt racism of the South. After disembarking from the ship which had brought the British team to Newport News, she had searched for seats for herself and her husband in a crowded train to Washington. She had found an almost empty car containing only 'two very nice negroes', only to be told that in the South 'transport was still segregated'.

The somewhat frenetic atmosphere of Los Alamos exactly suited the exuberant Genia, who, according to Laura Fermi, required 'incessant action'. She was soon busily organizing picnics to the ruins of old Indian pueblos. On one of these outings Laura found herself being driven by Fuchs. She thought him 'an attractive, young man, slim, with a small, round face and dark hair, with a quiet look through round eyeglasses'. She tried to make conversation but he answered her only 'sparingly, as if jealous of his words'. Genia, who had known and fussed over Fuchs since he had first come to work for her husband in Birmingham, laughingly nicknamed him ' "Penny-in-the-slot Fuchs" because talking to him was like putting a coin into a vending machine. You got only one response to each question.' Despite this reticence, Fuchs soon became popular at Los Alamos. He was a good dancer, enjoyed a drink and was ever willing to babysit. He was living a life which, in his later confession, he would describe as 'controlled schizophrenia'. It allowed him to 'establish in my mind two separate compartments. One compartment in which I allowed myself to make friendships ... the other compartment to establish myself completely independent of the surrounding forces of society.' He did this so successfully that Richard Feynman once joked with him about which of the two of them would be the most credible suspect as a spy. They agreed it was Feynman.

Feynman was an effervescent character whom

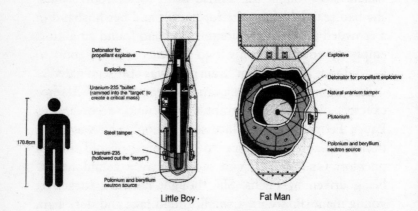

Little Boy, the bomb dropped on Hiroshima, was some 3 metres long, had a diameter of 0.7 metres and weighed about 4 tons. Fat Man, the bomb dropped on Nagasaki, was some 3.2 metres long, had a diameter of 1.5 metres and weighed some 4.5 tons

C. P. Snow later described as a cross between Groucho Marx and Einstein. He played the bongos and was once commissioned to paint a nude female toreador. He took delight in outwitting the increasingly sophisticated locking devices fitted to the Los Alamos filing cabinets. As he recalled in his memoirs, everybody thought their reports were safe, but, as he repeatedly demonstrated by presenting astonished colleagues with their own papers, the complex arrangements of steel rods, padlocks and, later, combination wheels 'didn't mean a damn thing'.

Feynman's wise-cracking boisterousness and passion for pranks masked a personal tragedy: his wife Arlene was dying of tuberculosis in a hospital in Albuquerque. Knowing that the end could come at any time, Feynman asked Fuchs whether he could borrow his car so he could get to Arlene's bedside quickly. Fuchs, always obliging to his friends, readily agreed. When the summons finally came, Feynman tore off in Fuchs' old blue Buick and, despite three flat tyres, reached the hospital in time to be with Arlene when she died.

With the work at Los Alamos focused on two completely different designs of atom bomb – the uranium device 'Little Boy' and the plutonium device 'Fat Man' – the scientists made their best estimates of how much uranium and plutonium respectively each would require to produce the necessary critical mass. They calculated that 'Little Boy' would need between 87 and 133 pounds of U-235 to cause an explosion equivalent to the detonation of between ten thousand and twenty thousand tons of TNT. Estimates of the necessary amount of plutonium for 'Fat Man' were even more uncertain. As everyone knew, having a bomb of either or both types available in time to influence the course of the war depended above all on whether sufficient fuel could be produced in time. Groves had always believed this to be the hardest part of the project.

Manufacturing uranium and plutonium was by then a massive effort. Oak Ridge and Hanford were the heart, but they were supported by factories and laboratories in thirty-nine states. Groves later estimated that by the war's end more than six hundred thousand people had contributed, directly or indirectly, to the Manhattan Project. The Y-12 electromagnetic uranium separation plant at Oak Ridge, where operators sat on high stools six feet apart, produced its first two hundred grams of U-235 in February 1944, barely a year after its construction began. However, production remained worryingly slow until the discovery that feeding the plant with uranium that had already been slightly enriched with U-235 significantly increased the yield. By late 1944, Y-12 was producing more substantial amounts of U-235. Meanwhile, K-25 – the plant using gaseous diffusion to separate U-235, built in sections by the Chrysler Corporation in Detroit and then assembled at Oak Ridge – was nearing completion. It would not become fully operational in time to make a major contribution. Still, the U-235 it began producing in April 1945, by pumping uranium gas against a porous membrane so that the lighter U-235 passed through more rapidly than the heavier U-238, could be used as feed for Y-12.*

The first plutonium-producing plant at Hanford, where workers had laboured in nine-hour shifts six days a week, was brought up to full power in late September 1944. Scientists observed the controlled chain reaction with satisfaction. However, after a while the power mysteriously began to drop and the reactor effectively shut itself down. Soon after, the power level began to rise again, only to be followed once more by a seemingly inexplicable shut-down. Scientists discovered the reason to be a rare isotope – Xenon-135, created during the fission process – which sucked up neutrons, thereby causing the chain reaction to peter out. They solved the problem by increasing the amount of fuel loaded into the reactor. Fortunately the Du Pont engineers had, with Groves' backing, designed in a larger number of slots for fuel than the scientists had thought necessary. The first plutonium was extracted from the reactor around Christmas 1944 and despatched

---

* After the war, as the gaseous diffusion technique was perfected, it would replace electromagnetic separation – rejected as too costly and cumbersome for mass production – and become the sole technique used by the United States. In 1991, Western scientists were surprised by evidence that Saddam Hussein was attempting to build a bomb using 'old-fashioned' electromagnetism.

to Los Alamos in early 1945, by which time the second and third plutonium plants at Hanford were also coming on line. The plutonium was transported by military convoy as air transport was considered too risky in case of a crash, while train connections from Washington State were too few.

The prospect that sufficient plutonium and U-235 would soon be available to build the bombs induced 'great pressure to be ready with all the necessary developments for making and detonating them', according to Rudolf Peierls. Oppenheimer drove his teams hard, determined, as he later wrote, 'to interpose no day's delay between the arrival of the material and the readiness of the bomb'. Scientists worked eighteen-hour days. Physicist's wife Ruth Marshak recalled how 'The Tech Area was a great pit which swallowed our scientist husbands out of sight, almost out of our lives. They worked at night, and often came home at three or four in the morning. Sometimes they set up army cots in the laboratories and did not come home at all.'

Oppenheimer was particularly anxious that everything for 'Fat Man', the plutonium-fuelled implosion bomb, should be in place – that the essential physics research had been completed, that the explosive lenses designed by James Tuck had been made and that an electric detonator system, developed by Luis Alvarez, was ready. Although the original plan had envisaged using uniform pressure waves to squeeze a thin, hollow shell of plutonium into a sphere, the necessary calculations had proved so complex that the idea had been abandoned for a simpler alternative. Robert Christy, by then working in Hans Bethe's division, had proposed using a solid sphere of plutonium comprising two fused hemispheres, together roughly the size of an apple. Christy calculated that the force of the implosion would at least double the plutonium's density, shortening the neutrons' route between nuclei and thereby swiftly accomplishing the required chain reaction. The device also included an initiator and a natural uranium tamper or shell. It was vital for achieving the chain reaction that the plutonium sphere remained spherical and did not distort, and the shell's purpose was to compensate for any asymmetrical effects during implosion.

As significant quantities of fissionable material began to reach Los Alamos, radiological protection measures were increased. Since the cavalier days of the 1920s, scientists and the public had become increasingly aware of the adverse effects of radiation on those exposed to it. The use of radium in tonics and potions and even face creams sold over the counter had become strictly controlled during the 1930s. The painful death of a wealthy Pittsburg industrialist, Ebert M. Byers, from the effects of a radium tonic, also advertised as an aphrodisiac, had been a particular catalyst for reform. In line with the instructions, he had daily consumed four doses, each containing two microcuries of radioactive material. The potential for radiation to cause genetic defects in unborn generations had also been recognized since 1927, following work on fruit flies by an American scientist named Herbert Muller. As a consequence, groups of experts had agreed internationally accepted limits of radiation exposure for both public and workers, albeit considerably more lax than those in force today.

As the Manhattan Project progressed, the American authorities deployed ever-increasing resources on research on health effects. Because of worries about security in the wide number of universities and other academic institutions involved, a new term, 'health physics', was adopted to embrace all radiation protection activities. The use of 100 hamsters, 200 monkeys, 675 dogs, 1,200 rabbits, 20,000 rats, 277,400 mice and 50,000,000 fruit flies in radiation experiments at one research establishment alone gives some indication of the scale of the work. Experiments were carried out on humans, too, sometimes without their knowledge or consent. The case of Ebb Cade, a black Oak Ridge worker, was one of the most shocking. After a car accident he was taken to hospital with broken limbs. Without his consent he was injected with plutonium. Later, fifteen of his teeth were extracted and bone samples taken to see how the plutonium had migrated around his body.*

Louis Hempelmann had been in charge of health safety at Los Alamos from the start, establishing limits for radiation exposure and developing ways of monitoring radiation levels. With the arrival of plutonium in 1944 his role intensified as he set up rules for the handling of plutonium and lectured teams about its extreme toxicity. Philip Morrison recalled, 'we had film badges, ionising gauges, counters. We were very seriously monitored.' There were no immediate radiological fatalities during the war, but there were serious incidents born of inexperience and carelessness. While leaning for a couple of seconds over blocks of U-235 in an assembly he nicknamed 'Lady Godiva' because of its unshielded nakedness, Otto Frisch

---

* The scale of experiments on humans came to light in 1996 in a report by the President's Advisory Committee on Human Radiation Experiments. Describing tests on some fifty people, the report concluded that 'patient subjects . . . were never told that the injections were part of a medical experiment for which there was no expectation that they [would] benefit, and [to which] they never consented . . .'

noticed out of the corner of his eye that the little red signal
lamps which flickered according to the number of
neutrons being emitted were 'glowing continuously'. His
body had 'reflected some neutrons back into Lady Godiva
and thus caused her to become critical'. Hastily he leaned
back and removed some of the uranium blocks. He calculated that during those two seconds 'the reaction had been
increasing, not explosively but at a very fast rate, by
something like a factor of a hundred every second'. The
radiation dose he had received was fortunately small, but
'if I had hesitated for another two seconds before remov-
ing the material . . . the dose would have been fatal'.

The Los Alamos teams were indeed engaged on dangerous work. According to Philip Morrison, 'we had the
temerity to "tickle the dragon's tail" by forming a supercritical mass of uranium. We made a much subdued and
diluted little uranium bomb that we allowed to go barely
supercritical for a few milliseconds. Its neutron bursts
were fierce, the first direct evidence for an explosive chain
reaction.' Such evidence gave Oppenheimer confidence
that the calculations for the uranium bomb were accurate
and that the gun-assembly method would work. He
advised Groves that, with a war on, the army should take
possession of the uranium bomb, Little Boy, untried.
Groves, convinced that the technology was straight-
forward and reassured by the exhaustive testing of the
actual gun mechanism by Deak Parsons' Ordnance Team,
readily accepted the advice.

Fat Man was a different matter. There was, as Peierls
recalled, 'much more room for doubt in the case of
plutonium, which depended on the very complex
implosion technique'. The scientists' advice was that the
plutonium bomb had to be tested. Groves at first objected,

fearing that if the test failed his precious plutonium would be scattered across the desert. However, as he later wrote, he was eventually persuaded of the need to check that 'the complex theories behind the implosion bomb were correct, and that it was soundly designed, engineered, manufactured and assembled – in short, that it would work'. What particularly swayed him was the argument that if the plutonium bomb failed to detonate when deployed the enemy would be presented with a fine gift of plutonium.

In March 1944 detailed planning began at Los Alamos for a test of a plutonium implosion bomb. Oppenheimer, recognizing the sombre significance of what would be mankind's first nuclear explosion, searched for a suitable codename. He chose 'Trinity' for reasons he never fully explained, although in a post-war letter to Groves he would suggest his inspiration derived in part from a devotional poem by the seventeenth-century metaphysical poet John Donne beginning, 'Batter my heart, three person'd God'.

As the bombs came closer to reality, so the misgivings of some scientists at Los Alamos grew. Joseph Rotblat's only motivation for working on the bomb had been the fear that the Germans would get there first. However, from his first days at Los Alamos he had been uneasy. As he recalled, 'When I saw the magnitude of the project at Los Alamos, how many people worked there, how no effort was spared, no money, I could see straight away that, even with all this, it would take a long time before the bomb was made. The Germans could never match it. In 1944, Germany was being bombarded day and night, industry was being destroyed. It would have been impossible for them to do anything like this in the conditions.'

Rotblat's anxiety heightened when in March 1944, during dinner at James Chadwick's house, he heard General Groves declare that 'the real purpose in making the bomb was to subdue the Soviets'. Rotblat was not a communist. Like Marie Curie, he was a Polish patriot highly conscious of Russia's long suppression of his native land. However, Groves' belligerence towards the Soviet Union was, as Rotblat later recalled, 'a terrible shock – I had been a bit naive, an idealist, I thought we are all fighting together against a mortal enemy, and here we are on the other hand doing something against the person who counts as our ally'.

Rotblat shared his worries with Niels Bohr, one of his closest friends at Los Alamos. As Rotblat recalled, 'we hated the US news – ten seconds of news then fifteen seconds of "Ex-Lax" ads'. Instead they listened to the BBC World Service on Rotblat's shortwave radio. Afterwards, they would talk long into the night. Bohr 'inspired' Rotblat with 'thoughts of scientists' responsibilities'. He also told him his ideas for a system of international control to head off a post-war arms race. He believed passionately that the three great powers – the United States, Britain and the Soviet Union – had to agree how atomic energy should be applied and controlled before the bomb was completed and deployed. This meant telling the Russians about the Manhattan Project and subsequently making arrangements for the internationalization of knowledge for the benefit of all.

Bohr's international stature meant that he had contacts on both sides of the Atlantic willing to help him take his views to the highest level. Indeed, when he had first visited the United States after fleeing from Denmark, he had discussed his fears of a nuclear arms race with US Supreme Court Justice Felix Frankfurter, an adviser to President Roosevelt, whom Bohr had known since 1933. Frankfurter had passed Bohr's comments on to the President, who eventually sent back a noncommittal message that he was interested to know Churchill's reactions to Bohr's views.

Bohr responded readily to the implied invitation to go to England, flying there in April 1944. Churchill was not, however, keen to meet him. Lord Cherwell, another old acquaintance of Bohr's, was unable to secure a meeting for Bohr with the Prime Minister until May, and then it was not a success. There was no empathy. At one point Churchill turned to Cherwell to demand, 'What is he really talking about? Politics or physics?' To Bohr there was no difference. To Churchill politics was strictly his and Roosevelt's sphere. He was left appalled by what Bohr had to say, believing his advocacy of openness to be highly dangerous. In his opinion, Bohr 'ought to be confined or at any rate made to see that he is very near the edge of mortal crimes'. He also took a strong personal dislike to the celebrated Dane, writing resentfully to Cherwell that 'I did not like the man when you showed him to me, with his hair all over his head'. Bohr, in turn, was shocked by Churchill's attitude: 'It was terrible. He scolded us [Bohr and Cherwell] like two schoolboys.'

After Bohr's return to the United States, Frankfurter arranged in August 1944 for the Dane to meet Roosevelt face to face. Bohr found the President seemingly more receptive than the British Prime Minister. However, as events proved, he had not convinced Roosevelt either. With the war in Europe entering a decisive phase after the successful invasion of Normandy, and with the Russians advancing briskly from the east, Roosevelt had no desire to share America's hard-won and expensive secrets with Stalin. In September 1944, he and Churchill agreed an aide-mémoire stating that 'the suggestion that the world should be informed regarding Tube Alloys with a view to an international agreement regarding its control and use, is not accepted'. They also agreed that 'Enquiries should be made regarding the activities of Professor Bohr and steps taken to ensure that he is responsible for no leakage of information, particularly to the Russians.'

Before Bohr was allowed to return to Los Alamos, Groves and Chadwick were instructed to interview him in Washington. Churchill had been particularly incensed by reports that Bohr had been in touch with a Russian professor. Bohr explained that the professor was Ernest Rutherford's former protégé, Peter Kapitza. Learning of Bohr's flight from Copenhagen, Kapitza had invited him to work in Moscow. Bohr satisfied Chadwick and Groves that he had behaved impeccably, reporting the contact to British intelligence and politely declining the offer. 'Mr Baker' was free to go back to Los Alamos, where he joined in the work on implosion.

While Bohr continued to fight his ideological battle from within, Joseph Rotblat chose another path – becoming the only key scientist to quit the Manhattan Project during wartime for reasons of conscience. He had only been waiting until he could be 'absolutely sure' that the Germans had no bomb. Sufficient proof came at the end of 1944 when Chadwick, who had by then moved to Washington, which was more convenient for him as head of scientific liaison for the British team, visited Los Alamos. Chadwick had access to high-level intelligence reports and he confided to Rotblat that there was no evidence of a German atomic bomb. On the strength of this, Rotblat resigned forthwith from the Manhattan Project. As Rotblat remembered, Chadwick 'didn't like it', knowing it would cause ructions with the Americans, but he forwarded Rotblat's request to the army authorities.

Their response was to present Chadwick with a hefty dossier purportedly proving Rotblat to be a Russian spy. The core of the evidence was that he had told a young woman in Santa Fe that he intended to go to England, join the RAF and then parachute into Russia or Soviet-occupied Poland to tell the Russians what was happening on 'the hill'. Rotblat recalled, 'within this load of rubbish was a grain of truth'. He had indeed talked to someone in Santa Fe, but only with Chadwick's knowledge and approval and the conversation had not concerned Los Alamos. Army intelligence backed down, conceding that the dossier on Rotblat was worthless. However, Chadwick wisely advised Rotblat that he should cite, as his formal reason for wishing to leave Los Alamos, his anxiety about the wife he had been forced to leave in Poland at the start of the war. Groves agreed that Rotblat should leave immediately.

Shortly before Christmas 1944, Rotblat travelled by train to the east coast to stay for a few days with the Chadwicks in Washington. He then caught a train to New York from Union Station and Chadwick helped him carry on board a large wooden box filled with research notes and personal papers. Curiously, by the time the train arrived in New York Rotblat's box had vanished, and he sailed to England on Christmas Eve without it. Despite many enquiries the papers, no doubt spirited away by US Army intelligence, have never been found. Their

disappearance symbolized an almost paranoiac and in-
discriminate approach to security as the bomb project
approached its final months. Even Chadwick was not
immune. When, in January 1945, he wished to visit the
British programme in Canada he was told there were
strong objections on grounds of security which, he
decided, 'it would be most impolitic to ignore'.

The Anglo-Canadian project was, in fact, giving Groves
particular security problems. With the Allied invasion of
Europe and the liberation of France in 1944, the refugee
French scientists involved as part of the British team
wished either to visit their homeland or, indeed, return for
good. Groves, who had done everything he could to
exclude them from the Manhattan Project, was appalled
at the thought of them taking even their limited knowledge home to France. Groves suspected Frédéric
Joliot-Curie was a communist – as indeed he had become
– and would pass whatever he learned from his
compatriots straight on to Moscow.

James Chadwick struggled to cool both French and
American tempers. He persuaded Bertrand Goldschmidt
to be patient and to remain in Canada a while longer.
However, Hans von Halban, leader of the French team,
insisted on being allowed to return to Europe, and the
British agreed. When they learned that he was in London,
the Americans demanded assurances that he would not be
permitted to travel on to France. They were outraged
when a few days later he flew to Paris where he was soon
briefing Joliot-Curie. Groves blamed British duplicity,
erroneously believing the British were trying to ensure a
greater nuclear role for France in the post-war world in
exchange for rights to certain French nuclear patents. The

affair soured Anglo-American relations. As a British diplomat wrote with tongue-in-cheek reference to Gallic culinary habits, 'the salad is heaped in a bowl permanently smeared with the garlic of suspicion'. When von Halban returned from France to New York, an embarrassed Chadwick asked Goldschmidt to visit von Halban 'to learn as much as possible about what he had said to Joliot-Curie' and then to report to Chadwick in Washington. When Goldschmidt, who was perfectly happy to oblige, duly arrived in Washington to report, Chadwick apologetically and naively admitted that he already had 'a transcript of everything'. The meeting between Goldschmidt and von Halban had been bugged.

However, unknown to either Groves or Chadwick, the French team in Canada had already succeeded in passing information direct to General de Gaulle. Having learned that the general was to visit the Free French delegation in Ottawa in July 1944, Goldschmidt and colleagues Pierre Auger and Jules Guéron had told the head of the delegation that 'there is something so secret we can't tell you. We must tell de Gaulle direct. We need fifteen minutes.' The man had agreed to set up a meeting but insisted that only one of them could meet the general and that he could have only three minutes. The trio decided that Guéron, who had already met de Gaulle, should be the messenger, and he rehearsed the words again and again. The gist, as Goldschmidt recalled, was: 'There's going to be a new weapon. It will be ready in a year. It will be used first on Japan. If Germany is still in the war they'll be told the second one is for them. It will revolutionize warfare. You must start work in France as soon as possible with Joliot-Curie.'

The brief encounter took place by design in the privacy

of the gentlemen's lavatory in the French delegation's villa. Goldschmidt worried that, in the circumstances, the general might not have grasped the full import of Guéron's whispered words. However, later that day, when Goldschmidt was formally introduced to de Gaulle, the general said meaningfully to him, 'Thank you, professor. I understood very well.'

Meanwhile, Los Alamos's real and most dangerous spy, Klaus Fuchs, was pursuing his activities undetected. The Russians had, with wry humour, nicknamed the Manhattan Project 'Enormous' and devised codenames for some of the chief associated cities: Washington DC was 'Carthage', New York City was 'Tyre' and San Francisco was 'Babylon'. Their name for Fuchs was 'Rest'. The bearded Igor Kurchatov, in charge of the revived Soviet fission programme, wanted information that would fill in the gaps in Soviet knowledge and was driving Fuchs' agenda. Kurchatov wanted to know about gaseous diffusion as a means of separating out U-235. However, he was especially excited about information, gained through espionage, suggesting that there was an alternative bomb fuel to uranium – plutonium. Such prospects were, he wrote, 'unusually captivating'.

On first arriving in the United States with the British mission, Fuchs had spent nine months in New York working on the theory of gaseous diffusion. During evening meetings, usually in Manhattan though sometimes in Brooklyn or Queens, he gave his handler 'Raymond' – the alias of Soviet agent Harry Gold – information about it. At first, Fuchs just talked; then he began handing over notes which, like everything he gave to Gold, he had written himself. Fuchs' information convinced Kurchatov to concentrate on gaseous diffusion for separating uranium.

Fuchs' transfer to Los Alamos in August 1944, especially his assignment to work on implosion and plutonium, was a major breakthrough for the Soviets. For the first time, Fuchs saw the scale of the American programme and understood the importance of plutonium as an alternative fuel to U-235. Somehow, the earnest young scientist, always ready to run errands for others in the beaten-up Buick he had loaned Richard Feynman, became a familiar figure around the site, whose movements, despite tight security, went unremarked. In February 1945 he passed Harry Gold a detailed report on the design of the plutonium bomb. It described the problems with spontaneous fission that had led the Los Alamos teams to develop implosion. Fuchs also explained that far less plutonium than uranium was needed to make a bomb – only five to fifteen kilograms.

Over subsequent months Fuchs handed over further details, including a sketch of the bomb. His reports, in conjunction with details about high-explosive lenses supplied by machinist David Greenglass, who was busily casting the explosives to be used in the lenses, were welcomed in Moscow as 'extremely excellent and very valuable'. They convinced Kurchatov to recommend to Stalin that the Russians too should pursue an implosion plutonium bomb. Fuchs also told Gold that, if the testing of the plutonium bomb was successful, there were plans to drop it on Japan.

Japan's own nuclear programme was struggling. Yoshio Nishina had reported to the Imperial Navy the scientists' conclusions that, although an atomic bomb was feasible, it might take ten years to build and would require immense resources. After a series of meetings culminating in March 1943, about which the naval representative reported that the more the scientists debated the issue 'the more pessimistic became the atmosphere', the navy, unsurprisingly, lost interest. Instead, they asked the scientists to focus on shorter-term projects such as radar.

However, just as the navy had taken the lead when the army's commitment to nuclear research had waned, so, in May 1943, the army intervened to fill the nuclear vacuum. It decided to subsidize what it called the 'N-Project', in tribute to Nishina, and left it to him to decide how best to direct his research. He decided to focus the work of his group at Tokyo University on the separation of the fissile U-235 from U-238 by the use of thermal diffusion. However, he made slow progress, and it was only with great difficulty that his group manufactured small quantities of uranium hexafluoride gas. In July 1944 they made their first attempts at isotope separation using a thermal diffusion column wherein they hoped the effect of heat would separate the U-235 in the hexafluoride gas from the U-238. The lighter U-235 would rise to the top of the column and the heavier U-238 would fall to the bottom. Yet however hard they tried and whatever modifications they made, Nishina and his team could not get the apparatus to work.

Japan's military situation had also not prospered. Two years earlier, on 5 June 1942, she had suffered her first major reverse when an attack by a large Japanese carrier task force on Midway Island, a stepping stone to the planned conquest of Hawaii, was beaten back by American naval air power. Japan lost 332 aircraft and four aircraft carriers, three of which had taken part in the attack on Pearl Harbor. Although the United States had also suffered losses, her industrial power had allowed her to replace them much more easily than Japan. Japanese expansion had reached its high watermark, and slowly the Allies began to push back its armies in the Pacific, in New Guinea and on the frontiers of India. In April 1943, naval commander-in-chief Isoroku Yamamoto died when the aircraft in which he was travelling on an inspection visit was shot down in the South Pacific by an American fighter as a result of an intercepted and decoded message.

In July 1944, US marines took Saipan. Thirty thousand Japanese troops and fifteen thousand Japanese civilians died, many by their own hands to avoid capture. Tinian fell quickly thereafter. Saipan was the first piece of what had been Japanese territory before the war to be lost. The Japanese did not admit the loss until twelve days later when they praised the garrison which had 'fought victoriously to the last man'. Tokyo Radio then continued, on behalf of the government, 'The American occupation of Saipan brings Japan within the range of American bombers but we have made the necessary preparations.'

That summer, too, the cinemas began to show a news-reel entitled 'The Divine Wind Special Attack Forces Take Off', glorifying the first kamikaze pilots as, before their one-way missions, the young suicide bombers vowed fealty to their Emperor and, smiling, climbed into their cockpits. The 'divine wind' or 'kamikaze' was a reference to the winds said to have been sent by the deities to protect their favoured country, Japan, at critical times in her history. In particular, in 1281 the 'divine wind' had destroyed an invading Mongol fleet.

In Hiroshima, neighbourhood associations began to organize air-raid drills and to give guidance on rallying areas in case of attack. The associations distributed little brown and white pottery cups with bracing inscriptions such as 'Neighbourhoods unite and resist'. Those whom neighbourhood leaders observed or overheard engaging in defeatist talk or activity were reported to the feared secret police, the Kempei-Tei, based in Hiroshima castle. School-children of thirteen years and older had already been conscripted to work for up to eight hours a day in weapons factories. Now, in their spare time, they were ordered to dig trench shelters in hillsides surrounding the city as protection against the bombers. Everyone, young and old, male and female, had to drill with bamboo spears.

The Japanese Steel Products Group organized a conference in Hiroshima to encourage increased productivity to retaliate against the Anglo-Americans. 'The beasts are desperate, we must strike back', the workers were warned. However, lack of the very raw materials, such as oil and iron ore, which the Japanese had gone to war to obtain meant that there were no longer private cars or taxis, only trams or bicycles. There were few trucks and much less shipping in the harbour. Nearly 80 per cent of the Japanese merchant fleet had been lost, as had nearly 50 per cent of Japanese naval tonnage. Lack of steel meant that replacements for the merchantmen were being made of wood; lack of fuel meant that pilots received less training, and that the coastal patrol boats were almost invariably in Hiroshima harbour rather than at sea.

Lack of fuel, coupled with American attacks on the few fishing trawlers that did find enough to sail, also meant that fish – a staple of the Japanese diet – was becoming scarcer. Like other food, it was tightly rationed and 'canned' in patent earthernware jars to save metal. In a single week the ration for a family might be a cake of bean curd, one sardine or small mackerel, two Chinese cabbages, five carrots, four aubergines, half a pumpkin and a little rice. Most meals consisted mainly of watery soup with a few vegetable shreds. The citizens grew and collected what extra food they could. Schoolchildren, digging the shelters on the hills, brought back the excavated earth which they laid in layers on flat roofs or piled in old containers to grow vegetables such as sweet yams. Bramble shoots were stripped of their prickles and chewed. Reeds from the city's rivers were boiled and eaten. Anyone who had the opportunity to leave the city took a heavy stick with them so that they could hunt down the few remaining wild rabbits. Their meat was a tasty supplement to the diet, and their fur was collected by the neighbourhood associations for use in lining pilots' flying jackets. When the rabbits were gone, worms, grubs and insects were spitted and barbecued over such small fires as the shortage of coal and coke allowed. Necessities such as real soap and toothpaste were only available on the black market, and there were continual reminders of severe penalties for those trading there. Most people made do with an ersatz soap made of rice-bran and caustic soda, and with a gritty, salty paste for their teeth. Now that the Japanese forces had mainly departed for overseas and their return was largely prevented by Allied aircraft and submarines, there was little trade for the kimono-clad prostitutes collected in Hiroshima's red-light districts' 'houses of joy'. Their few customers were asked to pay in food, not cash.

CHAPTER TWENTY

'THIS THING IS GOING TO BE VERY BIG'

ON 1 SEPTEMBER 1944, ON A US AIR FORCE BASE IN NEW Mexico, Lieutenant-Colonel Paul Tibbets climbed into the pilot-seat of one of the new B-29 Superfortresses, the world's first pressurized bomber, which had been designed for long-range conventional bombing attacks against Japan.* His destination was Colorado Springs.

The twenty-nine-year-old had taken his first flight seventeen years earlier in Florida. At that time pilots were hired to make low advertising runs over public gatherings, and a barnstormer had taken the eager eleven-year-old aloft with him to rain promotional chocolate bars on the crowds on a local racecourse. This, the greatest thrill of young Tibbets' life, inspired him with the wish to fly, and in 1937 he joined the United States Air Corps. After the United States' entry into the war, he flew twenty-five

---

* The prefix 'B' in 'B-29' did not, as often thought, stand for Boeing, its manufacturer. Under an American military naming convention introduced in 1924, 'B' was for 'bomber' and '29' meant that the plane was the twenty-ninth model of bomber. Fighters had the designation 'P' for 'pursuit', although they are now assigned the letter 'F'.

combat operations in a B-17 Flying Fortress over occupied Europe and North Africa. In so doing, he acquired a high reputation as an excellent and unflappable pilot before returning to the United States to spend nearly a year as one of the test pilots for the Superfortress.

That day in Colorado began with a grilling by a security officer about his personal life. Tibbets began to suspect that the possible new assignment, which was the subject of his visit, was 'considerably more important than I had imagined'. The last question was whether he had ever been arrested. Tibbets confessed that when he was nineteen a nosy policeman had interrupted what Tibbets called 'a love-making episode' in a parked car on 'a secluded beach in Florida'. With this confession Tibbets satisfied the security officer, who identified himself as a Colonel Lansdale.

Lansdale took Tibbets to meet another group of men, including Deak Parsons, who was introduced as an explosives expert. One of the men asked Tibbets whether he had ever heard of atomic energy, and then went on to tell him that 'the United States has now split an atom. We are making a bomb based on that. The bomb will be so powerful that it will explode with a force of 20,000 tons of conventional high explosive.' Tibbets had been chosen to command the air force operation to drop the bomb. He was told that although it had the potential to end the war, the weapon was an unknown quantity that might not be ready for twelve months. If it exploded successfully, the bomber might suffer structural damage or even be thrown out of control, unless it put at least eight miles between itself and the explosion.

It would be up to Tibbets to lead a team to modify the B-29 aircraft to carry the bomb, which might weigh as much as ten thousand pounds, and to develop tactics for the operation. He would be given the 393 Bomb Squadron as an operating nucleus but should recruit other men as he thought necessary. Tibbets was given a choice of three remote locations as his base. He chose Wendover Field in Utah, 'only' 125 miles from Salt Lake City and 'surrounded by miles and miles of salt flats' in a 'virtually uninhabited' part of the state. It was, however, within easy flying distance of Los Alamos, and of suitable test bombing ranges.

The next stage in Tibbets' briefing included a meeting with General Groves, whom he described as 'of bulldozing efficiency'. Tibbets summed up his own position as he began his task as 'we would be organised for the purpose of dropping a bomb that hadn't been built on a target that hadn't been chosen'.

Undaunted, Tibbets recruited further personnel. A key choice was that of Major Thomas Ferebee, a farmer's son from North Carolina, as his bombardier. They had flown in the same crew in Europe, and together they co-opted others from their previous crews. In particular, from their European tour of duty they chose navigator Theodore 'Dutch' Van Kirk, tail-gunner Staff Sergeant George 'Bob' Caron and flight engineer Staff Sergeant Wyatt Duzenberg, and from Tibbets' B-29 testing days pilot Robert Lewis from New Jersey. Among those from 393 Squadron whom Tibbets chose to involve closely in the project was the squadron radio officer, Jacob Beser.

Beser had enlisted the day after Pearl Harbor and had secured some of the top marks in his training class. He would be responsible for the sophisticated radio and electronic equipment being developed to forestall any Japanese attempt to detonate the bomb prematurely, or to confuse the aircraft's navigational systems. On 19 September 1944 he was at Los Alamos with Tibbets, being given further details of the project. To Beser, it was 'the most fantastic day in my life' being introduced to scientists such as Bethe and Oppenheimer and learning the importance of the work on which he was engaged. Leaving the offices for the guest house in the late evening, he took a wrong turn and walked straight into Oppenheimer's quarters where he found his wife Kitty alone and stark naked, lying on a sofa sipping a cocktail. Considerably less embarrassed than Beser, she gave him directions and continued with her drink.

As he made further visits to Los Alamos, Tibbets became increasingly at ease with the scientists, recalling that 'although some did indeed have their heads in the clouds, others had the same interests as the normal every-day citizen'. He was particularly impressed by Oppenheimer, whom he thought 'unpretentious . . . highly nervous' and so able he could do at least three things at once. Tibbets remembered how Oppenheimer glanced into a room and saw a puzzled scientist staring at a black-board covered in scribbled formulae. After going a few steps further, Oppenheimer turned back, entered the room, erased a few numbers from the board, inserted some more and left the scientist exclaiming, 'My God, how did you do it? I've been looking for that mistake for three days.'

General Groves had some private reservations that Tibbets was 'too young' for the job and he was un-convinced of his abilities as a commander, as distinct from as a pilot. He therefore took particular care to impress Tibbets with the need for strict security and provided him with a security group of about thirty men, led by Major

William 'Bud' Umana. His job, Tibbets recalled, was 'literally to spy on our people to be sure there was no information leak'. In addition to monitoring mail and phone calls and eavesdropping, Umana went so far as to deploy his men as agents provocateurs. They accosted airmen as they left the Wendover Field for leave, asking seemingly innocent questions about the base's work. Those who blabbed always received a severe dressing-down from Tibbets, and often a posting to Alaska.

Among Tibbets' first tasks was to supervise modifications to the B-29s chosen to carry the atomic bomb.* The modifications were codenamed 'Silverplate' and were accorded the highest priority by Material Command. To allow the aircraft to fly higher than the limit of about thirty thousand feet that anti-aircraft flak could reach, and to provide extra speed to outrun enemy fighters, Tibbets ordered all the guns except the two twenty-millimetre cannon in the tail-turret, as well as the armour plating, to be stripped from the planes. By so doing, he saved seven thousand pounds in weight and achieved his goals of increased height and speed.

The B-29's two bomb bays had already been turned into one, and two twenty-seven-foot-long pneumatically operated bomb doors replaced the four twelve-foot ones. Bomb hooks of the type used to hold the largest conventional bombs, the British Lancaster's ten-thousand-pound 'Tallboys', were then installed to hold the atomic bombs in the enlarged bay.

---

* Serious consideration had been given to the use of the British Lancaster bomber, which would have needed less modification, but the proposal was rejected by Groves, who found it 'beyond comprehension to use a British plane to deliver an American A-bomb'.

Tibbets specified to his pilots how best to make the 155° diving turn he thought necessary after they had released the bomb to get them away from the much-feared shock wave. He had carefully worked out that the bomb would take only forty-three seconds to fall from 31,000 feet to its explosion height of around two thousand feet. The shock wave would take about forty seconds to travel eight miles – the minimum distance at which the plane was advised to meet the wave. However, the B-29 would need two minutes, not one minute and twenty-three seconds, to fly eight miles, so the sharp diving turn was necessary to pull the plane eight miles diagonally from the detonation before the wave struck. (The plane was fastest in a dive.)

Tibbets then moved his crews on to practice bombing, using casings simulating the likely shape of the atomic bomb. Tibbets tried to hide his frustration as the scientists regularly varied the weight and shape, seemingly oblivious to the impact on the performance of the plane and the mode of delivery.

In December 1944, Tibbets' command was re-designated the 509th Composite Group to reflect that, in addition to 393 Squadron's B-29s, it contained a scientific group and other elements to make it self-sufficient in communications and supply. In January 1945, many of Tibbets' planes flew down to Cuba for four weeks to continue their training from Batista Field, twelve miles from Havana. In the better weather of the Caribbean they could practise long flights over water and the transition from flying over water to over land, which would be important on approaching enemy coasts. In March, Captain Deak Parsons visited Wendover to give Tibbets his first briefing on the detailed mechanics of the bomb and its fusing. The premature explosion of a dummy unit carrying only a small amount of black powder did not improve the crew's feelings about the safety of their mission.

At the end of June 1945, re-equipped with new, stripped-down planes with improvements such as fuel-injection engines and reversible pitch propellers to replace the older models worn out by months of testing, the group took off for the Pacific. Their destination was the airfield on Tinian, one of the Northern Marianas Islands, captured by US marines less than a year earlier. The field was only 1,300 miles from Japan. Tibbets had had the rare privilege of choosing his own aircraft off the assembly line at Omaha, Nebraska, on 9 May, the day after the German surrender. A foreman had assured him that the workmen had been so careful to check and re-check everything that 'even the screws on the toilet seat were given an extra turn'.

On 27 June, it was Robert Lewis who piloted Tibbets' as yet unnamed plane to Tinian. Tibbets, Ferebee and Van Kirk, as commander, group bombardier and group navigator, were not with him, as they did not fly frequently themselves, though when they did it was nearly always with Lewis. A nineteen-year-old radio operator named Richard Nelson had joined Lewis's crew at the end of April. He was both thrilled and scared when Lewis buzzed the Wendover base in a hair-raising goodbye gesture.

While Tibbets and his crew had been training, Allied bomber squadrons had undertaken major raids against both Germany and Japan causing massive devastation and heavy casualties. At the Yalta Conference in early February 1945, the Russian High Command asked for assistance from the British and American Bomber Commands to prevent the transfer of large numbers of German reinforcements to the Eastern Front. Roosevelt and Churchill agreed. The targets would be the transport hubs of Dresden and Leipzig.

Neither the British nor the Americans had previously targeted Dresden, a historic city with many fine baroque buildings. The British attacked first, in two waves on the night of 13/14 February, aiming at the marshalling yards and creating a massive firestorm with temperatures at its centre of above 1,800° Fahrenheit. American writer Kurt Vonnegut, a prisoner of war in the city at the time, wrote that bodies dissolved in 'the semi-liquid way that dust actually returns to dust'. In the morning, 450 American bombers arrived to add to the destruction. One survivor wrote, 'Dead, dead, dead everywhere. Some completely black like charcoal. Others completely untouched, lying as if they were asleep.' Another saw nothing but parts of bodies being shovelled up into a big heap, then burned. The casualties numbered at least sixty thousand and perhaps significantly more, since the city was filled with refugees in addition to its recorded inhabitants. The inscription on one of the mass graves reads:

> How many died?
> Who knows the number?

Some in the Allied Command thought that, over and above its tactical benefits, the destruction had given the Russians a salutary demonstration of Allied air power.

After a precision raid on a Tokyo aircraft factory on 4 March, the US Air Force, under its recently appointed commander Curtis LeMay, decided on the carpet-bombing of whole Japanese cities. The first target was Tokyo. On 9 March, more than three hundred bombers took off from Tinian, and at around eleven p.m. Tokyo time pathfinder planes dropped coloured target markers illuminating the city. Then came the bombers, flying lower than usual because of the lack of anti-aircraft fire. They dropped two thousand incendiary bombs, some containing for the first time a new American invention, 'sticky fire' – napalm. The flaming napalm ran down the city's buildings, most of which were made of wood. Fire was blown from one building to the next, creating a firestorm which destroyed sixteen square miles of Tokyo and killed more than a hundred thousand people. Tokyo residents followed government directions to form bucket chains, but many suffocated from smoke inhalation or from the deprivation of oxygen as it was burned from the air, even before the flames consumed them. One survivor saw piles of blackened bodies piled outside the Meiji Theatre, so burned and disfigured that she could not even identify their sex or anything else about them. Radio Tokyo condemned the attacks as 'slaughter bombing'.

Over the following three months, Kobe, Osaka and Nagoya were destroyed by fire and the death toll rose to at least a quarter of a million, but still the resistance of the Japanese government and its obedient, patriotic people did not seem to crack, although Emperor Hirohito was said to have known the war was lost when he saw charred corpses heaped by the side of the river in Tokyo. In a later raid on Tokyo on 13 April an incendiary bomb set fire to the laboratory in which Yoshio Nishina and his team were still trying unsuccessfully to persuade their thermal diffusion column to separate U-235 from U-238. The fire destroyed the laboratory and in the ashes perished any faint, lingering hope that Japan might progress towards a nuclear weapon.

Hiroshima remained intact, but the authorities were nervous. Most of the city's houses were timber-framed with wooden walls and paper partitions under a tiled roof and thus were highly inflammable. At the beginning of 1945, the government ordered the mayor of Hiroshima to begin demolishing buildings to construct fire breaks against incendiary bombing raids to supplement the natural barriers afforded by the city's rivers. Both adults and schoolchildren went at the demolition work with a will. Wood from the fallen buildings could be used for fuel in the winter cold or, if a few nails could also be salvaged, turned into wooden sandals for the many who by then lacked them.

In April 1945, the Japanese authorities cut food rations again for all citizens, including the inhabitants of Hiroshima. The rice ration of three bowls a day had for a long time been routinely mixed with soya, but rice was henceforth provided on only twenty days out of any month. However, one resident recalled that the continued availability of tea was consoling, providing 'comfort and a reminder of the rituals of pre-war life'. That same month, the evacuation of some of Hiroshima's school-children began. They were sent to rural temples and assembly halls. Older pupils of no more than sixteen or eighteen years of age supervised those who were left in order to free teachers for war work. They were given the briefest of training and instructed not to use the same toilets as their pupils since, as 'higher beings', they should not be seen by them to perform basic bodily functions. They were also told that in the event of an attack 'their first priority, even before the safety of the pupils, should be to protect the portraits of the Emperor and Empress' which hung in every classroom.

Hiroshima had still not, however, been attacked. Inhabitants speculated about why the much-feared 'B-San' or 'Mr B.', as the B-29 Superfortress bombers were known, had not visited their city as they had so many others. In earlier years lack of viable agricultural land had forced many people from the area around Hiroshima to emigrate to Hawaii and California. As a consequence, some Hiroshima residents accepted as true the comforting speculation that President Roosevelt had agreed to spare Hiroshima from attack in response to petitions from Japanese-Americans, many of whom still had relatives in Hiroshima. Others thought that the city was being saved to serve as America's headquarters when the Americans conquered Japan. Such defeatism was becoming more common, so the secret police, the Kempei-Tei, based in Hiroshima castle, began a round-up of dissidents and defeatists in early May. Among the more than three hundred people swiftly detained in Hiroshima was diplomat Shigeru Yoshida, later to become Prime Minister of Japan.

A few weeks earlier Hiroshima had welcomed a new arrival, Field Marshal Shunroku Hata. The sixty-five-year-old veteran of the wars in China had been given the task of defending Japan against invasion and he chose to make his base in the city, establishing the Second General Army Headquarters there. He immediately gave orders for further military drills for all ages and both sexes.

Scarce fuel was set aside so that children could make Molotov cocktails to be stockpiled for use against the invaders. Even the infirm and wheelchair-bound were put to work making booby traps to protect the beaches. The many Koreans transported from their homeland to undertake forced labour in Hiroshima's factories were compelled to work longer hours, despite reduced rations. In the dockyards, the Japanese began assembling suicide craft to defend Hiroshima Bay. They packed small boats with explosives and a motor sufficiently powerful to speed them on a one-way mission to explode against the invaders' landing craft. Suicide divers, known as 'Fukuryus' or 'crouching dragons', were trained to swim out to sea to attach limpet mines to ships. Experiments were made with concrete shelters in which squads of Fukuryus could lie concealed offshore for ten hours before rising to attack the incoming landing craft. Every day the newspapers, all of which were state-controlled and strictly censored, urged their readers to give thanks for imperial benevolence and to be ready to die for Hirohito. Many of those in Hiroshima would have little choice in the latter.

'GERMANY HAD NO ATOMIC BOMB'

IN JANUARY 1945, WALTHER GERLACH, THE NEWLY APPOINTED German 'plenipotentiary' of fission research, ordered Heisenberg and all remaining scientists to flee Berlin immediately. Otto Hahn had left a few weeks earlier. During 1944, Allied bombs had destroyed a wing of the Kaiser Wilhelm Institute for Chemistry and reduced his office to rubble; among the possessions he most regretted losing were letters from Ernest Rutherford. He had decided to send his team and whatever he could salvage to the small town of Tailfingen in south-west Germany, not far from Heisenberg's evacuated team under Max von Laue at Hechingen. He arrived there himself in late 1944.

Hahn watched uneasily as Allied bomber squadrons passed overhead, but no bombs fell on Tailfingen. In early 1945 he found himself in greater danger from the local Gestapo for trying to shield Frau von Traubenberg, the Jewish physicist wife of one of his team, when, after her husband died suddenly of a stroke, she was arrested. Hahn argued that the woman was vital to what he called 'our secret work on uranium' but failed to secure her release. However, she was sent to the Theresienstadt holding camp where she was given a small room in which to work, and she survived the war. Hahn himself was by now under increasing surveillance from the Nazi authorities to whom he had been denounced as hostile to the Third Reich and who subjected him to harsh interrogations.

Kurt Diebner had also despatched some members of his small, army-sponsored reactor project in Berlin to greater safety, choosing Stadtilm, near Weimar. Yet, like Heisenberg, he had chosen to remain in Berlin to continue working on his reactor model. Neither of their programmes had yet yielded significant results, and certainly no chain reaction. Paradoxically, Diebner, with the least resources, had made the most progress. The rivals had been experimenting with different configurations of uranium to see which produced more neutrons. Diebner's trials, using cubes of natural uranium suspended on wires in heavy water, had generated more neutrons than Heisenberg's use of uranium plates. However, Heisenberg clung stubbornly to his preferred plate design until, admitting defeat at last in late 1944, he ordered the plates to be re-made into cubes. But he had left it too late. In January 1945, just as he and his team had finished attaching hundreds of cubes of uranium to aluminium wires and submerging them in heavy water, came Gerlach's order to leave Berlin. The next day, Diebner also departed, fleeing in a convoy of trucks containing both his own and Heisenberg's equipment.

Dodging bombs and the strafing of low-flying Allied fighter planes, Heisenberg reached Hechingen safely, where he lodged directly opposite a house that had once belonged to Einstein's uncle. He was not, however, reunited with his uranium and heavy water until the end of February after a squabble with Diebner, who had tried to appropriate them for his own experiments at Stadtilm. Heisenberg spent the last weeks of the war reassembling his reactor in a wine cellar cut deep into rocks in the village of Haigerloch near Hechingen. He was joined by von Weizsäcker, who, as the Allies advanced, fled from the French city of Strasbourg where since 1942 he had held the physics chair of a new university set up by the occupying Nazis. In their cave, Heisenberg and von Weizsäcker managed to generate more neutrons than ever before, but in these desperate, dying days of the war they still could not make their reactor go critical.

Unaware of the small scale and technical failures of the German fission programme, General Groves had long feared that 'the Germans would prepare an impenetrable radioactive defense against our landing troops'. In late November 1943 he had argued forcefully for a scientific intelligence-gathering unit to be set up, and, as usual, got his way. The mission itself, without Groves' prior knowledge, acquired the name 'Alsos' – ancient Greek for 'grove'. No-one was quite sure how.

Groves chose as the unit's military and administrative leader Lieutenant-Colonel Boris Pash, an FBI-trained security officer whose Russian émigré father was the senior Eastern Orthodox Bishop of North America. Partly as a consequence of his background, Pash loathed communists. In 1943 he had investigated Oppenheimer's alleged communist leanings. Oppenheimer had admitted to Pash that he and scientists at Berkeley had been approached by a Berkeley academic acting for the Soviet Union. He refused to reveal the man's name but insisted he had not divulged any information. Unconvinced by

Oppenheimer's protestations of innocence, an alarmed Pash had told Groves that Oppenheimer could well be a spy. Only after Pash's departure to Alsos was the matter cleared up. Oppenheimer revealed to Groves that the mysterious academic was Haakon Chevalier, a left-leaning professor of French literature at Berkeley, who had been trying to recruit Oppenheimer's brother Frank to spy for the Soviet Union. Oppenheimer assured Groves that he had advised Frank to have nothing to do with Chevalier.

Sam Goudsmit, the multilingual theoretical physicist who had left Holland in the 1920s to work in the United States, was appointed head of the Alsos scientific team. As a student in Amsterdam he had studied scientific techniques for solving crimes. He had no detailed knowledge of the Allied bomb project so, as he later wrote, 'I was expendable, and if I fell into the hands of the Germans they could not hope to get any major bomb secrets out of me'. He was also 'personally acquainted with many of the European scientists, knew their specialities, and spoke their languages'. Werner Heisenberg had been his guest in Ann Arbor in July 1939 during his final visit to the United States before the war.

Since then, Heisenberg had, unknown to Goudsmit, been asked to help the latter's elderly Jewish parents. When the Nazis announced the deportation to concentration camps of Holland's Jews, Dutch physicist Dirk Coster, who had worked so hard to save Lise Meitner, asked Heisenberg to use his influence to aid the Goudsmits. In February 1943, Heisenberg sent Coster a letter for him to show the authorities. It pointed out that the Goudsmits' son was an eminent scientist in America and that their fate would attract attention abroad. It also emphasized Goudsmit's supposed admiration for Germany. Heisenberg concluded that he personally 'would be very sorry, if for reasons unknown to me' the Goudsmits suffered 'any difficulties'. However, the letter arrived too late to have any influence. In March 1943 in America, Sam Goudsmit received a note from his parents bearing the address of a Nazi concentration camp. After that he heard no more.

On 25 August 1944, Boris Pash and an advance team from Alsos entered Paris with the first Allied troops. Dodging rooftop sniper fire, they found Frédéric Joliot-Curie safe at the Collège de France and very grateful to see them. He told Pash he had been afraid for his life. Goudsmit followed two days later. The Frenchman claimed that German scientists had learned nothing of military value during their years working at his college, where, in the last days of the occupation, he had turned his hand to making Molotov cocktails.

By 7 September, the Alsos team were in newly liberated Brussels. Pash was shocked to see alleged Nazi collaborators, 'haggard and wild-looking' men and women, penned up in the zoo in cages whose original occupants had been destroyed or eaten. Finding their way through the shabby, war-sullied streets to the offices of the Union Minière, the Alsos team were not surprised to find evidence that most of the company's uranium stockpiled in Belgium had been taken by the Germans in 1940. Searching through the paperwork for clues to where the uranium had gone, Goudsmit found references to a chemist employed by the German Auer Company but based in Paris. Following the trail back to Paris, Goudsmit discovered little about uranium but unearthed papers showing that, shortly before the liberation, the chemist had ordered a large stock of thorium to be sent to Germany. Since the Alsos team knew that thorium could be used to make fissionable material for an atom bomb, as Goudsmit later recalled, 'this really scared us'. Some weeks later, Goudsmit discovered the farcical rather than sinister reasons for spiriting away the thorium. As he later wrote, Auer still had a patent on thoriated toothpaste, as used by James Chadwick, and 'were already dreaming of their advertising for the future. "Use toothpaste with thorium! Have sparkling, brilliant teeth – radioactive brilliance!"'

While awaiting the moment when the Alsos team could enter the Reich itself, Sam Goudsmit, at last, had the opportunity to visit his childhood home in The Hague. He wrote:

Driving my jeep through the maze of familiar streets . . . I dreamed that I would find my aged parents at home waiting for me just as I had last seen them . . . The house was still standing but as I drew near to it I noticed that all the windows were gone. Parking my jeep round the corner so as to avoid attention, I climbed through one of the empty windows. The place was a shambles. Everything that could possibly be burned had been taken away by the Hollanders themselves to use as fuel that last cold winter of the occupation . . .

Climbing into the little room where I had spent so many hours of my life, I found a few scattered papers, among them my high-school report cards that my parents had saved so carefully through all these years . . . As I stood there in that wreck that had once been my home, I was

gripped by that shattering emotion all of us have felt who
have lost family and relatives and friends at the hands of
the murderous Nazis – a terrible feeling of guilt. Maybe I
could have saved them . . . Now I wept for the heavy feeling of guilt in me. I have learned since that mine was an
emotion shared by many who lost their nearest and
dearest to the Nazis. Alas! My parents were only two
among the four million victims taken in filthy, jampacked
cattle trains to the concentration camps from which it was
never intended they were to return.

The world has always admired the Germans so much
for their orderliness. They are so systematic; they have
such a sense of correctness. That is why they kept such
precise records of their evil deeds, which we later found in
their proper files in Germany. And that is why I know the
precise date my father and my blind mother were put to
death in the gas chamber. It was my father's seventieth
birthday.

Four months later, the Alsos mission reached Strasbourg
in Alsace and went at once to the university to look for
von Weizsäcker. In his hurry to flee he had left a stack of
revealing paperwork including letters showing that the
Kaiser Wilhelm Institute for Physics had moved to
Hechingen. They even gave the address and phone
number, making Goudsmit wish 'we could fly to
Switzerland and call them up from there!' He and his
colleagues 'studied the papers by candlelight for two days
and nights until our eyes began to hurt'. By the end of
January 1945 Goudsmit felt confident enough to inform
Washington that, while the Germans were clearly investigating the military applications of nuclear fission, their
work was still at an experimental stage and the immediate

focus appeared to be nuclear power rather than weapons. In other words, 'Germany had no atom bomb'.

General Groves and American military intelligence were reassured, but wanted conclusive proof. With the war fast drawing to a close, Groves had an additional worry – how to prevent key German fission scientists and facilities falling into the hands of the Russians. In February 1945 the Allies had agreed which zones of a defeated Germany each would occupy. Groves was especially concerned that the Auer Company uranium processing factory at Oranienburg outside Berlin would be in the Russian zone. He successfully arranged for it to be destroyed from the air, but keeping individuals out of Russian hands was more difficult. The Alsos team was detailed to locate and take into custody Germany's most important atomic scientists.

In late March 1945, following swiftly in the wake of Allied troops, the Alsos team crossed the Rhine, entered Heidelberg and seized Walther Bothe's institute, home of Germany's only functioning cyclotron. Bothe, the scientist whose mistaken conclusions had convinced his colleagues that they needed heavy water not graphite as a moderator, was the first enemy scientist to be apprehended whom Goudsmit knew personally. Bothe shook Goudsmit's hand warmly but refused to talk about his military work.

Moving on to Göttingen, Goudsmit met Morris 'Moe' Berg, former catcher for the Washington Senators and Boston Red Sox baseball teams, and now an American secret agent who had been involved in a scheme to capture, even to assassinate, Heisenberg. The idea of kidnapping Heisenberg had first been mooted in October 1942 when news of Heisenberg's appointment as director of the Kaiser Wilhelm Institute of Physics some months earlier first trickled through to refugee scientists in the United States. Their respect for Heisenberg's abilities had made them highly nervous of what he might achieve. After discussing the problem with Hans Bethe, Victor Weisskopf had sent a three-page letter to Robert Oppenheimer stating their concerns. 'By far the best thing', Weisskopf had suggested, 'would be to organize a kidnapping.' He had even volunteered to undertake the mission himself. Oppenheimer thanked him for his 'interesting' letter which he had submitted 'to the proper authorities', but told Weisskopf, 'I doubt whether you will hear further of the matter'.

However, the idea had not gone away. In December 1944 US special operations had sent Moe Berg to Zurich, where Heisenberg was to lecture, with a gun in his pocket and orders that if Heisenberg said anything suggesting that German scientists were close to making an atomic weapon Berg was to shoot him dead in the auditorium. Berg had hung on Heisenberg's words but heard nothing to convince him to fire his gun. Later, Berg engineered an introduction to Heisenberg and accompanied him on a long walk through ill-lit streets back to Heisenberg's hotel, during which he badgered him with questions. Berg was a good linguist and spoke German well. Heisenberg, who had no idea his life was hanging on what he said, assumed the pushy stranger was Swiss. Unsurprisingly, he responded guardedly and again revealed nothing implying that Germany possessed a war-winning weapon. Indeed, Heisenberg seemed regretfully resigned to Germany losing the war. Berg allowed him to leave Switzerland unharmed.

Berg's reassuring view of the relative poverty of the German nuclear capability did not distract Alsos from urgently tracking down German scientists and facilities.

High on their list was Diebner's German Army Ordnance fission project operating at Stadtilm, to which papers found in Heidelberg had alerted them. Immediately on hearing that Stadtilm was in Allied hands, Goudsmit flew there from Paris only to discover that the Gestapo had whisked away Kurt Diebner and several trucks of equipment two days earlier. However, he got his first look at part of the German fission programme: 'It was located in an old school-house. The cellar of that place looked almost like a natural cave and seemed quite bombproof. It was there that our men found the few remaining physicists huddled together with their families.'

A few days later at the town of Celle, north of Hanover, the Alsos team discovered an isotope separation laboratory hidden away by Paul Harteck in a parachute silk factory. Harteck himself had fled, but a brief examination of the centrifuge he had been developing satisfied Goudsmit that it 'would have taken a hundred years' to produce sufficient quantities of U-235 for a bomb. Even more importantly, soon afterwards an Anglo-American strike force located the bulk of the uranium taken by the Germans from Belgium and seized it from under the noses of advancing Russian troops near Magdeburg. On 23 April 1945, Groves told Army Chief of Staff General George Marshall categorically that the risk of a German nuclear weapon was over.

That same day, Colonel Pash, rushing to get there in advance of French troops, reached Haigerloch. He feared attack by the 'Werewolves', a fanatical Nazi resistance group, but as he drove in, white pillowcases, sheets and towels fluttered from every window. His men quickly found Heisenberg's and von Weizsäcker's secret German laboratory. Pash later wrote that it was an 'ingenious set-up' camouflaged and protected by 'a church atop a cliff'. He found 'a box-like concrete entrance to a cave' in the side of the cliff. Inside was 'a concrete pit about ten feet in diameter. Within the pit hung a heavy metal shield covering the top of a thick metal cylinder. The latter contained a pot-shaped vessel, also of heavy metal, about four feet below the floor level.' It was, Pash realized, 'the Nazi uranium "machine"'. In another chamber containing a series of cylinders he found a blackboard on which was chalked, 'Let rest be holy to mankind. Only crazy people are in a hurry.'

Leaving a team to photograph and dismantle the contents of the cave, Pash moved on to nearby Hechingen. The first thing he saw on entering Heisenberg's office in the woollen mill chosen to house the Kaiser Wilhelm Institute for Physics was the surreal sight of a photograph of Heisenberg and Goudsmit taken in 1939 at Ann Arbor. Heisenberg, in Goudsmit's view 'the brains of the German uranium project', was gone. However, Goudsmit, who arrived there soon afterwards from Haigerloch, was able to begin interrogating some twenty-five captured scientists and technicians. These included von Weizsäcker, who had quickly been apprehended; Karl Wirtz, one of Heisenberg's key assistants in Berlin; and Erich Bagge and Horst Korsching, who had been working on isotopic separation. Von Weizsäcker objected that the latter were too junior and insignificant to detain. According to Goudsmit, he remarked, 'What kind of selection is this?' Goudsmit had also taken Max von Laue into protective custody. He knew that von Laue had no direct connections with the nuclear work but he respected him deeply and believed he should play a part in the re-construction of science in post-war Germany.

Goudsmit learned from von Weizsäcker that his final contribution to the German war effort had been to lower a sealed metal drum containing key research notes into a stinking cesspit. The drum was retrieved. So were three drums of heavy water and the uranium used in the Haigerloch reactor, which Heisenberg had ordered to be buried. The Alsos team celebrated their success by consuming the contents of von Weizsäcker's wine cellar, which they had also discovered.

The Alsos team withdrew just before French Moroccan troops swept into Hechingen. Pash headed next for Tailfingen and Otto Hahn. He found Hahn and his entire staff assembled calmly in their laboratory. Hahn was extremely co-operative. As Pash recalled, he 'unhesitatingly' produced a pile of scientific reports and volunteered his view that a nuclear bomb could not be built. Goudsmit took him into custody.

In early May in Bavaria, the Alsos team caught up with Walther Gerlach and Kurt Diebner. They also finally captured their primary target, Werner Heisenberg. At three o'clock on the morning of Friday, 20 April, he had cycled out of Hechingen into the darkness, determined to reach his wife and children 120 miles away at Urfeld. He completed the journey, evading low-flying Allied aircraft and equally dangerous groups of hard-line Nazis roaming the countryside shooting or hanging anyone they took it into their heads to suspect of disloyalty to the Fatherland. When, a few days later, Pash arrived to arrest him, Heisenberg's initial reaction was relief. He recalled that he felt 'like an utterly exhausted swimmer setting foot on firm land'. However, knowing that he had to leave his wife and children behind and anxious how the locals would treat them if he were seen to co-operate, he begged

Pash to make it look as if he was being arrested against his will.

Heisenberg was taken to Heidelberg to be interrogated by Sam Goudsmit. Face to face with his old friend after six years, Goudsmit's overwhelming impression was that Heisenberg was 'actively anti-Nazi but strongly nationalistic'. Heisenberg was openly curious about Allied progress on fission and asked Goudsmit whether there was any programme in America like the Germans'. Goudsmit implied there was not, prompting Heisenberg's cheerful suggestion that 'If American colleagues wish to learn about the uranium problem I shall be glad to show them the results of our researches if they come to my laboratory.' The German's misplaced and bumptious confidence struck Goudsmit as pathetic.

Goudsmit's last major target, Paul Harteck, was arrested in Hamburg, bringing the total Goudsmit considered worth keeping in special detention to ten – Heisenberg, Diebner, Gerlach, von Weizsäcker, Hahn, von Laue, Wirtz, Bagge, Korsching and Harteck. On 3 July they were flown by Dakota to England for internment at Farm Hall, the elegant country house in Cambridgeshire where in 1942–3 Norwegian commandos had trained for their seemingly suicidal attack on the heavy water factory at Vemork.

At the beginning of August 1945, Sam Goudsmit was surprised to be recalled suddenly from Berlin, where relations with the Russians were already tense, to the safety of the United States military headquarters in Frankfurt. A few days later he would understand why.

'A PROFOUND PSYCHOLOGICAL IMPRESSION'

ON 12 APRIL 1945, PRESIDENT ROOSEVELT DIED, AGED SIXTY-three, from a cerebral haemorrhage which had struck reputedly while he was in bed with his mistress. Among his papers was found a draft of a speech in progress containing the following sentence: 'More than an end to war, we want an end to the beginning of all wars – yes, an end to this brutal, inhuman and thoroughly impractical method of settling the differences between governments.' His successor, the sixty-year-old Harry S. Truman, scarcely knew of the Manhattan Project and its war-winning potential. However, within the first twenty-four hours of his presidency he was briefed by Secretary of War Henry Stimson, who told him of 'the development of a new explosive of almost unbelievable destructive power', which was 'so powerful as to be potentially capable of wiping out entire cities and killing people on an unprecedented scale'. James Byrnes, an adviser to Roosevelt and soon to be designated by Truman as his Secretary of State, replacing Edward Stettinius, told him the bomb might well put the United States in a position to dictate her own terms at the end of the war.

In this and other briefings, the new President does not seem to have queried the underlying assumption that, when available, the bomb should be used and that the key questions were how and where. At the end of April, German capitulation was clearly only a short while away (VE Day was on 8 May). Japan was therefore the only remaining target for the bomb which would not be ready for some weeks yet.

At the suggestion of General Groves, a Target Committee was established chaired by his deputy, General Thomas Farrell. Its purpose was, in Groves' words, 'to make plans for the bombing operation itself, even though we still had no assurance that the bomb would be effective'. Among the members were five scientists, including John von Neumann and William Penney, air force officers and other project staff. Groves addressed the initial meeting of the group on 27 April and in his usual blunt style first reminded all present of the need for secrecy. He then went on to suggest, before departing, that four potential targets in Japan should be identified for attack in July, August or September. They should be within the B-29's range of 1,500 miles. An air force meteorologist gave the bad news that the summer months were the least likely to provide the clear weather required for the bombing. Of the three months specified, August was relatively the best.

The group went on to consider basic targeting criteria using guidelines given to Farrell by Groves, who recalled them in his autobiography: 'I had set as the governing factor that the targets chosen should be places the bombing of which would most adversely affect the will of the Japanese people to continue the war. Beyond that, they should be military in nature, consisting either of important headquarters of troop concentrations, or centres of production of military equipment and supplies. To enable us to assess accurately the effects of the bomb, the targets should not have been previously damaged by air raids.' By the end of the first meeting, the committee had chosen seventeen targets for initial study. William Penney was asked to consider 'the size of the bomb burst, the amount of damage expected, and the ultimate distance at which people will be killed'.

The committee next met at Los Alamos on 10 and 11 May, two days after the German surrender. Robert Oppenheimer, Deak Parsons, Hans Bethe and several other project staff also attended. Bethe gave his latest guesstimates of yields from the bombs – five thousand to fifteen thousand tons of TNT equivalent for the uranium bomb Little Boy and, with less confidence and subject to the forthcoming Trinity test, seven hundred to five thousand tons for the plutonium bomb Fat Man. A detailed discussion followed of the best height at which to detonate the bombs to produce maximum impact from the blast, since this was governed by their yield.

Moving on to the targets themselves, the committee refined their criteria: '(1) they be important targets in a large urban area of more than three miles diameter, (2) they be capable of being damaged effectively by a blast, and (3) they are likely to be unattacked by next August'. The committee also agreed 'that psychological factors in the target's selection were of great importance. Two aspects of this are, (1) obtaining the greatest psychological effect against Japan, and (2) making the initial use sufficiently spectacular for the importance of the weapon to be internationally recognised when publicity on it is released.'

Five cities were selected. First was Kyoto, 'an urban industrial area with a population of 1,000,000 . . . the former capital of Japan . . . from the psychological point of view there is the advantage that Kyoto is an intellectual centre for Japan and the people there are more apt to appreciate the significance of such a weapon as the "gadget" [the atomic bomb]'. Second on the list was Hiroshima and its 350,000 inhabitants, 'an important army depot and port of embarkation in the middle of an urban industrial area. It is a good radar target and it is such a size that a large part of the city could be extensively damaged.' Next was the port city of Yokohama near Tokyo, followed by the Kokura arsenal, and finally the port of Niigata on the north-west coast of Honshu. The committee considered but rejected a direct strike at the apex of the Japanese power structure: 'the possibility of bombing the emperor's palace was discussed. It was agreed that we should not recommend it but that any action for this bombing should come from authorities on military policy.'

The third meeting was held in the Pentagon on 28 May, with Paul Tibbets present to report on the operational readiness of his crews. The committee finally recommended three cities as targets and to be exempted from conventional air attack. They were Kyoto, Hiroshima and Niigata. Aiming instructions were much simplified, 'to endeavour to place first gadget in centre of selected city . . .' and significantly 'to neglect location of industrial areas as pin point target, since on these three targets such areas are small [and] spread on fringes of cities'. This recommendation was not entirely in line with the thinking of a more senior committee which met three days later.

At the end of April, Secretary of War Henry Stimson

had agreed with President Truman to chair a committee to advise on nuclear energy policy while the topic was entirely secret and before it could be put to Congress for decision. It thus became known as the Interim Committee. Stimson recommended that the President appoint 'a personal representative of himself'. He chose James Byrnes. Among the other members were Vannevar Bush and James Conant. The committee quickly spawned a scientific advisory panel comprising Arthur Compton, Enrico Fermi, Ernest Lawrence and Robert Oppenheimer. Although the topic did not fall within its initial terms of reference, the Interim Committee discussed the deployment of the bomb on 31 May and 1 June, with its scientific advisory panel in attendance.

As often, some of the most important discussions were informal ones. On this occasion they took place not in the men's washrooms but around the lunch tables. Lawrence, in discussion with Byrnes, proposed that the bomb 'ought to be demonstrated to the Japanese in some innocuous but striking manner, before it should be used in such a way as to kill many people'. In a discussion which lasted no more than 'perhaps ten minutes', according to Lawrence, much cold water was poured on Lawrence's idea. Oppenheimer could not envisage a demonstration that would be 'sufficiently spectacular to convince the Japs that further resistance was useless'. Byrnes was concerned that if the Japanese were warned of impending nuclear attacks they might move Allied prisoners of war into the target areas. Certainly their air defences, such as they were, would be activated. Others worried what would happen if a supposedly imposing demonstration were to fail. Stimson suggested that casualties from an attack would be unlikely to differ significantly from those resulting from the massive fire raids on cities such as Tokyo.

When the meeting formally reconvened after lunch, the minutes record the following decision: 'After much discussion concerning various types of targets and the effects to be produced, the Secretary [Stimson] expressed the conclusion, on which there was general agreement, that we could not give the Japanese any warning; that we could not concentrate on a civilian area; but that we should seek to make a profound psychological impression on as many of the inhabitants as possible . . . the most desirable target would be a vital war plant employing a large number of workers and closely surrounded by workers' houses.' The target of a vital war plant, even closely surrounded by workers' houses, differed from the city centre suggested by the Target Committee and was later adjusted towards the latter's views.

Stimson's diary reveals that he was concerned not only about the effects of the use of the atomic bomb but also about the morality of the growing civilian casualties caused by conventional bombings. Prior to the Interim Committee meeting he had acted decisively to remove one city, Kyoto, from the list of potential targets for the atomic bomb. According to Groves, 'the reason for his objection was that Kyoto was the ancient capital of Japan, a historical city, and one that was of great religious significance to the Japanese . . . the decision should be governed by the historical position that the United States would occupy after the war. He felt very strongly that anything that would tend in any way to damage this position would be unfortunate.' Although Groves pursued the matter several times, this was one of the few occasions he failed to win. After the war, he conceded, 'I was very glad that I had been over-ruled.' Groves noted that Kyoto benefited by being retained by him on the reserved target list for some time in the hope of changing Stimson's mind. The city was thus protected from conventional bombing and survived the war virtually intact. Over the next few weeks, Nagasaki replaced Kyoto on the target list.

Immediately after the Interim Committee meeting, on 1 June, Byrnes reported the strategy to the President, who, according to Byrnes, told him 'he could think of no alternative and found himself in accord'. This, and an entry in Truman's diary for 24 July in which he likened the bomb to the fiery destruction prophesied in the Bible to follow Noah's Flood but noted his agreement to its deployment, are the nearest to recorded decisions by President Truman to proceed with the use of the bomb.

The Interim Committee also discussed at its 31 May meeting the international dimension of nuclear energy, and in particular relations with the Soviet Union. Oppenheimer argued that the United States should offer 'to the world free interchange of information with particular emphasis on the development of peace-time uses. The basic goal of all endeavours in the field should be the enlargement of human welfare. If we were to offer to exchange information before the bomb was actually used, our moral position would be greatly strengthened.' However, Byrnes quashed the proposal, as the minutes record:

Mr Byrnes expressed a fear that if information were given to the Russians, even in general terms, Stalin would ask to be brought into the partnership ... particularly ... in view of our commitments and pledges of cooperation with the British. In this connection Dr. Bush pointed out that even the British do not have any of our blueprints on plants. Mr Byrnes expressed the view which was generally agreed to by all present that the most desirable program would be to push ahead as fast as possible in production and research to make certain that we stay ahead and at the same time make every effort to better our political relations with Russia.

Byrnes had been impressed by evidence from scientists and industrialists that the Soviet Union would take from four to ten years to catch up with the United States and that this would give the US a major diplomatic advantage throughout the period.

General Groves introduced a further topic at that meeting – the presence within the project of 'certain scientists of doubtful discretion and uncertain loyalty'. This was a reference above all to the activities of Leo Szilard. Groves was still deeply suspicious of the Hungarian. He had never allowed him to set foot on the Los Alamos site and had for a time had him removed from the project. Although Szilard had been reinstated to work as a consultant at the Met Lab in Chicago, Groves had ordered the FBI to keep him under strict surveillance. Their reports threw up only the obvious: 'The subject is of Jewish extraction, has a fondness for delicacies and frequently makes purchases in delicatessen stores, usually eats his breakfast in drugstores . . . usually is shaved in a barber shop, speaks occasionally in a foreign tongue, and associates mostly with people of Jewish extraction. He is inclined to be rather absent-minded and eccentric . . .'

By early 1945, Szilard had become preoccupied by the potential risks of a nuclear arms race leading to a first strike motivated by fear or perceived danger rather than an actual threat. He was convinced that only an international system of controls could forestall such a danger. He decided that because he was so far out of the Manhattan Project's mainstream, and because his nemesis Groves had so compartmentalized the project on security grounds, he should approach Roosevelt direct. Because 'I didn't suppose that he would know who I was', he approached Albert Einstein for a letter of introduction, which the latter gladly provided. Believing that Eleanor Roosevelt might be a useful conduit to the President, Szilard sent the introductory letter to her in late March. She agreed to meet him in early May, but the President's death intervened.

Szilard then managed to secure a meeting at the White House on 25 May, where he was told by Truman's appointments secretary that the President had suggested he should meet Byrnes at his home in South Carolina to discuss the issue. The meeting to which Harold Urey and another scientist accompanied him two days later was not a success. (Urey had become another of Groves' *bêtes noires*. He thought him completely ineffectual: '. . . he was not a doer himself and could never make decisions. At heart he was a coward.') Szilard lectured Byrnes about the dangers of even testing a bomb if an alarmed Stalin was not to start an arms race. Szilard was 'flabbergasted' by Byrnes' contrary 'assumption that rattling the bomb might make Russia more manageable' including in relation to how the Soviet Union controlled Eastern Europe and Szilard's Hungarian homeland.

Following this mutually unsatisfactory conversation, Byrnes was only too ready to take a firm position on the dissident scientists. The Interim Committee agreed that 'nothing could be done about dismissing these men until after the bomb has actually been used or, at best, until after the test has been made', but then 'steps should be taken to sever these scientists from the program'.

Nevertheless, when he returned to the Met Lab after the committee, Arthur Compton told his colleagues that the committee would be prepared to listen to their views about the future of the atomic project, including by implication the concerns of those who had doubts about the deployment of the bomb. The scientists set up committees on topics ranging from research programmes to social and political implications. James Franck, whom Groves regarded as 'a babe in the woods' when it came to national and international affairs, chaired the latter, and Szilard was an enthusiastic member. Among the recommendations of their thirteen-page report was that the bomb should be demonstrated before it was used against Japanese civilians. Compton sent the report to Stimson 'at the request' of the laboratory 'for the attention of your Interim Advisory Committee'. He noted that the committee's scientific panel had not yet considered the report. When they did so on 16 June, the four men do not seem to have been in full agreement. Lawrence, with perhaps some support from Fermi, persisted in favour of the demonstration, but Compton and particularly Oppenheimer were strongly opposed. Oppenheimer's report to Stimson – 'we can propose no technical demonstration likely to bring an end to the war; we see no acceptable alternative to direct military use' – let Stimson, Byrnes and the other politicians off the hook.

Meanwhile, Leo Szilard wrote secretly to Edward Teller and other colleagues at Los Alamos urging support for a petition to Truman advocating a demonstration and the avoidance of an arms race. One paragraph of his letter read:

Many of us are inclined to say that individual Germans share the guilt for the acts which Germany committed during this war because they did not raise their voices in protest against these acts. Their defence that their protests would have been of no avail hardly seems acceptable even though these Germans could not have protested without running risks to life and liberty. We are in a position to raise our voices without incurring any such risks even though we might incur the displeasure of some of those who are at present in charge of controlling the work on 'atomic power'.

Teller felt that before replying 'he had to talk to Oppenheimer'. To his surprise, an impatient Oppenheimer spoke harshly and vehemently about both Szilard and Franck, questioning, 'What do they know about Japanese psychology? How can they judge the way to end the war?' He suggested that the decision should be left to 'the leaders in Washington and not individuals who happen to work on the bomb project'. Teller was relieved 'at not having to participate in the difficult judgements to be made'. He wrote back to Szilard a six-paragraph letter, concluding, 'I feel I should do the wrong thing if I tried to say how to tie the little toe of the ghost to the bottle from which we just helped it to escape.'*

Szilard continued to work on how the dissenting scientists' views could be got to the President. However, Truman, Stimson and Byrnes, confirmed as Secretary of State, had chosen their path. In addition to the ongoing

---

* In his autobiography, Teller reflected that the scientists should have done more to understand what would have been involved technically in a demonstration to the Japanese, and then to have informed the politicians.

conduct of the war against Japan, they were focusing on the forthcoming conference with Churchill and Stalin in July at Potsdam in defeated Germany. Churchill too had given Britain's formal consent to the use of the bomb against Japan, as required by the Quebec Agreement, by simply initialling a note requesting him to do so. British agreement was noted in the minutes of a UK/US Combined Policy Committee meeting in Washington on Independence Day, 4 July. Among the topics for discussion at Potsdam would be the future of Eastern Europe, the programme of the United Nations – whose charter had been signed on 26 June – and potential Russian participation in the war on Japan.

Thinking on the latter had gone through a number of stages. For a considerable time American policy had been that the entry of the Soviet Union into the war against Japan was highly desirable. By invading Manchuria, Soviet troops would tie down Japanese divisions and prevent them being returned to Japan to defend it against American forces.

The basic terms for Soviet entry into the war against Japan had been settled at Yalta in February 1945. These included the preservation of Outer Mongolia as independent from China and the restoration of the concessions made by Russia at the end of the Russo-Japanese War of 1904–5, including the return of the southern half of Sakhalin Island. In addition, Russia was to annex the Kurile Islands. Some of the provisions, such as the status of Mongolia, needed the consent of China, and this was left for discussion between the Russian and Chinese governments. Stalin had promised to join the war against Japan no more than two or three months after Germany's defeat. However, by the spring American planners realized that their naval and air forces could prevent Japanese troop ships sailing from Manchuria to the defence of the home island. They began to hope that Soviet entry into the war alone might be sufficient to force the Japanese leadership to surrender without an invasion of the Japanese home islands.

The US administration was desperately concerned about the cost in Allied lives of such an invasion. Japanese resistance was unrelenting. Kamikaze planes zeroed in on the US carrier fleet protecting the invasion forces off Okinawa. On 11 May, an attack on the *Bunker Hill* killed 396 men – three times more than the number of revolutionary forces who had died in 1775 in the engagement after which the carrier was named. The Allied servicemen could not understand the mentality of those prepared to undertake suicide attacks; thus they proved highly disturbing while at the same time reinforcing stereotypes of the Japanese as a race apart. So strong was this sense of distance that when one engineering officer on a US warship hit by a kamikaze later found the decaying leg of the pilot, he gave it to his comrades 'to make some souvenirs out of it'. He recalled 'the guys actually sliced up the bones into cross-sections. They made necklaces out of that pilot.'

Okinawa was not conquered until 21 June. Over twelve thousand American servicemen lost their lives on the island or in related operations, together with around eighty thousand local people and upwards of a hundred and twenty thousand Japanese. Three days earlier, the chairman of the US Joint Chiefs of Staff, William Leahy, had told the President that the front-line marine divisions had suffered 35 per cent casualties on Okinawa and that if, as seemed likely, a similar percentage were lost in the attack on the first of the Japanese main islands, Kyushu, planned for the autumn, casualties would be over one quarter of a million. Truman replied that he 'hoped that there was a possibility of preventing an Okinawa from one end of Japan to the other'.

If the Trinity test of the plutonium bomb succeeded, the new weapon might offer such a possibility, and also avoid the complication of Russian involvement in Japan and China. Stimson noted in his diary on 14 May that America's wealth and possession of the bomb were 'a royal straight flush and we mustn't be a fool about the way we play it'. The next day his diary described the bomb as a 'master card'.

The best way of preserving American lives while defeating Japan quickly and limiting Soviet influence preoccupied Truman, Stimson and Byrnes as they crossed the Atlantic in the cruiser USS *Augusta* to the Potsdam Conference. So convinced were they of the value of a successful atomic bomb test to the strength of their negotiating position with Russia that Truman had delayed the conference from the originally proposed date of mid-June until mid-July – the scientists' estimate of the earliest they would be ready to conduct the Trinity test. Churchill had been concerned that this might allow the Russian hold on Eastern Europe to consolidate, so Truman had sent one of his advisers, Joseph Davies, to London to tell him Truman 'didn't want to go to Potsdam to meet Stalin until he knew the outcome of the test'.

The US delegation was also aware that Emperor Hirohito had asked his ministers to put out diplomatic feelers about means to end the war. Both British and American intelligence had decoded subsequent signals by the Japanese foreign minister to Japan's ambassador in

Moscow to ask the Soviets to act as an intermediary. According to one of the intercepts it was 'His Majesty's heart's desire to see the swift termination of the war'. On the basis of their detailed intercepts, the two allies dismissed the initiative as not offering the unconditional surrender they demanded, which, indeed, it did not. There was disagreement among the Japanese leaders as to the wisdom of opening any negotiations. The militarists believed, in line with samurai tradition, that surrender was dishonourable in any form. The more liberal faction, which included the Emperor's Keeper of the Privy Seal, Koichi Kido, saw the need to end the war but feared a military coup if they proceeded too quickly or overtly. Neither faction could contemplate the abdication of the Emperor or the loss of the monarchy, which unconditional surrender might imply.

In Los Alamos, Oppenheimer and his team, planning the Trinity test of the plutonium bomb, were, in his words, 'under incredible pressure to get it done before the Potsdam meeting'. They succeeded. At 5.30 a.m. on Monday, 16 July, the sound of an explosion awoke a New Mexico storekeeper. He rushed into the street, where he found another man 'just standing there' looking 'dumbfounded'. When the storekeeper asked him what had happened, the man replied, 'Look over yonder, the sun blowed up.'

By early July, construction workers at the Trinity test site, which lay on a ninety-mile tract of high desert, the Jornada del Muerto, had completed their task, peppering the arid landscape with bunkers and shelters. The Jornada del Muerto, which translates roughly as 'Dead Man's Way', was located in central New Mexico, south of Los

Alamos. The area designated for the test lay largely within the Alamogordo Air Base, where five hundred miles of cable connected highly sensitive instruments, some of which would have just a split second to relay their data before the blast vaporized them. More than fifty high-speed cameras, some tied to adapted machine-gun mountings, would capture the image of the expected mushroom cloud.

On the night of 12 July scientists began assembling the two hemispheres of plutonium that made up the core of the test bomb. Final assembly took place on 14 July: this entailed fitting a small beryllium initiator between the two hemispheres, then placing the resulting solid sphere into the encircling tamper or inner shell – a hollowed cylinder of natural uranium. The next and highly tricky task was to position the core of the bomb within the shell of high explosives that would trigger the implosion process. There were anxious moments when the core would not fit into place. Oppenheimer lost his temper, then paced the ground in seeming silent despair. However, the problem was minor – the plutonium had expanded a little from heat. When, after a few minutes, it cooled, the core snapped easily into position. Later that day engineers slowly hoisted the 'gadget', as they called it, onto its platform atop a hundred-foot-high steel tower after piling army mattresses beneath the tower in case the bomb slipped and fell. On 15 July, scientists fitted the detonators. Oppenheimer climbed the tower to gaze for himself on the world's first nuclear bomb.

One crucial element lay outside the scientists' and the politicians' control – the weather. Chief meteorologist for the Trinity site Jack Hubbard had identified the optimum conditions for the test, which included visibility greater than forty-five miles, humidity below 85 per cent and clear skies. But the test team did not have the luxury of waiting for optimum conditions. General Groves was insisting that the test take place on 16 July, unless the weather made it quite impossible. Yet more weathermen arrived to help with the complex and vital task of forecasting, including Colonel Ben Holzman, who had helped select the date of the D-Day landings.

Zero hour for the test was four in the morning on 16 July, but at two a.m. violent thunderstorms erupted unexpectedly. Winds of 30mph scoured the desert, while heavy rain battered the shelters and the Trinity base camp. The scientists even feared that the bomb might 'be set off accidentally', as Isidor Rabi recalled. Hearing 'an unbelievable noise', Emilio Segrè went to investigate but found it was only the sound of 'hundreds of frogs in the act of making love in a big hole that had filled with water'. The presence of a highly agitated Groves added to the pressure on the three men, all by then exhausted, who had the power to cancel the test: Robert Oppenheimer, Kenneth Bainbridge, a Harvard experimental physicist appointed test director in March 1944, and Jack Hubbard. They debated what to do. Hubbard advised that the test could not go ahead at four a.m. Yet, somewhat to Groves' surprise, after careful scrutiny of the data he predicted that by dawn conditions would be acceptable. The men set a new time of 5.30 a.m.

At 4.15, Hubbard's final weather forecast confirmed that the test could indeed proceed. Shortly after five, Bainbridge ordered the bomb's timing device to be activated. Meanwhile, Groves, fearful of the possible effects on the local community, alerted the governor of New Mexico that he might have to declare a state of emergency and evacuate people from the neighbourhood. He gave no details as to why.

At 5.29, a rocket streaked across the sky to signal that there was only a minute to go before detonation. At their observation point on Compañia Hill, twenty miles north-west of ground zero, Hans Bethe and Edward Teller applied suntan lotion to protect themselves from the coming flash. Bethe had calculated that the Trinity bomb could not ignite the earth's atmosphere. Fear that the explosion might trigger an unstoppable, catastrophic sequence that turned the earth into a burning star had long troubled the Los Alamos team. Their concern was that this might result not from fission but from fusion. If a nuclear explosion produced by atomic fission of heavy elements, such as uranium or plutonium, generated sufficient heat and pressure, this might in turn fuse together the light atoms of hydrogen, helium and nitrogen in the atmosphere and cause them to ignite – the process which causes the sun and stars to burn. In the early days of the project, Oppenheimer had been so worried by some figures produced by Teller that he had consulted Arthur Compton, who told him that if the risk exceeded three in a million the bomb project must halt. Teller had sub-sequently reworked his figures showing the risk to be much less than he originally supposed and justifying the project's continuation. He had also confirmed Bethe's more recent calculations. Nevertheless, both Teller and Bethe spent a long, fraught sixty seconds.

It happened as planned. James Chadwick, also on Compañia Hill, remembered, 'the first grey light of dawn was appearing as we lay or sat on the ground. Except for the faint twitterings of a few early birds there was com-plete silence. Then a great blinding light lit up the sky and earth as if God himself appeared among us'; then came the 'explosion, sudden and sharp as if the skies had cracked . . .' Beside him, Otto Frisch, afraid of being dazzled, had turned his back. He was watching the landscape take substance in the pale dawn when 'suddenly and without any sound, the hills were bathed in brilliant light, as if somebody had turned the sun on with a switch'. He turned to the source of the light but it was still too bright to focus on. Stealing a few brief, tantalizing glances, he gained 'the impression of a small very brilliant core much smaller in appearance than the sun, surrounded by decreasing and reddening brightness with no definite boundary . . .' After a few seconds he was able to look at it properly and saw 'a pretty perfect red ball, about as big as the sun, and connected to the ground by a short, grey stem. The ball rose slowly, lengthening its stem . . . A structure of darker and lighter irregularities became visible, making the ball look somewhat like a raspberry. Then its motion slowed down and it flattened out, but still remained connected to the ground by its stem, looking more than ever like the trunk of an elephant. Then a hump grew out of its top surface and a second mushroom grew out of the top of the first one . . .'* The whole was surrounded by 'a purplish blue glow'. Frisch waited, fingers in ears, for the expected blast. The noise, when it reached the man who six years earlier, with his aunt Lise Meitner, had proved the reality of nuclear fission, was in his view 'quite respectable'. A long rumbling followed,

---

* The emblematic mushroom effect resulted from the thermal updraught created by the explosion and the heat it produced, which sent debris up into the sky where it flattened out as it reached the stratosphere and the energy dissipated.

'not quite like thunder but more regular, like huge noisy wagons running around in the hills'.

To Rudolf Peierls, the explosion had a symbolic as well as a scientific significance: 'To us that trial explosion had been the climax . . . The brilliant and blinding flash . . . told us . . . we had done our job. In that instant . . . still awed by the indescribable spectacle . . . we thought more about the work successfully completed than about the consequences.' Robert Oppenheimer's first reaction was also a surge of relief, though, as he later told reporters, he was 'a little scared of what we had made'. A line from his beloved *Bhagavadgītā* raced through his brain: 'I am become Death, the shatterer of worlds'. To others, though, Oppenheimer's mood seemed close to euphoria. Rabi recalled that Oppenheimer's 'walk was like [the film] *High Noon*, I think it's the best I could describe it – this kind of strut'. General Groves' reaction was unequivocal satisfaction. During the final seconds he had thought 'only of what I would do if, when the countdown got to zero, nothing happened'.

The important question now was how big the blast had been. Enrico Fermi, who to Groves' irritation had the night before been taking bets on the chances of the explosion igniting the atmosphere, conducted a simple but ingenious experiment. Just after the flash, Groves saw him 'dribbling' some torn fragments of paper 'from his hand toward the ground. There was no ground wind, so that when the shock wave hit, it knocked some of the scraps several feet away.' Fermi measured precisely how far the blast wave had carried them, then, using his slide rule, he calculated the force of the explosion. It was equivalent, he reckoned, to some ten thousand tons of TNT. His improvised 'paperchase', given how much was unknown, was surprisingly accurate: the blast had, in fact, been equivalent to twenty thousand tons of TNT.

Only a few hours after the Trinity test of the plutonium bomb, the heavy cruiser the USS *Indianapolis* left San Francisco. On board were a gun assembly in a fifteen-foot crate and a lead bucket containing a uranium core – the key components for Little Boy, the uranium-fuelled bomb which the scientists had decided need not be tested before its use in the field.

The message announcing the success of Trinity reached Stimson at Potsdam in the following terms: 'Operated on this morning. Diagnosis not yet complete but results seem satisfactory and already exceed expectations . . .' Stimson informed Truman and Byrnes. That night, Truman wrote in his diary, 'I hope for some sort of peace, but I fear that machines are ahead of mortals by some centuries, and when mortals catch up perhaps there'll be no reason for any of it.'

The next day, Stimson passed Churchill a cryptic note, 'babies are satisfactorily born', which Churchill failed to understand. Stimson then told him explicitly. His diary records Churchill's subsequent reaction: ' "Now I know what happened to Truman yesterday. I couldn't understand it. When he got to the meeting after having read this report he was a changed man. He told the Russians just where they got on and off and generally bossed the whole meeting." Churchill said he now understood how this pepping up had taken place and that he felt the same way.' According to the diary of his own top general, Lord Alanbrooke, Churchill was 'completely carried away. It was now no longer necessary for the Russians to come into the Japanese war; the new explosive alone was sufficient to settle the matter. Furthermore we now had something in our hands which would redress the balance with the Russians. The secret of this explosive and the power to use it would completely alter the diplomatic equilibrium which was adrift since the defeat of Germany. Now we had a new value [said Churchill], pushing his chin out and scowling, now we could say if you insist on doing this or that, well we can just blot out Moscow, then Stalingrad, then Kiev, Kharkov, Sevastopol, etc., etc. Then where are the Russians!' A note from Churchill to Foreign Secretary Anthony Eden confirmed that his elation and disdain for Russia were shared by his American counter-part: 'It is quite clear that the United States do not at the present time desire Russian participation in the war against Japan.'

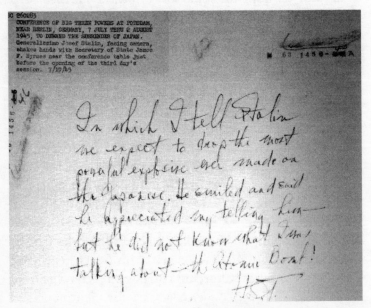

Truman's note recording his discussion with Stalin at Potsdam

With Churchill's agreement, Truman told Stalin about the bomb. At the end of one day's meetings he simply wandered over to the Soviet leader and 'casually mentioned to Stalin that we had a new weapon of unusual destructive force. The Russian premier showed no special interest. All he said was that he was glad to hear it and hoped we would make "good use of it against the Japanese".' Their mutual nonchalance concealed not only Truman's understanding of the bomb's potential, but also Stalin's prior knowledge of the bomb from Klaus Fuchs' detailed reports. Stalin was already pressing his generals to hasten their plans for Soviet entry into the war. Nikita Khrushchev later wrote, 'Stalin had his doubts about whether the Americans would keep their word . . . What if Japan capitulated before we entered the war? The Americans might say, "We don't owe you anything." '

The Soviets had, immediately prior to the conference, received renewed peace feelers from the Japanese to which they had given a noncommittal reply. Fortified by their knowledge of the Trinity test, aware of the Japanese peace approaches and without consulting Stalin, on 26 July Truman, Churchill and Chinese leader Chiang Kai-shek issued the Potsdam Declaration, offering Japan what they called 'an opportunity to end this war' on the basis of 'unconditional surrender'. The declaration ended, 'The alternative for Japan is prompt and utter destruction.'

Two days earlier, General Marshall and Henry Stimson had approved a directive drafted by General Groves authorizing the atomic bombing of Japan. Although they must have consulted President Truman, his formal consent does not appear in the surviving documents. The first part of the directive to General Carl Spaatz, the newly appointed commander of the Strategic Air Force, reads:

1. The 509 Composite Group, 20th Air Force will deliver its first special bomb as soon as weather will permit visual bombing after about 3 August 1945 on one of the targets: Hiroshima, Kokura, Niigata and Nagasaki . . .
2. Additional bombs will be delivered on the above targets as soon as made ready by the project staff . . .

On 26 July, at the end of her ten-day voyage from San Francisco, the USS *Indianapolis* reached Tinian. Two days later, the Japanese Prime Minister rejected the Potsdam offer. His government would 'ignore it [and] press forward resolutely for the successful conclusion of the war'. The Tokyo newspaper *Mainnichi* dismissed the Potsdam Declaration as 'a laughable matter'.

'AN ELONGATED TRASH CAN WITH FINS'

TINIAN IS A SMALL ISLAND ABOUT TWELVE MILES LONG AND only five miles wide at its broadest. Fringed by pale sand and coral reefs, Tinian was the centre of the American strategic bombing offensive against Japan, which lay some 1,300 miles directly to the north. The four parallel runways of North Field, one of the two airfields on the island, were, at 8,500 feet, said to be the longest in the world at the time and could launch four B-29s simultaneously at forty-five-second intervals, or more than three hundred planes per hour. Many of the streets on the base were named after those in Manhattan, such as Park Avenue and Riverside Drive. The area reserved for Paul Tibbets' 509 Composite Group was known as 'the Columbia University District'.

Their compound was surrounded by a high fence topped with barbed wire with an inner compound of windowless huts surrounded by more barbed wire and guards. Here, a thirty-seven-man-strong technical team, including Luis Alvarez and William Penney, under the command of Deak Parsons worked on the assembly of the two bombs. Conditions in general on Tinian were fairly primitive, and the 509 Officers' Club was made of plywood from recycled packing crates, canvas and mosquito netting. Many of the officers and men lived in tents. One of their occupants wrote, 'the bathroom is a pipe in the open connected to a waterpipe; the tank itself is made out of salvaged bomb containers; there is no hot water. Beside the coral path leading to the bathroom is a derelict Japanese foxhole. Beyond it by the sea is the graveyard of 1,100 marines who died in the battle for the island.' Despite such discomforts, the 509th had the highest priority rating on the island and the best of everything going. Their priority status provoked jealousy among the base's other personnel, who were entirely ignorant of the mission of their pampered neighbours. Some disgruntled spirits threw stones over the wire onto the corrugated iron roofs of the group's accommodation huts to keep the occupants awake at night.

Each night also, US marines went out to hunt the small numbers of Japanese soldiers still holding out in the luxuriant jungle, or in caves in the hundred-foot-high cliffs overlooking the air base, and its surrounding sugar-cane fields. One of them, Chief Warrant Officer Kizo Imai, later claimed to have noticed the special compound set up for the 509th and pondered how best to get out a message for it to be attacked. Paul Tibbets recalled that the Japanese propagandist Tokyo Rose mentioned the distinctive arrow marking on the tails of 509's bombers in one of her radio broadcasts. Perhaps an unknown Japanese soldier did get a message through.

The crews of 509 spent July 1945 in more training, including flying daylight missions of two or three bombers over Japan during which they dropped orange, pumpkin-shaped practice bombs containing only sufficient powder to show where detonation occurred. These practice sessions confirmed to Tibbets the crews' competence and the inability of Japanese anti-aircraft fire to reach thirty thousand feet. Fighters were only rarely seen, and Tibbets hoped that the insignificant effects of the practice bombs might lull the Japanese into a false sense of security about lone, high-flying bombers. Tibbets himself went on none of these flights; he was forbidden to fly over enemy territory in case he was shot down and captured. Instead he worked out with senior officers the operational plans for the final bombing mission. Because August was often cloudy over Japan and the mission needed clear weather, his bomb-carrying plane would be preceded by three weather aircraft, one to check the weather over the primary target and the others over each of the alternatives. The bombing plane would be accompanied by a plane carrying scientists and scientific instruments and another to take photographs. A spare aircraft would fly as far as Iwo Jima, halfway along the six-hour route to Japan. In the event of any mechanical problems aboard the lead plane, it would land at Iwo Jima and the bomb would be transferred to the spare plane, which would complete the mission.

The *Indianapolis* had anchored a thousand yards offshore because there was no quay deep enough for her to berth alongside. Her secret cargo was transferred to a tank-landing craft, brought ashore and carried to the windowless huts. The *Indianapolis* then sailed unescorted for Guam. C-54 Green Hornet transport aircraft delivered some final components for the uranium bomb – 'Little Boy' – on 28 July, and the bomb's assembly was virtually complete by 31 July. Little Boy was 120 inches long,

28 inches in diameter and weighed about 9,700 pounds. To one observer it looked like 'an elongated trash can with fins'.

The final components – the bomb casings – for the plutonium implosion bomb Fat Man, the same type as the one tested in the Jornada del Muerto, did not arrive until midday on 2 August. Among the other cargo in one of the B-29s transporting them was a ten-foot-high statue of Christ being taken to Tinian at the request of one of the chaplains. The components of Fat Man were hurried to the secure inner area where Luis Alvarez began quickly to assemble them.

One of Britain's most experienced bomber pilots, Captain Leonard Cheshire, holder of Britain's highest decoration, the Victoria Cross, and a former leader of the famous 617 'Dambusters' squadron, was on Tinian expecting to be an observer on the first atomic bomb mission. He shared a tent with William Penney, who also expected to fly on the mission as a British scientific observer. In early August Cheshire dropped in on the assembly of Fat Man. He recalled how Luis Alvarez 'straightened up and without much formality began explaining the basic functions of the gadgetry . . . little of which I grasped despite his obvious efforts to keep it simple. Then . . . he walked across to a yellow box lying on the floor and casually kicked it open with his foot. Inside I saw what appeared to be a metallic sphere about the size of a football . . . it did not strike me as anything very special.' Then Alvarez told him it was the plutonium core of the bomb.

I must have looked startled, for he told me not to worry; it was perfectly harmless and I was quite free to touch it if

I wanted, provided I wore a pair of gloves ... Disbelief
that the new monster bomb could be lying haphazardly on
the floor ... was followed by a sense of awe. Then I pulled
myself together, accepted the gloves that Alvarez offered
me and touched it. The sensation was rather like that of
the first time you touch a live snake: you recoil from what
you know will feel slimy and repulsive, and then to your
surprise find that it is warmish, almost friendly ...
Hitherto the bomb had conjured images of devastating,
unimaginable power ... True, there was a potentially
lethal side to it: but equally an inert side that left it totally
subservient to man's will.

Just after midnight on the morning of 30 July, the cruiser
Indianapolis was torpedoed en route to Guam by a
Japanese submarine and sank before being able to get a
distress call away. However, US Signals Intelligence
routinely intercepted and decoded a message from the
Japanese submarine reporting the sinking of a 'battleship
of Idaho class'. Intelligence passed the decode to naval
headquarters on Guam on the morning of 30 July but no
action was taken because no battleships were known to be
in the region. It was not until 2 August that a plane on a
routine patrol spotted survivors from the air. Only 318
sailors out of the crew of 1,169 were still alive to be
rescued; of those who survived the initial attack, 484
had died in the water of their wounds or of exposure, or had
been eaten by sharks. It was the greatest loss at sea in the
history of the US Navy and the last major warship to go
down in the Second World War. When news of the
Indianapolis's sinking reached Tinian, it darkened men's
moods, particularly that of Jacob Beser who, before the
cruiser left, had enjoyed a convivial reunion with an old

schoolfriend serving aboard her who was now among the dead.

On 3 August, Tibbets received the formal targeting order. The operation was to be codenamed 'Centerboard' and Hiroshima was the primary target; if it was covered with cloud the secondary target was Kokura, with Nagasaki as the fall-back.

The main briefing for the crews of the seven Superfortresses to take part in the operation was held on 4 August. Many smoked cigarettes nervously as in the afternoon heat they filed into a tin hut with closed curtains and under armed guard. According to one of those present, 'it was so hot and sticky just breathing was difficult'. Soon, everyone was wet with perspiration. Deak Parsons was flying on the raid as senior weapons officer. Pale and tight-lipped, and pausing frequently to wipe the sweat from his brow and bald head, he gave the lead presentation. He spoke slowly and softly: 'The bomb you are going to drop is something new in the history of warfare. It is the most destructive weapon ever devised.' He paused and cleared his throat. 'We think it will wipe out almost everything within a three-mile area, maybe slightly more, maybe slightly less.' Next he showed a film of the Trinity test.

Afterwards, Parsons told his visibly stunned audience that no-one knew what the exact effect of such a bomb dropped from the air would be as it had never been done. However, a flash of light much brighter than the sun was expected, against which crews would need to protect their eyes. For that purpose he distributed sets of goggles like those worn by welders. They had adjustable lenses, and the crews were told to switch them to the darkest setting.

Paul Tibbets told the crews, 'he was personally honoured, and he was sure all of us were, to have been chosen to take part in this raid which, he said – and all the other bigwigs nodded when he said it – would shorten the war by six months'. Everyone was told, 'no talking . . . no talking, even among yourselves. Loose lips sink ships. Be quiet. Say nothing. And over again, each phrase half a dozen times. And no letters. No writing home. Not to anyone, our wives, our mothers . . . The news, when it was released, would come from Washington, from President Truman himself.'

Neither Tibbets, Parsons nor anyone else mentioned the word 'nuclear' or 'atomic' in connection with the bomb. A subdued Leonard Cheshire and William Penney sat at the back, excluded at this late stage from taking part as observers almost certainly by General Groves, who wanted the first use of the atomic bomb to be an all-American affair.

Deak Parsons had seen a number of bombers crash and burn on Tinian's runways on take-off. On 5 August, the day before the scheduled first mission, he decided it would be unwise to arm the bomb until the aircraft was in flight in case it exploded in a crash on take-off and destroyed the island and the twenty thousand service personnel on it. A similar proposal had previously been rejected by General Groves. Although Parsons consulted Groves' deputy, General Farrell, the senior Manhattan Project officer on Tinian, who agreed that the bomb should be armed in flight, neither informed Groves of the change of plan. Groves later wrote, '. . . they just didn't have the nerve that was required, that was all. There had been quite a few crashes, but after all we had probably the best pilot in the air force, Colonel Tibbets . . . If I had known about it in advance they would have had a very positive order over there.' Parsons practised the necessary manoeuvres in the bomb bay, cutting his hands as he worked in the restricted space amid much sharp metal, but was able to satisfy himself that he could arm the bomb in flight.

On the afternoon of 5 August, in what to Leonard Cheshire resembled 'a military funeral cortege', a tractor moved Little Boy, painted a dull gun-metal grey, on a trailer covered with a tarpaulin under armed guard from the technical area to be winched aboard Tibbets' plane. Several messages to the Japanese had been scrawled on the bomb's casing, including one of vengeance for those lost on the *Indianapolis*. Like the bomb, the plane had gained a name – *Enola Gay* – after Tibbets' mother's first names. He had consulted some of his crew, but not Bob Lewis. Lewis had flown the plane much of the time in training and on 4 August had had the difficult task of telling his regular bombardier and navigator that they had to make way for Thomas Ferebee and Dutch Van Kirk respectively. When, a day later, he saw the name *Enola Gay* freshly painted on the plane he was, he recalled, 'very angry' and confronted Tibbets, but the name stayed. Less controversially, Tibbets had also had the arrow on the tail-fin noted by Tokyo Rose painted over so that *Enola Gay* no longer looked any different from any other plane on the base. A little later the Los Alamos scientific team was ordered to move to another part of the island in case of an accident on take-off. Knowing the power of the bomb, they realized how futile this would be and stayed put.

During the evening, the crew made their personal preparations. Some prayed or, if Catholics, went to confession. Others, including Ferebee, played poker and blackjack. Whatever else they did, most ate. Tibbets shared several plates of his favourite pineapple fritters with Van Kirk and Ferebee. However, he did not share with them that in his pocket he had a tin of cyanide capsules so that he and the crew could, if necessary, choose to die rather than face capture and torture. Tibbets, whose own 'tightly-wound nerves vetoed the idea' of sleep, held a short final briefing around midnight at which the Lutheran chaplain said a special prayer. Included were the words, 'Guard and protect them . . . May they as well as we, know thy strength and power, and armed with thy might bring this war to a rapid end.' Jacob Beser, who was Jewish, reflected that in his religion it was more usual to give thanks after coming through than 'to ask a special favour beforehand'.

One of the crew of the plane which would carry the scientific monitoring equipment remembered how the last hours felt: 'It's a little difficult to explain the emotions experienced just before a mission, when you know you're going and at what time and how far it is and what opposition is expected and when (*if*, of course, always *if*, although you never admit that, even to yourself, especially to yourself) you'll return. It's a little like going to the dentist's office. Once you've made the date, you relax a little . . . You know it's just a matter of sweating out the patients ahead of you, and you can't (or won't) run away; everything's set. It's irrevocable, and you accept it.' Bob Caron, the *Enola Gay*'s rear gunner, recalled:

It was about 1 a.m. . . . when we piled out of the trucks that drove us to the flight line. The *Enola Gay* was bathed in a flood of lights and the hardstand looked like a Hollywood movie set. A crowd was on hand, consisting of

military brass, other interested military personnel, and
some civilians whom we knew to be scientists.
Cameramen – still and newsreel – seemed to be everywhere. Frequently our preparations for take-off were
interrupted to have our pictures taken. Even the
photographers did not know why they were taking
pictures; they were just following orders. I recall
wondering whether we were being photographed for
historical interest – or because they didn't think we were
coming back.

After some time and many photographs, the crew
climbed the ladder into the plane and strapped themselves
in. Caron took his place in the tail-turret, not for fear of
attack, rather because 'there was a marginally better
chance of survival in the tail' in the event of a crash on
take-off.

At 2.45 a.m. Tibbets let go the brakes and opened the
throttles. *Enola Gay* moved down the mile-and-a-half-
long chopped-coral runway lined by fire and rescue
vessels. Loaded with the 9,700-pound Little Boy and the
7,000 gallons of aviation fuel necessary for the long flight,
she weighed about 65 tons, around seven more than the
usual take-off weight for a B-29. Consequently, she picked
up speed only slowly. To those on the ground, it seemed
that Tibbets was never going to pull back the stick and
take off. It seemed so too to Bob Lewis, who, from the
adjacent co-pilot's seat, urged Tibbets to lift off. But
Tibbets wanted maximum speed to lift his heavy load
with a 'cushion of safety in case we lost an engine at this
moment of maximum strain'. Tibbets was, in his
words, 'little more than 100 feet from the end of
the pavement' when he eased *Enola Gay* from the

ground safely and steadily into her climb before she vanished from the view of onlookers into the velvet northern sky.

'IT'S HIROSHIMA'

TEN MINUTES AFTER TAKE-OFF, THE *ENOLA GAY* WAS PASSING over Saipan and, flying at an air speed of 213 knots (247mph), was climbing to 4,700 feet, the initial cruising altitude for the first leg of the six-hour journey, three hours north-north-west to Iwo Jima.

Two minutes earlier, at about the time rear gunner Bob Caron tested his guns, Parsons and his assistant Morris Jeppson had, after securing Tibbets' permission, made their way to the bomb bay. Parsons climbed down through the hatch in the floor, squeezing himself into the small place behind Little Boy's tail to begin his delicate task of arming the bomb. Using only spanners and a screwdriver, he had to remove a series of protective shields and then insert an explosive charge. When triggered, the charge would propel a slug of uranium down the gun barrel into the uranium rings fitted into the nose of the weapon to achieve the critical mass necessary to begin the explosive chain reaction.

While Jeppson shone a torch from above, Parsons worked quickly in the confined, cold and unpressurized space of the bomb bay, trying not to cut himself on the sharp steel casings, as he had done in practice. As Parsons worked, Jeppson used the intercom to tell Tibbets of his progress, and Tibbets in turn informed Tinian over his low-frequency radio. Only twenty-five minutes after starting, Jeppson reported the task complete. Because of static interference Tibbets could not get this final message through to Tinian, but, he recalled, 'Progress was such by this time, they had no doubts of Parsons' success.'

Parsons left three green safety plugs in position which he would have to replace later with three red arming plugs to unlock the weapon's fusing circuits, which he would then carefully monitor using a bank of electronic equipment. The detonation of the live bomb would then depend on a series of triggers. The primary one was a kind of proximity fuse, a simple radar unit built into the bomb which closed a switch firing the explosive charge when the bomb fell to a predetermined height of some two thousand feet above ground.* The second was back-up clocks activated mechanically on the bomb's release and preventing detonation for at least fifteen seconds after that time. Finally, there was a barometric pressure switch which would not close until the air pressure had increased to that found at a maximum of seven thousand feet above ground. Both the back-up systems would give some protection to the aircraft if the primary system were to be activated too early for any reason. All three triggering systems contained duplicates to overcome an individual instrument failure.

---

* The unit had been adapted from an instrument designed to alert pilots to the approach of enemy aircraft to the rear. Instead of bouncing signals off an approaching hostile plane, it would respond to the approaching ground.

Tibbets checked with the plane carrying the scientists and their instruments and then with the photographic plane. Receiving confirmation that all was well aboard them, he made a quick tour of inspection of the *Enola Gay*, crawling back along the communication tunnel to talk to Caron and others. Satisfied that all was in order, and 'having had little sleep in the past forty-eight hours', he sensed he was 'operating on nervous energy alone' and so, making himself as comfortable in his seat as he could with the help of his lifejacket and parachute pack, he dozed for about an hour. Co-pilot Bob Lewis took a bite to eat while he kept an eye on the green-lit instrument panel and on the automatic pilot known in this and in other aircraft as 'George'.

Soon, Iwo Jima was in sight; according to the official log, they reached it at 5.55 a.m. Tinian time. In the soft pink light of dawn Tibbets circled the island's highest peak, Mount Suribachi, at 9,300 feet so that his instrument and photographic planes could take closer formation.

As they left Iwo Jima at 6.07 a.m., there were still three prospective targets – the primary target, Hiroshima, and the secondaries, Kokura or Nagasaki. The final choice would depend on reports from the three weather planes which had taken off from Tinian about an hour earlier than the *Enola Gay*, each assigned to a particular city. At 7.30 a.m. Deak Parsons and Jeppson made their way back to the bomb bay and carefully removed each of the green safety plugs, substituting the red plugs which activated Little Boy's internal batteries. Bob Lewis, who was keeping some authorized notes for a *New York Times* journalist, wrote, 'The bomb is now alive. It is a funny feeling knowing it is right in back of you. Knock wood.'

He worried that if they hit bad weather or turbulence, the bomb might detonate. Tibbets calmed himself by smoking his pipe 'with', in his words, 'a little more intensity than usual' as the *Enola Gay* climbed slowly to the bombing altitude of 30,700 feet.

At just after eight o'clock Tinian time, seven o'clock Japanese time, the weather plane assigned to Hiroshima, *Straight Flush*, piloted by Claude Eatherly, made a run towards the city. The plane approached, bumping through cloud cover, but then, directly over the city, came a large break in the clouds through which shafts of sunlight illuminated Hiroshima. At his request, Eatherly's radioman sent a signal consisting of the numbers and letters 'Q-3, B-2, C-1'. Aboard *Enola Gay*, young radio operator Dick Nelson picked up the transmission, decoded it and reported the result to Tibbets. The cloud cover at all altitudes was less than three tenths. 'Advice to bomb the primary target.' Tibbets recalled, 'over the intercom I gave the word to members of our crew, "It's Hiroshima" '.

Soon the *Enola Gay* crossed the first of the Japanese islands. Deak Parsons tested the bomb's electrical circuits with his instrument console one last time. Jacob Beser reported that he could detect no Japanese radio countermeasures. Tibbets recalled, 'we were eight minutes away from the scheduled time of bomb release when the city came into view. The early-morning sunlight glistened off the white buildings in the distance.' Tibbets reached the initial point of the bomb run. Surrounded by plexiglass panels in the exposed nose of the plane, Tom Ferebee crouched over the bombsight as Tibbets began the three-minute bomb run. There was no anti-aircraft fire. Soon the aiming point, the 'T'-shaped Aioi Bridge in the central

Salugakucho district, was clearly visible in Ferebee's bombsight. The bombardier activated a sixty-second radio tone to alert the *Enola Gay*'s crew and the two accompanying planes to the imminent release of the bomb. Bob Lewis scrawled on his notepad, 'there will be a short intermission while we bomb our target'. Tibbets remembered that at the end of the tone, at 9.15 Tinian time and 8.15 local time in Hiroshima, the bomb doors opened automatically and 'out tumbled Little Boy'.

Tibbets immediately pushed the *Enola Gay* into the 155° turn required to take her to safety. Bob Caron recalled, 'The manoeuvre felt like being on the cyclone rollercoaster ride at the Coney Island amusement park.' In making the diving turn, the plane lost 1,700 feet in height.

Tibbets was focusing so hard on the controls that the flash of the explosion did not have the effect he expected, but at the instant of the blast he recalled 'a tingling sensation in my mouth and the very definite taste of lead upon my tongue'. According to Tibbets, scientists later told him that this was caused by an interaction between the fillings in his teeth and the radiation released by the bomb. Among the *Enola Gay*'s crew, only Bob Caron, holding a camera alone in his tail-turret as the plane raced away, saw the explosion direct – bright even through the very dark goggles Tibbets had ordered him to don, like the rest of the crew, a minute before the attack. He saw the shock wave develop and seemingly rise towards the *Enola Gay* as if 'the ring around some distant planet had detached itself and was coming up towards us'. He yelled to warn the pilots. As he did so, the shock wave engulfed the *Enola Gay*, throwing her about and creating a massive noise. Both Ferebee and Tibbets thought the effect was that of an anti-aircraft shell exploding, while

Japanese propaganda cartoon showing caricatures of Roosevelt and Churchill huddling on the deck of an aircraft carrier.

Left: Paul Tibbets in front of the *Enola Gay* on Tinian.
Main picture: 'Little Boy', the first atomic bomb, explodes over Hiroshima on 6 August 1945.

Left: The Hiroshima Prefectural Industrial Promotion Hall after the explosion.

Right: An old woman, her daughter and grandchild, the only survivors of their family, flee Hiroshima with all their belongings on a cart.

Main picture: Japanese victims await help in the hours following the atomic bombing of Hiroshima.

Main picture: The Hiroshima Prefectural Industrial Promotion Hall as it is today.

Below, left: Klaus Fuchs checks his ticket as he leaves for East Germany after his release from prison in 1959.

Below, centre: Edward Teller after the Second World War.

Below, right: Peter Sellers portrays Dr Strangelove.

Lewis compared the shock to a giant striking the plane with a telephone pole.

Almost immediately, Caron saw another wave approach. This wave was the reflection of the first from the ground and its impact was less dramatic, but still sufficiently violent to propel Dick Nelson half out of his seat. Once the second wave had passed the air grew still once more and Tibbets circled the stricken city as Caron photographed the mushroom cloud, which seemed to rise sixty thousand feet into the air. Tibbets asked Beser to take round to each man a portable wire recorder he had been given and to ask them to record their impressions. The recordings have been lost, but to Bob Caron the scene was 'beautifully horrible'. He recollected his description in the following words:

A column of smoke rising fast. It has a fiery red core. A bubbling mass, purple-grey in colour, with that red core. It's all turbulent. Fires are springing up everywhere. Like flames shooting out of a huge bed of coals. I am starting to count the fires. One, two, three, four, five, six . . . fourteen, fifteen . . . It's impossible. There are too many to count. Here it comes, the mushroom shape that Captain Parsons spoke about. It's coming this way. It's like a mass of bubbling molasses. The mushroom is spreading out. It's maybe a mile or two wide and half a mile high. It's growing up and up and up. It's nearly level with us and climbing. It's very black, but there's a purplish tint to the cloud. The base of the mushroom looks like a heavy undercast that is shot through with flames. The city must be below that. The flames and smoke are billowing out, whirling out into the foothills. The hills are disappearing under the smoke.

Beser himself recorded, 'what a relief it worked'. Lewis wrote on his notepad, 'My God, what have we done?'

After completing three circuits, Tibbets headed *Enola Gay* for Tinian. Deak Parsons radioed back a report of success: 'Clear cut, successful in all respects. Visual effects greater than Trinity. Hiroshima. Conditions normal in airplane following delivery. Proceeding to regular base.' The chief weaponer then relapsed into what other members of the crew recalled as 'a withdrawn and meditative' mood. Bob Caron could still see the mushroom cloud from his tail turret until they were more than 350 miles from Hiroshima. Tibbets handed the controls to Bob Lewis while he napped for a while. As he flew the plane home, Lewis was assessing events. Later that day he told a reporter, 'Even though we had expected something terrific, what we saw made us feel that we were Buck Rogers twenty-fifth-century warriors.' More soberly, he wrote on his pad, 'I had a strong conviction that it was possible, by the time we landed, that the Japs would have thrown in the sponge. Because of the total destruction I didn't feel there was room for anything but complete surrender.' Tom Ferebee wondered whether the radiation to which they had been exposed might make him sterile.* Deak Parsons tried to reassure him.

Tibbets took the controls again to land on Tinian. As they came to a standstill they were greeted by a large crowd, the formal welcoming party many times out-numbered by well-wishers. When, pipe in mouth, Tibbets led his men down onto the tarmac through the hatch to the rear of the B-29's nose wheel, he was surprised to see General Carl Spaatz, Commanding General US Army

---

* Ferebee need not have worried. He later fathered four sons.

Strategic Air Force, approaching. Tibbets barely had time to palm his pipe before the general pinned the Distinguished Service Cross on his creased flight overalls. Then a scrum of well-wishers surrounded him, slapping his back, rejoicing in the mission's success and eager to hear more about it.

'MOTHER WILL NOT DIE'

TOWARDS MIDNIGHT ON 5 AUGUST, HATSUYO NAKAMURA, THE widow of tailor turned soldier Isawa Nakamura who had been killed more than three years earlier on the day Singapore fell, heard on her radio a warning to all inhabitants of Hiroshima that two hundred American bombers were approaching. Everyone should go to the safe areas. She woke her three children. With their sleeping rolls, they set out from their small wooden house in the part of the city known as Nobori-cho for their appointed safe area on the north-east side of the city near Hiroshima railway station. They returned at 2.30 a.m. after the all-clear signalled that the bombers had passed. Air-raid sirens woke Mrs Nakamura again at around seven a.m. as Claude Eatherly's weather plane approached, but she decided not to disturb her children to take them back to the safe area. The all-clear soon sounded once more. Her children were beginning to wake, so she gave them a few peanuts for breakfast and told them to try to sleep, soon made difficult by the noise of a neighbour knocking down his own house to comply with the order to make fire-breaks.

Mrs Nakamura was standing at her kitchen window watching him when, just after a quarter past eight, a white flash enveloped her. Her house, which was only three quarters of a mile from the Aioi Bridge, collapsed about her, burying her under the debris. As she struggled to free herself, she heard her youngest daughter Myeko cry for help and, twisting around, saw her buried up to her chest. She could see no sign of her ten-year-old son Toshio or eight-year-old daughter Yaeko. Then, from beneath the collapsed beams and tiles, came separate disembodied cries. Both were alive. Mrs Nakamura frantically pulled the wood aside to free them. Afterwards she released Myeko too.

All four were dusty, dirty, frightened and confused. Toshio and Yaeko said nothing, but young Myeko kept asking, 'Why is it night already? Why did our house fall down?' Mrs Nakamura got them out into the street where they saw the neighbour who had been demolishing his house lying dead. The authorities had previously designated Asano Park, a wooded area along the nearby Kyo River, as the evacuation place for Mrs Nakamura's neighbourhood. Now, with her children and with a hastily gathered bundle of clothes on her back, she and a neighbour almost instinctively hurried towards the park, past ruined houses. From under the rubble of some came cries for help half-muted by the debris. Mrs Nakamura felt compelled to ignore them as, in her determination to save her children, she pressed on.

About half an hour after the explosion, as the intensity of the fires consuming the city grew ever stronger, heavy rain began. At first it fell in large, sticky, black drops – 'black rain' – water intermingled with soot, muddy dust and debris flung into the air by the explosion. The rain also contained radioactive material released by the bomb. Some desperately thirsty, traumatized people instinctively opened their mouths and let the contaminated water cool their parched throats.

Mrs Nakamura's family were among the first to arrive at Asano Park – a private riverside estate with rock gardens, ornamental trees and bamboo groves. The Nakamuras had not been affected by the black rain. However, once in the park they went down to the river to drink from it. Immediately they began to vomit from the effects of the polluted water, as did all those around them who were doing likewise. The Nakamuras lay prostrate on the ground until, during the afternoon, the fires from the city raging out of control began to catch the trees of the park. Luckily, a storm combining black rain and strong winds held the fires back and the Nakamuras spent the night where they were. The next day a German Jesuit priest from their neighbourhood brought them to safety outside the city, with the children riding on a handcart. On 12 August, they moved in with Mrs Nakamura's sister-in-law in a nearby town.

Futaba Kitayama, the young mother who had seen the bomb fall from 'an airplane as pretty as a silver treasure' and explode 'with an indescribable light', made up her mind that, despite her severe burns, she must survive to see her three children again. Fortunately, she had sent them to the countryside a little while earlier for safety. She recalled:

Suddenly, driven by a terror that would not permit inaction I started to run for my life. I say 'run', but I had no idea where the road was. Everything was covered in wood and tiles so I had no idea which way to go. Such a bright morning until a moment ago, what in the world could have happened? Now we were under a thin cover of darkness, just like dusk. The dull haze, as if my eyes were covered with mist, made me wonder if I was losing my mind. Looking around unsteadily, I saw something that looked like people running on the bridge. 'That's Tsurumi Bridge. If I don't get over it right away, I'll lose my chance to escape,' I thought. Jumping over trees and rocks like a crazy person, I ran towards the Tsurumi Bridge. When I arrived I saw a horrifying spectacle. Countless bodies squirming and writhing in the flow of people and water under the bridge. Their faces were grey and so swollen I couldn't tell male from female. Hair stood straight up. Arms waved in the air. Voices groaned wordlessly. They were jumping into the river one after another. The strong ray had burned my work pants to rags, and my whole body was in agony, so I was preparing to jump in with them when I remembered that I couldn't swim.

When she turned around, she saw that the whole city was 'a solid sheet of flames. Calling out the names of my three children in turn, I encouraged myself over and over, saying, "Mother will not die. Mother will be all right." Looking back, I simply cannot remember where or how I ran. The many pitiful sights I saw are etched in my brain.' Eventually she came to another bridge, where

corpses were floating by like dead dogs and cats, their shreds of clothing dangling like rags. In the shoals near the banks I saw a woman floating face up, her chest gouged out and gushing blood. Could such terrifying sights be of this world? Suddenly I lost strength and [after crossing the bridge] had to sit right in the middle of the [neighbouring] drill ground. All around me, junior high school girls and boys from another volunteer corps writhed on the ground. They seemed crazed, crying 'Mother, mother.' As my eyes took in the cruel sight of their burns and gaping wounds so horrible I couldn't bear to look at them, an enormous rage welled up from deep within, but I didn't know where to direct it. Even these innocent children . . . crying for their mothers, first one, then another, breathed his or her last. All I could do was look at them.

I gathered all the strength in my flickering body and soul and fell in line with people heading towards the mountains. Probably about three p.m., having been utterly lethargic for some time, I sat down. As I gazed around with what was left of my eyesight I could tell that the station and all of Atagocho had become a sea of fire. I felt lucky to have escaped. Gradually my face grew stiff. Gently touching my cheeks with both hands, I measured with my eyes the distance between my hands as I took them away and saw that my face had swollen to about twice its normal size. My vision was more and more restricted. 'Oh no, soon I won't be able to see. Could I have come this far only to die here?' Stretcher after stretcher came by carrying the injured. Carts and trucks drove by full of injured people and corpses that looked more like monsters. On both sides of the road, many people wobbled this way and that, as if sleepwalking.

I realized that while I could see a little, I needed to find a safe place where I would not be hit by a truck and could quietly trust myself to fate. Peering here and there through barely open eyes, I saw my own sister squatting and resting. 'Sister, sister, help me.' Without thinking I ran towards her. My sister at first looked at me doubtfully.

Finally, she recognized me. 'Futaba-chan, you look . . .' She couldn't say any more and just held me. 'Sister, I can't see any more. Please take me to my children.'

Her sister put Futaba on a vegetable cart and took her to a relief station. After a couple of days the doctors said her case was hopeless. Futaba persuaded her sister to take her to her home.

That evening my children arrived at last. When I heard their voices scream 'mummy' I felt rescued from the depths of hell. 'I'm OK. These burns are nothing much.' And I cried as the children I had missed so much came and clung to me. On the 11th I was quietly preparing myself to give up and die when my husband arrived, having tracked me down. At that time my suffering was so bad I found brief solace thinking, 'Ah, good. Now if they lose their mother they'll still have their father,' and I was happy. Then, three days after finding us, my husband, who had no serious injuries at all, began vomiting blood. Then he was gone, leaving behind a wife, unsure she would see another day, and his three beloved children. Our little boy sat near my pillow crying, 'Mummy.' I almost bled with grief and, even now, as I recall that time, the tears flow. 'My poor children, I can't die now, I can't leave them orphans.' With all my heart I prayed to the spirit of my husband, asking help. Over and over I was told I had no hope, but miraculously I lived.

Down in the port area, two and a half miles from the centre of the blast, twenty-eight-year-old army doctor Hiroshi Sawachika remembered:

I had just entered [my office] and said good morning and I was about to approach my desk when outside it suddenly turned bright red. I felt very hot on my cheeks. I felt weightless as if I were an astronaut. I was then unconscious for twenty or thirty seconds. When I came to I realized that everybody including myself was lying at one side of the room. I went to the windows to find out where the bombing had taken place. And I saw the mushroom cloud over the gas company. I still had no idea what had happened. I realized that my white shirt was red all over. I thought it was funny because I was not injured at all. I looked round, then realized that the girl lying nearby was heavily injured with lots of broken glass stuck all over her body. Her blood had splashed and made stains on my shirt.

He was told that 'injured citizens were coming towards us for treatment. [There were] big hospitals in the centre of the town, so why should they come here, I wondered, instead of going there? With lots of injured people arriving, we realized just how serious the matter was. As they came to us they held their hands aloft. They looked like they were ghosts. We made the tincture for their treatment by mixing edible peanut oil and some other things. We had to work in a mechanical manner in order to treat so many patients.' After a while he went into a nearby room where patients were waiting. 'I found the room filled with a smell that was quite similar to the smell of dried squid when it has been grilled. The smell was quite strong. It is a sad reality that the smell human beings produce when they are burned is the same as that of the dried squid when it is grilled – the squid we like so much to eat.'

Afterwards he walked back through the rows of people awaiting treatment.

I felt someone touch my leg, it was a pregnant woman. She said that she was about to die in a few hours 'but I can feel that my baby is moving inside. It wants to get out of the womb. I don't mind if I die. But if the baby is delivered now it does not have to die with me. Please help my baby live.' There was no obstetrician there. There was no time to take care of her baby. All I could do was tell her that I would come back later when everything was ready for her and her baby. Thus I cheered her up, and she looked so happy. The image of that pregnant woman never left my mind. Later I went to the place where I had found her before; she was still there, lying in the same place. I patted her on the shoulder but she said nothing. The person lying next to her said that a short while ago she had become silent.

* * *

The atomic bomb had exploded with a temperature at its centre of one million degrees Celsius, generating a white-hot fireball. Immediately beneath the explosion the ground reached more than 3,000° Celsius (iron melts at half that temperature). Over a mile away, clothing on people outdoors spontaneously burst into flames. Pressure rose to over six tons per square metre (several hundred thousand times normal atmospheric pressure). Nearly all wooden houses within a mile and a half had, like Mrs Nakamura's, collapsed. Of the energy released, 35 per cent was in the form of heat, 50 per cent as blast and 15 per cent as radiation.

According to the Hiroshima city government, the death toll by December 1945 was 140,000 plus or minus 10,000 of the 350,000 people estimated to have been in the city that day. It included those who had died from the immediate effects of the bomb and later from radiation.

Deaths from radiation-related diseases continue.* The dead are estimated to include up to ten thousand Koreans, nearly all forced labourers, and about ten American airmen being held prisoner in Hiroshima castle after being captured when their bombers were shot down.

Because of the destruction of all means of communication, news of the attack did not reach Tokyo until around midday. Not until 7 August did the Japanese authorities send Professor Yoshio Nishina to Hiroshima to confirm that the bomb had indeed been an atomic one. His plane developed engine trouble and had to turn back, and he did not arrive until the next day, 8 August, forty-eight hours after the explosion. The rapid clicking of his Geiger counter, together with the evidence of high temperatures provided by melted clay roof tiles and the obvious radiation injuries suffered by the victims, left him in no doubt. He later expressed awe at 'the product of pure physics', but his greatest sensation was horror.

Pride or refusal to contemplate defeat led Field Marshal Shuntoku Hata, who had survived in his headquarters near Hiroshima, to play down the effects of the bomb. He reported that in his view the bomb was 'not that powerful a weapon'. However, during the evening of 8 August Nishina telephoned the Prime Minister's office in Tokyo with confirmation that it was indeed an atomic bomb

---

* Later deaths from cancer can be attributed to the effects of the bomb only statistically. This is done by assessing how many more deaths from cancer occur in the Hiroshima population than would be expected in a similar population not exposed to radiation. The calculations are fraught with difficulty in identifying comparable populations, and estimates vary widely. Official figures suggest fewer than a thousand such deaths since the end of 1945.

with, he calculated, a force of around twenty thousand tons of TNT that had obliterated Hiroshima.*

Fewer than six hours after Nishina's call, the Soviet Union implemented her declaration of war against Japan, communicated earlier on 8 August, and more than one and a half million of her troops crossed into Manchuria, pushing back the Japanese forces in front of them. Just a few hours later, at about eleven a.m. on 9 August, before the Japanese government had had time to consider either Nishina's report or the implication of Russian entry into the war, Fat Man destroyed Nagasaki.

After Hiroshima, General Spaatz had been ordered 'to continue operations as planned' in the original directive to him that additional bombs be 'delivered on' the target cities as soon as made available by the project staff. The second bomb had originally been scheduled to be dropped on 11 August, five days after Hiroshima, but the project team on Tinian had brought the date forward by two days in discussion with Paul Tibbets since good weather was forecast for 9 August and the five succeeding days were expected to be bad. General Farrell later explained, 'We tried to beat the bad weather. But secondly, there was a general feeling among those in the theatre [Pacific] that the sooner this bomb was dropped, the better it would be for the war effort.'

Nagasaki was the secondary not the primary target for the delivery aircraft *Bock's Car*, piloted by Charles Sweeney. The primary was Kokura, but Sweeney found it cloud-covered and diverted to Nagasaki. Leonard Cheshire, whom Groves had this time allowed to fly as an

* The actual force of the explosion is now generally accepted to have been equivalent to around fifteen thousand tons of TNT.

observer together with William Penney, described a boiling blackness with a mushroom cloud 'the colour of sulphur' with an 'evil kind of luminous quality'. The immediate death toll, at around forty thousand, was lower than Hiroshima, but the devastation was still immense.

That afternoon, 9 August, in Tokyo, an imperial conference convened at which Hirohito's ministers expressed differing views on the wisdom of surrender. The meeting went on through the evening and into the early morning of the next day. At two a.m. on 10 August, Hirohito gave his view that the Japanese should accept the Potsdam Declaration and surrender. The sole precondition his ministers appended before informing the Allies was that their acceptance was on the understanding that the Potsdam Declaration did 'not comprise any demand which prejudices the prerogatives of His Majesty as a sovereign ruler'.

Secretary of State Byrnes' response to the offer did not address the condition either way. Nevertheless, Japan formally surrendered on 14 August at the Emperor's express command. On the afternoon of the 15th the Emperor broadcast to his people for the first time. His thin, high-pitched voice told them in archaic court language that he had agreed to the surrender to save mankind 'from total extinction'. His government had conducted the war for self-defence and to preserve the nation's existence. Listeners then heard a cabinet announcement denouncing the United States for the use of atomic bombs in contravention of international law. Hirohito would die over forty-three years later, on 7 January 1989, still Emperor of Japan.

In America, President Truman had on 10 August given

the order to suspend further atomic bombing. He spoke to the American people in a nationwide radio broadcast after his return from Potsdam: 'Having found the bomb, we have used it. We have used it against those who attacked us without warning at Pearl Harbor, against those who have starved and beaten and executed American prisoners of war, against those who have abandoned all pretence of obeying international laws of warfare. We have used it in order to shorten the agony of war, in order to save the lives of thousands and thousands of young Americans.'

'A NEW FACT IN THE WORLD'S POWER POLITICS'

AMERICAN NEWSPAPERS SUPPORTED THE USE OF THE ATOMIC bomb. The *New York Times* said on 12 August, 'By their own cruelties and treachery our enemies had invited the cruelty.' An academic study later showed that of nearly six hundred American editorials fewer than 2 per cent opposed the bomb's use. In Britain it was much the same. The *Times* editorial pondered the consequences had Germany been first to the bomb and then rejoiced that 'in the intellectual sphere as on the battle field the discipline of free minds has its inalienable advantage. Pre-eminence in the pursuit of knowledge must belong to a social system in which men, whatever their origin, are free to follow whithersoever the argument may lead.' However, in both countries any concerns expressed stemmed less from worries about the morality of the bombing of Hiroshima than, as the American magazine *New Republic* put it, 'thoughts of its future use elsewhere and specifically against ourselves and our children'.*

* Not all journalists were preoccupied with grim reflections. Within hours of the announcement of the bomb, the Washington Press Club bar was selling 'atomic cocktails' – a blend of Pernod and gin.

The immediate response in the rest of the world's press was not so unanimous as in Britain and the United States. On 7 August, the Vatican's newspaper, *L'Osservatore Romano*, condemned the dropping of the bomb and juxtaposed its use with Leonardo da Vinci's reputed withdrawal of his invention of the submarine for fear of its misuse. The Swedish *Aftenblader* on 9 August reflected, 'although Germany began bomb warfare against open towns and civilian populations, all records in this field have been beaten by the Anglo-Saxons. The so-called rules of warfare . . . must brand the bombing of Hiroshima as a first-class war crime. It is all very well if atom raids can shorten the war but this experiment with the population of an entire city as guinea pigs reflects no martial glory on the authors.' In France, *Le Figaro* remarked that it was 'not probable that the Anglo-Saxons will long remain the sole possessors of this thunderbolt' and looked forward to French participation in nuclear projects.

Among Allied troops, including those preparing to move from the now quiet European front to take part in the invasion of Japan, there was only relief. On Okinawa, American forces fired so many weapons into the air in joy that shell splinters falling back to earth killed seven men. A British doctor wrote of the atomic bombings, 'We were packing for the invasion of Penang Island. None of us wept for the victims. Perhaps we were wrong, but on the night the war ended I don't think any of us gave a damn. Reprieve is sweet. I was home six months later.' American soldiers thought similarly. A twenty-one-year-old second lieutenant wrote, 'when the bombs dropped and the news began to circulate that [the invasion of Japan] would not, after all, take place, that we would not be obliged to run up the beaches near Tokyo assault-firing while being mortared and shelled, for all the fake manliness of our facades we cried with relief and joy. We were going to live. We were going to grow up to adulthood after all.'

Klaus Fuchs was not able to get the details of the Trinity test to his contact until September. Therefore, despite his previous forewarnings, the explosion of the bomb and its massive power were still a surprise to the Russian leadership. The London *Sunday Times*' experienced Moscow correspondent wrote that the news had had an 'acutely depressing effect' . . . 'it was clearly realised . . . that this was a new fact in the world's power politics, that the bomb constituted a threat to Russia, and some Russian pessimists I talked to that day dismally remarked that Russia's desperately hard victory over Germany was now "as good as wasted" '. Stalin told Kurchatov, 'Hiroshima has shaken the whole world. The balance has been destroyed.' Another of those working on the Russian bomb project, the physicist and future human rights activist Andrei Sakharov, wrote that when he saw the newspaper headline, 'I was so stunned that my legs practically gave way . . . Something new and awesome had entered our lives, a product of the greatest of the sciences, of the discipline I revered.'

Stalin redoubled efforts on the Soviet Union's own weapons programme, telling Kurchatov, 'if a child doesn't cry the mother doesn't know what he needs. Ask for whatever you like. You won't be refused.' The initial Soviet test would closely resemble Trinity, and the main Soviet separation plant near Sverdlovsk would be configured almost identically to the plant at Oak Ridge. Bolstered by further information from Fuchs, the Russians would test their first nuclear weapon on 29 August 1949, beating the British to become the world's second nuclear power.

The news of Hiroshima reached the German scientists rounded up by the Alsos mission and interned at Farm Hall in the late afternoon of 6 August. Farm Hall was a pleasant country house surrounded by gardens with a tennis court, a good library and a piano. Each scientist had a German POW as his orderly. When they had first arrived in July, Kurt Diebner had asked Werner Heisenberg, 'I wonder whether there are microphones installed here?' Heisenberg had laughed and replied, 'Microphones installed? Oh no, they're not as cute as all that. I don't think they know the real Gestapo methods; they're a bit old fashioned in that respect.' He was entirely wrong. Microphones had been installed in every common room and bedroom when the house was used to train agents so that the British spy masters could check on their morale and loyalty. Thus the officer in charge, Major T. H. Rittner, and his staff were able to listen to all the German scientists' discussions, the content of which confirmed that they did not suspect they were being recorded. He had translated transcripts made of the more interesting points and transmitted them to his own superiors. Other copies went to the American Embassy, whence they reached General Groves.

In his report for 6 August, Rittner recounted how Otto Hahn was the first at Farm Hall to be told of the bomb.

Hahn was completely shattered by the news and said he felt personally responsible for the death of hundreds of thousands of people, as it was his original discovery which had made the bomb possible. He told me that he had originally contemplated suicide when he realised the terrible potentialities of his discovery and he felt that now these had been realised and he was to blame. With the

help of considerable alcoholic stimulant, he was calmed
down and we went down to dinner where he announced
the news to the assembled guests . . . The announcement
was greeted with incredulity.

In a BBC interview, Otto Hahn would recall the events of
that evening slightly differently. Major Rittner 'told me
about the dropping of the bomb and I of course was
frightened, or should I say very sad about it and
depressed, and I told the man, "Couldn't you have not
done it another way?" Then the major answered me,
"Well I don't care about 100,000 or 150,000 Japs if we
can save a couple of our British and American people,
therefore we dropped the bomb." The man was very
satisfied about it.'

Major Rittner was certainly right about the scientists'
incredulity and conceit that anybody else could have done
something they had not achieved. Werner Heisenberg
insisted that it was a bluff; 'it's got nothing to do with
atoms'. Carl-Friedrich von Weizsäcker said, 'I don't
believe it has anything to do with uranium.' Otto Hahn,
perhaps braced by the alcoholic stimulant provided by
Major Rittner, responded, 'If the Americans have a
uranium bomb then you are all second-raters. Poor old
Heisenberg . . . you might as well pack up.' A little later,
another exchange took place between the three:

Von Weizsäcker: I think it is dreadful of the Americans to
have done it. I think it is madness on their part.
Heisenberg: One can't say that. One could equally well
say, 'That's the quickest way of ending the war.'
Hahn: That's what consoles me.

At nine p.m. they listened to the official announcement of the bombing on the BBC and appeared to be convinced of its authenticity. There was some squabbling as to why their own programme had failed. One of the junior scientists praised the co-operation between the American scientists and contrasted this with the disharmony within their own programme, wherein 'Each one said the other was unimportant'. Erich Bagge wrote in his diary that day that one reason for the failure was that some of the scientific leaders had looked down on isotope separation 'and only tolerated [work on] it at the margins'. Von Weizsäcker rationalized their failure another way: 'I believe the reason we didn't do it was because all the physicists didn't want to do it, on principle. If we had all wanted Germany to win the war we would have succeeded.' Otto Hahn replied, 'I don't believe that but I am thankful we didn't succeed.' Heisenberg took up von Weizsäcker's theme: 'At the bottom of my heart I was really glad that it [the object of their research] was to be an engine [reactor] and not a bomb.' Hahn said, 'I thank God on my bended knees that we did not make the uranium bomb' – which he called 'an inhuman weapon'. He later said that he would have 'sabotaged the [German] war [effort] if I had been in a position to do so'.

After further discussion together and in small groups they retired to bed. None of them got much sleep. Hahn was clearly very depressed and agitated. According to Bagge's diary, 'At 2 a.m. there was a knock on our door and Mr. von Laue entered. "We must do something; I am very worried about Otto Hahn. This news has shaken him horribly, and I fear the worst." We stayed awake a long time, and only when we were able to tell from the next room that

Mr Hahn had finally fallen asleep did we all go to bed.'

The transcribed discussions on both 6 and 7 August reveal fundamental misunderstandings on the part of the German scientists about how a bomb could be made to work. Heisenberg stated on 6 August that 'a ton' of U-235 would be required to produce the critical mass necessary for an explosion. In his explanation of why this was so he used an irrelevant (and arithmetically incorrect) calculation. His description also omitted any discussion of the effect of heat gasifying the material and causing only some 2 per cent of it to be consumed. Had Heisenberg realized this, his calculation would have produced an even higher figure of fifty tons. While inaccuracy in calculation can be forgiven in the heat of discussion and the stress of just having heard about Hiroshima, Heisenberg's account seems to miss too many fundamental points for there to be any doubt that the German project was some way from understanding even the theory of an atomic bomb, never mind the engineering practicalities of separating enough fissile material and of constructing one.

Over the next few days the transcripts reveal that the scientists, under the leadership of von Weizsäcker, began to develop a rationale of why their work had failed, perhaps designed to protect them against three kinds of criticism: from Germans who thought they should have done better to protect their fatherland; from Allied scientists who could not understand how they could have worked for Hitler on an atomic bomb; and self-criticism based on doubts about their own scientific abilities and moral values. Von Weizsäcker's statement that they did not do it because they did not want to formed one of the two key strands of what Max von Laue called their 'version' ('Lesart' in German) of events, from which he distanced himself. The other justification, not entirely compatible with the first but more pragmatic, was that it was impossible to produce a bomb during the expected duration of the war with the resources available to them. Von Laue noted that during their discussions Heisenberg was 'mostly silent'.

The sense of purpose that had fuelled Robert Oppenheimer ended with the war. He confessed that there was 'not much left in me at the moment'. Determined to return to academe, he resigned from Los Alamos. In the immediate post-war years he remained an influential adviser to the United States Atomic Energy Commission (AEC), the successor to the Manhattan Project. However, he did not support the AEC's plans to build the world's first hydrogen bomb based on the release of energy caused by the fusing of hydrogen atoms, believing the fission bomb quite powerful enough for America's military needs.

Some other scientists, especially Edward Teller, resented Oppenheimer's attitude. The passionately anti-communist Teller feared the Russians would soon acquire the capability to build an atomic bomb and had devoted himself to what he called the 'sweet technology' of the hydrogen bomb. His supporters included Ernest Lawrence, and the project went ahead. On 1 November 1952, the United States conducted the H-bomb equivalent of the Trinity test over the Pacific; the device destroyed an island a mile in diameter, exploding with a force five hundred times greater than Little Boy. To an observer it seemed like 'gazing into eternity, or into the gates of hell'.

The news depressed Oppenheimer deeply and convinced him he had lost all influence. Before long he fell victim to the prevailing mood of anti-communist hysteria centred around Republican senator Joe McCarthy. On 12 April 1954 the *New York Times* reported that the AEC had suspended Oppenheimer's security clearance and planned a hearing that day to consider charges that Oppenheimer's left-wing contacts and activities in the 1930s made him unfit to have access to classified information.

The AEC hearing was held in private and lasted more than three weeks. Oppenheimer was the first witness to appear before the three-man board. Many others followed, and, as the transcripts show, most gave him their wholehearted support, but some, including Edward Teller, did not. Under cross-examination, Teller stated, 'If it is a question of wisdom and judgment, as demonstrated by actions since 1945, then I would say one would be wiser not to grant clearance.' By 'actions' Teller was no doubt referring to Oppenheimer's overt opposition to the H-bomb, which he interpreted as unpatriotic. The board recommended by two to one that Oppenheimer's security clearance should not be renewed. The AEC endorsed that view but emphasized, somewhat ambiguously perhaps, that though they considered him a security risk, Oppenheimer's personal loyalty was not in question.

Oppenheimer was deeply wounded but refrained from public denunciation of his detractors. He continued as director of the Institute for Advanced Study in Princeton – a post he had taken up in 1947 – and in 1963 the AEC, in a gesture of rehabilitation, awarded him the Enrico Fermi Award for outstanding contributions to atomic energy. Throat cancer prompted the chain-smoking Oppenheimer's resignation from the institute in 1966, and he died in his elegant Princeton home the following year, aged sixty-two. Enrico Fermi himself had returned to

Chicago University but had died in 1954, aged just fifty-three, seven months after testifying in Oppenheimer's defence.

Oppenheimer had always remained on good terms with Leslie Groves. Groves retired from the army in 1948 when General Eisenhower, then chief of staff, who thought Groves insensitive, arrogant and ruthless, made it plain both that Groves would no longer exercise the same influence on nuclear policy as he had during the war and that he would not be appointed the army's next chief of engineers. Instead, the fifty-one-year-old Groves joined the Remington Rand Corporation. At the 1954 AEC hearings Groves did his best to support Oppenheimer, the man he had always considered a genius, asserting that he 'would be astounded' if Oppenheimer had ever committed a disloyal act. However, under cross-examination he admitted that, under a strict interpretation of the AEC's security rules, Oppenheimer should not be given security clearance. Groves died of a heart attack in 1970.

Ernest Lawrence excused himself from testifying at the Oppenheimer hearings on health grounds. Some colleagues claimed this was simply an excuse: he had been intending to testify against Oppenheimer but could not bring himself to go through with it. Others believed the excuse was genuine: he was suffering from severe ulcerative colitis. Lawrence spent his post-war career raising ever larger sums for ever larger cyclotrons. He finally overreached himself with plans for a device that contravened the special theory of relativity and was physically unachievable. His futile strivings to make it work undermined his already frail health. He died in 1958.

Edward Teller was longer lived. He became the inspiration behind President Reagan's 'Star Wars' strategy – the building of a defensive shield in space to ward off missile attack. Some also thought him and his views the inspiration for the film *Dr Strangelove*. He died in 2003, aged ninety-five.

Among the many young American scientists who contributed to the Manhattan Project, the mercurially brilliant, safe-cracking, wise-cracking Richard Feynman stands out. He became a highly influential figure in many areas of post-war science. He was awarded the Nobel Prize for Physics for his work on the theory of quantum electrodynamics and played a decisive role in diagnosing the fatal flaw that destroyed the *Challenger* space shuttle in 1986. Feynman died two years later.

Most of the scientists who found refuge in the United States before the war made their permanent home there. Albert Einstein continued to live quietly in Princeton, walking slowly each morning to the Institute of Advanced Study, then under Oppenheimer's direction. Suggestions that his letter to President Roosevelt was the catalyst for the bomb troubled him. He told his secretary, 'Had I known that the Germans would not succeed in producing an atomic bomb, I would never have lifted a finger. Not a single finger!' Einstein died in 1955.

Hans Bethe returned from Los Alamos to the Cornell Physics Department. In 1967 he won the Nobel Prize for Physics for his work on energy production in stars. Leo Szilard focused his energies on trying to halt the arms race. He urged the sharing of technology as a way of fostering peace and devised methods for checking that nuclear arms control agreements were being honoured. During the Cuban missile crisis in 1962 he fled to Geneva for safety. From there, he typically tried to contact President Khrushchev to urge dialogue with the United

THE BRITISH MISSION

INVITES YOU TO A PARTY IN CELEBRATION OF

THE BIRTH OF THE ATOMIC ERA

FULLER LODGE

SATURDAY, 22ND SEPTEMBER, 1945

DANCING, ENTERTAINMENT,
PRECEDED BY SUPPER AT 8 P. M.

Mr & Mrs C. Critchfield

R.S.V.P. TO MRS. W. F. MOON
ROOM A-211 (EXTENSION 250)

States. Szilard died two years later.

Of the British team that contributed to the bomb, James Chadwick was knighted in 1945 and returned to Liverpool University. Chadwick's role had been mentally and physically stressful. A perceptive colleague observed that he 'had plumbed such depths of moral decision as more fortunate men are never called upon to peer into'. In 1948 he moved back to Cambridge University as Master of Gonville and Caius College. On his recommendation, William Penney took charge of the British atomic weapons programme, effectively his successor. The first British atomic bomb was exploded in 1952. Chadwick died in 1974.

The authors of the catalytic Frisch-Peierls memorandum returned to Britain to become university professors at Cambridge and Oxford respectively. Frisch's much-loved aunt, Lise Meitner, moved to Cambridge in old age to be near him. She died in 1968, just a few weeks after Otto Hahn and shortly before her ninetieth birthday. The inscription on her gravestone in an English country churchyard reads, 'A physicist who never lost her humanity'. In 1994 a new element, 109 in the Periodic Table, was named 'Meitnerium' in her honour.

After the war, Joseph Rotblat tried in vain to learn the fate of his wife in Warsaw and concluded she must have perished in the Holocaust; he never remarried. Rotblat returned to Liverpool University to work once more with James Chadwick before becoming Professor of Physics at St Bartholomew's Hospital in London, studying nuclear medicine and campaigning for nuclear non-proliferation. He worked with Bertrand Russell, Albert Einstein and others to found the Pugwash Conferences on Science and World Affairs, whose aim was to bring scientists of the rival nuclear powers closer together. In 1995, aged eighty-six, Rotblat was awarded the Nobel Peace Prize, was knighted three years later and died in 2005.

Australian Mark Oliphant, as horrified as Rotblat by the bombing of Hiroshima, was also a founder member of the Pugwash movement. Describing himself as 'a belligerent pacifist', for the rest of his long career he refused all work of a military nature. After a further period at Birmingham University he returned to academic life in Australia where, in the 1950s, he spoke up for Robert Oppenheimer and was consequently refused a visa to enter the United States. Oliphant became Governor of South Australia in 1971 and died in 2000.

Niels Bohr went home to Copenhagen, where he resumed control of his institute. He continued to campaign for scientific openness and the peaceful applications of nuclear power, and against the arms race, remaining, until his death in 1962, one of the most respected senior statesmen of the scientific community.

Klaus Fuchs was finally arrested as a Soviet spy in 1950. By then he was a senior scientist at the Harwell atomic research establishment near Oxford. He was unmasked following the FBI's cracking of the Soviet codes. For the first time they were able to decipher messages between the United States and the Soviet Union which they had intercepted during the war. One of these was a report on the Manhattan Project by Fuchs. This, in itself, did not mean Fuchs was a Soviet agent, but detailed correlations between Fuchs' movements and the passage of information revealed the truth. The evidence of Fuchs' spying thoroughly alarmed the United States authorities, who feared that he might have passed H-bomb technology to the Soviets. This evidence of espionage in the heart of Los Alamos fuelled the suspicion that would fall on Oppenheimer, though no connection between Oppenheimer and Fuchs' spying was ever established by the FBI.* Alerted by the FBI, British counter-intelligence coaxed a confession out of Fuchs. And a confession was important: had Fuchs denied the charges at his trial, the British and Americans would have had to produce evidence revealing to the Russians that their codes had been broken.

Colleagues and friends were shocked. Rudolf Peierls learned of Fuchs' arrest from a journalist. It seemed 'quite unbelievable'. He hurried to Brixton prison to ensure that Fuchs had proper legal representation. There, Fuchs confessed to Peierls that he now regretted his actions as he had since 'learned to appreciate [the British] way of life

---

* FBI investigations also led to the unmasking of Fuchs' handler Harry Gold, David Greenglass and the Rosenbergs.

and values'. When Peierls expressed surprise that Fuchs, 'as a sceptical scientist . . . had been willing to accept the Marxist orthodoxy', Fuchs replied, 'You must remember what I went through under the Nazis. Besides, it was always my intention, when I had helped the Russians to take over everything, to get up and tell them what is wrong with their system.' Peierls was 'shaken by the arrogance and naivete of this statement'.

Fuchs was sentenced to fourteen years in prison. His model conduct earned him remission so that he served only two thirds of that time. To his regret, the British government revoked his citizenship and on his release in 1959 Fuchs went to East Germany to become deputy director of a nuclear research laboratory. By a strange irony, his boss, Heinz Barwich, later defected to the West. Fuchs' powers of self-delusion remained undiminished. One visiting Western scientist wrote, 'I have never before known a person who possesses such a marvellous ability to think in abstract terms who is at the same time so helpless when it comes to either observe or evaluate reality.' Fuchs died in 1988.

The interned German scientists were eventually returned to Germany in early 1946, and soon after released back to their scientific work. As he had hoped and anticipated, Werner Heisenberg became an influential figure in West German post-war science. He promoted the peaceful uses of nuclear power, opposed nuclear weapons and helped launch the European centre for nuclear research, CERN. From 1946 until his retirement in 1970 he headed the Max Planck Institute for Physics and Astrophysics.*

---

* The Max Planck Institutes were the post-war successors to the Kaiser Wilhelm Institutes.

Nevertheless, his wartime behaviour, especially his visit to Niels Bohr in Copenhagen, dogged him and he became involved in a heated dispute with Sam Goudsmit over allegations in the latter's book about the nature of Germany's wartime atomic programme. He died in 1976.

Heisenberg's close colleague Carl-Friedrich von Weizsäcker also had a successful post-war career in physics and philosophy. His younger brother Richard – who had defended their father at the Nuremberg war crimes trials, where he was convicted of war crimes and sentenced to prison – became President of Federal Germany from 1984 to 1994.

Otto Hahn, after rebutting unfounded allegations that he had been a Nazi, also helped shape West Germany's science policy, having survived an assassination attempt in 1951 by a frenzied, frustrated inventor. He remained in touch with Lise Meitner but, despite the changed political circumstances, never publicly acknowledged her contribution towards the discovery of fission, or Otto Frisch's. His Nobel Prize acceptance speech made no mention of either, and in his autobiography he gave himself the full credit for the discovery. Hans Bethe thought his attitude 'very nasty'. Meitner remained fond of Hahn and grateful to him for helping save her from the Nazis, but she believed him guilty of 'suppressing the past', recognizing with bleak clarity that 'I am part of that suppressed past'. Hahn died in 1968.

Fritz Strassmann, who had always considered Lise Meitner the intellectual leader of their team, refused to accept Hahn's offer of 10 per cent of his Nobel Prize money, made to him alone. He encouraged Meitner to return to Berlin after the war but said he would understand if she refused. They remained friends until her death.

Max Planck was the only German scientist invited to London in 1946 for the belated celebration of the tercentenary of Newton's birth. He died the following year having never recovered from the execution of his last surviving son for plotting to assassinate Hitler in 1944.

Of the French atomic scientists, Irène Joliot-Curie died in 1956 of leukaemia at the age of fifty-eight. Her husband Frédéric became High Commissioner for Atomic Energy under Charles de Gaulle but was dismissed for his connections with the French Communist Party and for his opposition to the military uses of nuclear science. He died in 1958, also aged fifty-eight, of cirrhosis of the liver induced, a doctor friend believed, by exposure to polonium.

After the war, Paul Tibbets served as a senior officer in the United States Air Force, including a tour of duty with NATO in France. He also worked with Boeing on the development of the B-47 – the first successful jet bomber. He retired from the air force as a brigadier general to continue what he called his 'love affair' with aeroplanes by running an aviation company. He repeatedly stated that he felt no personal guilt that as a member of the armed forces he had planned and carried out the mission assigned to him to the best of his ability. However, in his memoirs he wrote, 'Let it be understood that I feel a sense of shame for the whole human race, which through all history has accepted the shedding of human blood as a means of settling disputes between nations.' He added, 'Let those who honestly desire peace among nations also condemn all forms of international terrorism that are meant, by their perpetrators, to set the stage for war.'

Tibbet's co-pilot, Robert Lewis, helped raise money for medical treatment of the so-called 'Hiroshima maidens' –

young girls disfigured by the atom blast. In 1971 he sold his 'log' of the flight of the *Enola Gay* at auction for $37,000 and used some of the money to buy marble to sculpt images with a religious theme. These include a phallic 'mushroom statue' symbolically leaking blood. Lewis called it 'God's Wind'.

Hiroshima is again a vibrant city with a population more than three times as large as in August 1945, a symbol of the unquenchable human spirit. Citizens bustle to work over the many bridges that link the fingers of land separated by the river delta. Yet Hiroshima remembers. The area beneath the hypocentre of the bomb – the vanished district of Salugakucho – is now the site of the Peace Memorial Park. Memorials such as the bronze 'Statue of Mother and Child in the Storm' recall the lost people of Hiroshima; another honours the Korean forced labourers, brought to Hiroshima against their will, who also perished. The dome of the former Hiroshima Prefectural Industrial Promotion Hall by the T-shaped Aioi Bridge has simply been left, a shattered icon. Every morning at 8.15 a bell rings out by the dome, and, for just a moment, passers-by pause and reflect.

EPILOGUE

QUESTIONS OF 'WHAT IF?' LITTER HISTORY. ANSWERS TO THEM can inherently never be certain and therefore they attract historians like bees to honey. Nevertheless, they provide a useful way of analysing some of the key facets of the fifty-year story beginning with Marie Curie's pioneering work on radium and culminating in the destruction of Hiroshima by Little Boy.

The question what if Heisenberg, Bohr and von Weizsäcker had all died in the avalanche of 1933 which buried Heisenberg leads easily into a discussion of how much a single individual's input advances the course of science. Shortly after his death, Ernest Rutherford's colleagues debated how much difference it would have made if he had not lived. How much delay would there have been in understanding the nucleus? Some answered 'ten years', some answered 'more likely only five'. Even five is probably at the very top end of any realistic scale. Ideas have their time, and if not discovered by one person they will be by another. Robert Hooke claimed that some of his ideas about natural science predated some of Isaac

Newton's work; Charles Darwin and Alfred Wallace worked entirely independently and at the same time on theories of evolution, as did William Swann and Thomas Edison on the lightbulb. Erwin Schrödinger's wave theory was published within months of Heisenberg's quantum mechanics work and was equally illuminating of the movement of atomic particles.

In the early 1930s, Ernest Lawrence at Berkeley, not James Chadwick at the Cavendish, could easily have been first to the neutron, and could also have beaten the Joliot-Curies to 'artificial' radioactivity and Fermi to the production of radioactivity using neutrons. Insights into the atom were shared and built on the work of others, as can be seen by the usually high number of references in scientific papers. Good examples of co-operation include Rutherford's and Bohr's work on atomic structure and Bohr's debates with Heisenberg on uncertainty and complementarity. Perhaps a particularly gifted individual can make two or three years' difference. There are of course exceptions. Einstein's paper on the special theory of relativity, which 'quietly amalgamated space, time and matter into one fundamental unity', has no references and cites no authorities. How long it would have taken for someone else with his genius to emerge can only be guessed.

What if Britain had not cajoled America to become involved in a bomb project by divulging the secret Maud Report and its trigger, the Frisch–Peierls memorandum, and by sending high-level scientific missions such as that of Henry Tizard with his 'black box' of secrets and committed advocates such as Mark Oliphant to the United States? What if Britain had not sent scientists to Los Alamos? Because even a few months' delay could have meant that the war was ended before Little Boy was ready, this question deserves investigation. Even the Anglophobic General Groves wrote privately in 1949, 'The main British contribution was in arousing and maintaining the interest and enthusiasm of President Roosevelt in the project. This was of real value. Among other things it was probably the major factor in our keeping top priority throughout the war.' One American scientist suggested that the British saved a year in making the bomb. Hans Bethe thought the British contribution to his theoretical team's work 'essential'. Robert Oppenheimer agreed. 'I think that the fact that the British were convinced very early by Simon and Peierls probably was the greatest single factor in getting the job done when it was . . . If the British Government had not been committed we might have been very much slower in this country to put the necessary resources into it . . . The British at Los Alamos were very valuable. If they hadn't been there it is hard to know who would have taken their place.'

Other historians have, like Groves, emphasized the Frisch–Peierls memorandum, the Maud Report and Churchill's constant pressure on Roosevelt to work faster as more important than the British work on the ground. What is clear overall is that without the British contribution a bomb would not have been ready until at least very early 1946, after the planned invasion of Japan had gone ahead. It is also true that without Klaus Fuchs' involvement in the Los Alamos project as part of the British team and his transmission of key secrets to the Russians, the first Soviet test would have been delayed a year or two.

A related and intriguing question is what if an experienced, eloquent and charismatic British prime minister had tried to persuade a new American president, in-experienced in foreign affairs, to stay his hand rather than use the A-bomb as soon as it was available? Britain, as junior partner, had to give consent under the Quebec Agreement to the bomb's use; it would have been difficult politically for the United States to go it alone. This did not happen and, given Churchill's determined purpose and view that the Axis was reaping the whirlwind, was virtually inconceivable. Britain and America had come too far together and had too many joint interests to pursue. Churchill gave his consent, and throughout his life stuck to the position that the bomb was necessary to save Allied lives, both British and American, despite having privately suggested to Truman at Potsdam that the Japanese should have been given clarification of the unconditional surrender terms.

Another question is what if Churchill's old ally, President Roosevelt, had lived a little longer? Would history have followed the same course? This question pre-supposes that Roosevelt's health would not have deteriorated further (ill health made him less politically adroit at Yalta than at previous Great Power meetings). He was the architect, along with Churchill, of the un-conditional surrender policy, designed to show that victory was unequivocal and to give the victors the paradoxical ability to impose democratic institutions. He had not been swayed by his meeting with Bohr about the internationalization of knowledge about atomic energy. However, he was a more skilled diplomat than Truman and confident in his ability to manipulate both Churchill and Stalin, being more inclined to cajole Stalin into co-operation than to threaten him. Had Roosevelt lived, Edward Stettinius would have remained Secretary of State and James Byrnes, the most committed proponent of immediate action against Japan, and of the diplomatic power of the bomb against the Soviet Union, would not have been in a position of such influence. Therefore Roosevelt may well have listened to those, including Churchill, Stimson and the joint chiefs of staff, arguing for the issue of clarification of what unconditional surrender might mean for the future of the Japanese dynasty. Roosevelt may have allowed more time than the inexperienced Truman to see the effect of Japanese diplomatic manoeuvres through the Soviet Union.

What if Roosevelt and Churchill had accepted the proposals from Bohr, Szilard and others to internationalize the project? Would an arms race with Russia still have resulted? The answer is, probably, yes. Bohr's idealistic concept was essentially a free exchange of information internationally. All nations would pool scientific knowledge rather than keep it secret, and an international body consisting mainly of scientists would oversee its exploitation. These ideas harked back to the free flow of information about physics in the fifty years before the Second World War, a period Bohr regarded as a golden age. However, not only times but nuclear physics had changed. Nuclear physics was by then perceived as having not only massive military potential but real commercial value for power generation. Both these factors conferred great diplomatic, economic and political power. For Stalin, possession of nuclear capability had immense importance, both symbolically and practically. Generation of electricity from nuclear power had the potential to achieve his long-stated aim to 'catch up and overtake' the West in terms of industrialization. Nuclear weapons would give him the ability to rule securely over his increasing empire in Eastern Europe while allowing him to appear as, and to act as, the equal or better of the West elsewhere. Western lack of trust in a totalitarian regime made a race inevitable.

However, one of Bohr's other pleas – for politicians not to view the bomb as 'just another military weapon' – seems to have been heeded. No nuclear weapon has been used since 1945, but this is probably due to the immense destructive power demonstrated by these bombings rather than to Bohr's words or to the misgivings expressed by himself and other scientists after it had been dropped. The fact that Hiroshima was destroyed by a single weapon dropped from a single plane and that survivors could appear healthy but then succumb months or years later to radiation effects set Hiroshima and Nagasaki apart. Conventional bombings, such as the attacks on Dresden and Tokyo, although costly in human life and property were inflicted by greater numbers of planes and their physical if not mental effects, however ghastly, were fully evident within a day or two.* The silent, unseen and deadly effects of radiation, which may not appear for decades and can be passed to future generations, attract a unique revulsion.

Leo Szilard believed that General Groves had delayed the Manhattan Project significantly because of his obsessive desire for compartmentalization. Compartment-alization to protect security may have imposed small delays, but, to compensate, Groves had single-minded drive, great project management skills and the ability to

---

* About three thousand B-29s would have been needed to drop conventional bombs carrying the amount of TNT equivalent to the bomb *Enola Gay* dropped on Hiroshima.

focus on essentials. Edward Teller recollected that 'between 1943 and 1945 General Groves could have won almost any *un*popularity contest in which the scientific community in Los Alamos voted'. Nevertheless, most scientists who worked with him thought that without him the project would have been severely set back rather than advanced. Both Teller and Bethe agreed with James Chadwick's assessment that 'without Groves the scientists would never have finished anything'.

However, if we accept the Szilard view, we are left with the intriguing question, what if the bomb had been ready in, say, February 1945? Would Roosevelt and Churchill have used it against Germany? The answer is almost certainly yes. General Groves recalled that 'Mr. Roosevelt told me to be ready to do it'. Arguments about saving Allied lives would have been stronger. There would have been considerable political advantage in forestalling Soviet occupation of parts of Eastern Europe, in addition to demonstrating to Stalin the West's military power. The Allies' agreed priority was to defeat Germany. They showed no compunction in bombing an untouched German city, Dresden, in February 1945. Knowledge of the effects of radioactive fall-out did not deter the Trinity test from being carried out on the mainland of the United States and so would have been unlikely to stop a bomb being dropped on Europe. Knowledge of Germany's treat-ment of Jews and other minorities was already leaking out, following the Russian liberation of Auschwitz at the end of January 1945. Many of those involved in the bomb project had seen it as primarily directed at forestalling a German atomic attack, and there was at that time still a fear of final German vengeance weapons, for example improved V-2s.

The only argument that the decision to drop a bomb on Germany may have been different assumes that racism was a factor in deciding to bomb Japan. In Allied countries there was certainly racism. It was generally acceptable to attribute characteristics to a whole race and to make judgements about individuals against these stereotypes, as in the case of the academic reference given to Oppenheimer which commented on his Jewish background. There was segregation in the southern states of America. Undeniably there was racist sentiment against the Japanese in the United States. Japanese-Americans were interned en masse, German-Americans were not. Some 13 per cent of the American public surveyed in a Gallup poll in December 1944 favoured the extermination of all Japanese. A US Marine publication described the Japanese as lice and said that the Marine Corps had been 'assigned the gigantic task of [their] extermination . . . but before a complete cure may be effected the origin of the plague, the breeding-grounds around the Tokyo area, must be completely annihilated'. *Life* magazine carried a photo of the girlfriend of an American sailor gazing wistfully at a gift he had sent her from the Pacific – a Japanese skull signed on the top by him and his friends. British and American journals regularly portrayed the Japanese as monkeys. American admiral William Halsey spoke of the Japanese as 'yellow monkeys' and before one operation publicly proclaimed he was 'rarin' to go get some more monkey meat'.

President Roosevelt himself was not without prejudice, as a note by a British diplomat about a 1942 conversation with him reveals: 'It seemed to him [Roosevelt] that if we got the Japanese driven back within their islands, racial crossings might have interesting effects. For instance

Dutch-Javanese crossings were good ... Japanese-European thoroughly bad, Chinese-European not at all bad.' The diplomat summed up, 'as far as I could make it out, the line of the President's thought is that an Indo-Asian or Eurasian or (better) Eurindasian race could be developed which could be good and produce a good civilisation and Far East order to the exclusion of the Japanese, languishing in Coventry within their original islands'. Churchill, too, had colonialist prejudices against Asian races and battled to retain Britain's right to rule over Indians and others. However, there is no evidence that either adopted different military policies under the influence of such views, nor is there any evidence of any racist element in the decision to drop the bomb on Hiroshima.

Another intriguing question is what if German chemist Ida Noddack's views on fission in 1934 had been taken seriously? As well as her previous failure to substantiate her claimed discovery of masurium, the reasons they were not have something to do with anti-feminism. It is easy to forget that when Marie Curie was making her discoveries the only country in which women had the right to vote was Ernest Rutherford's homeland of New Zealand, gained in 1893. (Perhaps this is one of the reasons why he had a relatively enlightened attitude to women students and supported Marie Curie.) Two years after Marie Curie received the Nobel Prize for Physics, former American President Grover Cleveland could still write that 'sensible and responsible women do not want to vote. The relative positions to be assumed by man and woman in the working out of our civilisation were assigned long ago by a higher intelligence than ours.' The early years of Lise Meitner's career were held back by her gender rather than by racism.* The frantic attacks on Marie Curie after the Affaire Langevin contained a strong line of anti-feminism as well as of xenophobia. It is, perhaps, symptomatic of the predominance of males in the Manhattan Project that, when a series of code words relating to birth was agreed to report on the Trinity test, the birth of a boy was to stand for success, that of a girl for the failure of the bomb to detonate.

The sidelining of Ida Noddack also had something to do with the elitism of physicists. She was a chemist, and there was the perception among some physicists that physics was at the top of a hierarchy of science and that those lower down, such as chemists, could not be relied upon for original thought. Rutherford once said that 'All science is either physics or stamp collecting.' When Austrian physicist Wolfgang Pauli's wife left him for a chemist, he told a friend, 'Had she taken a bull-fighter I would have understood . . . but a *chemist* . . .' Both Otto Hahn and Bertrand Goldschmidt indicated that they too faced such prejudices, and as late as the 1970s entrants into the nuclear industry could be told, semi-humorously, of a hierarchy with theoretical physicists at the top and engineers at the bottom.

If Noddack's ideas had been followed up in 1934, fission might have been discovered a year later – four years earlier than it actually was. This does not mean that the nuclear bomb would have been available in 1941, though. It took wartime pressures for the Manhattan

---

* Discrimination may have been subliminal. In 1922, Lise Meitner's inaugural university lecture about 'The Significance of Radioactivity for Cosmic Processes' was reported in the academic press as a talk about the significance 'of Radioactivity for Cosmetic Processes'.

Project to receive the massive funding it required. However, the uses of fission would have been more widely debated and information more widely pooled before the outbreak of war. For example, the use of graphite as an alternative moderator to heavy water might have been publicized and become known to the Germans. The German programme would have gone into wartime isolation more advanced and posing much more immediate moral dilemmas for the scientists involved than was, in fact, the case. There might have been much more French information and many more facilities for the Germans to capture. On the Allied side, the British might have begun to work harder on nuclear issues earlier as, in the late 1930s, arguments for rearmament made by Churchill began to be heeded. But, on balance, it was better for the world that the scientific community, including her fellow women such as Lise Meitner, dismissed Noddack's work as, in the words of Noddack's fellow chemist Otto Hahn, 'really absurd'.

The 'what if?' question that has most preoccupied historians is what if the bomb had not been dropped? On the very big assumption that no new diplomatic initiative would have been launched, Russia would still have entered the war against Japan. Her plans to do so to secure a share of the spoils were highly advanced and Hiroshima brought them forward by only a week. As the Russians' initial progress in fighting the Japanese in Manchuria demonstrated, they would have advanced swiftly into China and occupied much of the country, perhaps hastening by a year or two the fall of Chiang Kai-shek and the communist take-over.

When Japan did surrender, the Russians would have

played a greater part in determining the peace and might have asked for a role in the occupation of Japan, perhaps even seeking an occupation zone of their own. This would not have been to the Allies' liking. Truman wrote, 'the experience at Potsdam now made me determined that I would not allow the Russians any part in the control of Japan'; Secretary of State Byrnes wrote, 'in the days immediately preceding the dropping of that bomb his [Truman's] views were the same as mine – we wanted to get through with the Japanese phase of the war before the Russians came in'. Washington insider Walter Brown recorded that Byrnes believed 'after the atomic bomb Japan will surrender and Russia will not get in so much on the kill'. Clearly, Truman and Byrnes preferred to end the war quickly without the added difficulties that full Soviet involvement would entail.

The Japanese would undoubtedly have continued to resist for some time. Their defeat would probably have required an invasion of their home islands with the heavy loss of Allied lives so much feared by Truman and Churchill. This was despite the fact that, as Allied military leaders such as Marshall, LeMay, Arnold, Eisenhower and Alanbrooke agreed, they were militarily already thoroughly beaten. Their cities were defenceless against US air attack and their supply routes to and from their homeland and between their armies were severed. Admiral Leahy wrote that by the beginning of September 1944 'Japan was almost defeated through a practically complete sea and air blockade'. There would have been further deaths of Allied airmen, of sailors in kamikaze attacks and of soldiers fighting in Burma and elsewhere. More Japanese cities would have been destroyed and many lives lost. General LeMay told his superior, General

Arnold, in June 1945 that by September or October of that year his pilots would have run out of industrial targets to bomb.

The argument made by Major Rittner to Otto Hahn that he did not care about 100,000 or 150,000 Japanese if a couple of Allied lives could be saved cannot be defended morally, even if one can understand how, in wartime, it might have been made. However, when the number of deaths reaches closer to parity, its force increases.

Some have argued that by dropping the bomb more Japanese lives were saved than lost; they claim that Hiroshima gave the peace faction in the Japanese government the grounds for pressing for a surrender. In the words of General Marshall, the bomb provided an opportunity 'to shock them into action' and 'out of their determination to sacrifice great numbers of their people in futile further defence'. Keino Kido, the Emperor's confidant, said in an interview in 1966, 'there was also a plus aspect to the atomic bombs and the Soviet entry into the war. I assumed at the time that if there had been no atomic bombs and the Soviet Union hadn't joined in, we might not have succeeded in [making peace].' A senior Japanese officer saw the same two events as 'in a sense heaven-sent blessings. This way we didn't have to say we quit the war because of domestic circumstances.' Even Taro Takemi, Japan's leading practitioner of nuclear medicine, who accompanied Professor Nishina to Hiroshima on 8 August to investigate the explosion, thought that the bomb might have had a beneficial effect. He later wrote, 'When one considers the possibility that the Japanese military would have sacrificed the entire nation if it were not for the atomic bomb attack, then this bomb might be described as having saved Japan.'

All the above discussion assumes that no change was made in the Allies' diplomatic position, and in particular that they would have continued to insist on unconditional surrender. The reasoning behind their unconditional surrender demand included avoiding any future claims (akin to those made by some Germans after the First World War) that Japan had not been defeated and thus allowing militarism to rise once more. Such fears were, however, a reason for insisting on occupation to make defeat unequivocal. They were not arguments against modifying the term 'unconditional surrender' to make clear that the Japanese ruling house could be preserved in some form – as, in fact, happened.

The importance of the monarchy to the Japanese position was appreciated in Washington. Under Secretary of State Joseph Grew, US Ambassador to Tokyo until Pearl Harbor, advised Truman on 28 May, 'the greatest obstacle to unconditional surrender by the Japanese is their belief that this would entail the destruction or permanent removal of the Emperor and the institution of the Throne'. His opinion was backed up by several other government members. In a memo to Truman on 2 July about the drafting of a statement on surrender terms for the Japanese, Stimson advised, 'I personally think . . . we should add that we do not exclude a constitutional monarchy under the present dynasty. It would substantially add to the chances of acceptance.' Winston Churchill also suggested to Truman that he might consider whether unconditional surrender 'might not be expressed in some other way, so that we got all the essentials for future peace and security, and yet left the Japanese some show of saving their military honour and some assurance of their national existence'. The joint chiefs of UK/US staff were also sympathetic to such a clarification, suggesting the inclusion in the Potsdam Declaration of the following addition: 'Subject to suitable guarantees against further acts of aggression the Japanese people will be free to choose their own form of government' – which, by implication, would include the continuation of the monarchy. Stimson had a final meeting on 24 July with Truman about the issue of the Potsdam Declaration and recorded the upshot in his diary as follows: 'I spoke of the importance which I attributed to the reassurance of the Japanese on the continuance of their dynasty and I had felt that the insertion of that in the formal warning was important and might be just the thing that would make or mar their acceptance . . . I heard from Byrnes that they [Truman and Byrnes] preferred not to put it in.'

Because so many of the key Japanese documents – including the diary Emperor Hirohito is said to have kept since the age of eleven, and his family correspondence – are still kept secret by the Imperial Japanese Household Agency, the issue of whether the Japanese liberal faction would have felt strong enough to promote surrender before the atom bomb was dropped remains clouded. They were fearful of a military backlash against any premature initiative leading to their own murder and the virtual imprisonment of the Emperor. At the same time they were concerned not to allow domestic conditions to deteriorate to the extent that there was a popular revolt against the throne.

Whether the Japanese accepted it or not, there seems no reason why, based on the knowledge they had at the time, the Allies should not have included in the Potsdam

Declaration a concession on the ruling dynasty, if not the continuance of Hirohito's own rule. If the Japanese had rejected it, the case for the deployment of the bomb would have been strengthened, not weakened, at no cost in human life.

The available sources contain no substantive information as to why Byrnes and Truman chose not to include such a concession, although it seems clear that the strongest opponent of doing so was Byrnes. It may be that the two men feared public criticism if they did so. An unpublished Gallup poll in June 1945 showed that 77 per cent of the US public wanted the Emperor severely punished. However, after the bombs were dropped, Truman and Byrnes were prepared to face the outcry at allowing the Emperor to remain.

Perhaps conscious of the power of the new weapon, the two men disregarded other options to end the war. The Manhattan Project had a momentum of its own. General Groves described Truman as 'like a little boy on a toboggan'. Some have suggested that the subsequent use of the bomb in combat validated in the minds of officials the amount of government resources spent on its development without specific Congressional approval. They cite Stimson, who in 1947 wrote in *Harper's* magazine, 'At no time, from 1941 to 1945, did I ever hear it suggested by the President . . . that atomic energy should not be used in this war . . . on no other ground could the war-time expenditure of so much time and money be justified.' Truman and Byrnes were certainly also both conscious of the diplomatic advantage the deployment of the bomb would give them in their difficult relationship with the Soviet Union, and that its deployment might prevent Russian involvement in the occupation of Japan and the dictation of peace terms. This suggests that both momentum and strategic diplomatic benefits played a part.

Neither reason is, however, sufficient to explain fully why Truman and Byrnes took no action to prevent a second bomb being dropped so quickly on Nagasaki. The Manhattan team and the US Air Force personnel on Tinian brought the drop forward by two days for operational reasons. Fat Man fell on Nagasaki on 9 August, before the Japanese had time to respond to the Soviet invasion launched at about the same time the B-29 *Bock's Car* took off from Tinian. The American authorities in Washington, but presumably not the team on Tinian, had been aware of the Russian declaration of war, issued at five p.m. Moscow time on 8 August, and of the Soviet intention to invade Manchuria.

Another major 'what if?' relates to the conduct of the German atomic bomb programme. What if the key German scientists had been more committed to their work on an atomic bomb for Hitler? The question contains a major assumption that these scientists *could* have been more committed, for there is considerable evidence for varying degrees of lack of commitment. Many – von Weizsäcker, Heisenberg, Hahn and Strassmann, for example – were not members of the Nazi Party when it was politic to be so. Hahn helped Lise Meitner to flee and met and corresponded with her thereafter. Strassmann, with Hahn's knowledge, risked his life and that of his family to hide Jewish pianist Andrea Wolffenstein. Houtermans got word of the German work on the atomic bomb out to America through Fritz Reiche. Even after being warned off by the Gestapo, Heisenberg made tentative attempts to help some of his colleagues.

Edward Teller saw a continuing 'conflict between Heisenberg's patriotism and Heisenberg's thorough unwillingness to help the Nazis'. He also recalled, 'I could not imagine that he would support the Nazis willingly, much less do so enthusiastically.' Teller's remarks are typically clear-cut, the reality perhaps less so. Heisenberg and von Weizsäcker made compromises with the authorities when, as von Weizsäcker stated in a private conversation with Heisenberg at Farm Hall, they were aware that 'the right [correct] position [for us] would really have been in a concentration camp [as protesters], and there are people who chose that'. Heisenberg remained involved because, like many others, he was a German patriot who feared the consequences of defeat, particularly by the Russians, at least as much as those of victory. Heisenberg was also naive, tactless and entirely unable to appreciate the perspectives of others, and hence the impact of his words and actions upon them. In the words of Sam Goudsmit to Rudolf Peierls, long after the war, the problem was that 'this great physicist, our idol, wasn't any better [morally] than we are'.

Levels of commitment fluctuated with the progress of the war. When Heisenberg visited Niels Bohr in Copenhagen in autumn 1941, when the Germans seemed to have nearly won the war, one of Heisenberg's messages was that Bohr must accept the reality of a German victory. At around that time too, Houtermans, despite previously leaking information to the Allies at great personal risk, told his superiors and others of his work on plutonium instead of attempting to conceal it. Perhaps scientific conceit played a part, but there was also an element of accommodating to the likely outcome of the war. Such pragmatism seems to have influenced most of the Germans who remained, with the notable exception of Max von Laue, who did not work on the German bomb project and gave moral support to Jewish former colleagues. According to Houtermans, when he told von Laue that he was worried about the possibility of Germany making a bomb during the war, von Laue's ambiguous reply was, 'An invention is not made which one doesn't want to make.'

The two writers of the memorandum which spurred Britain and America to action, Otto Frisch and Rudolf Peierls, both commented in 1965 about the attitude of the German scientists they had known well before the war. Peierls suggested, 'in the West, we all felt that this was our war and that for Hitler to acquire complete domination of the world would be a disaster. The German scientists, I think, were not so identified with their own regime . . . it may be that as a result they were less active in thinking about these possibilities.' Peierls, however, added a sentence suggesting plausibly that scientific curiosity might have kept them involved despite themselves: 'the possibility was a very exciting thing for any physicist, whatever you decide to do with it'. Frisch thought similarly: 'In the first place I think it is true to say that all the best scientists did not wish to have such a frightful weapon in Hitler's hands. I think that many of them hoped that once the war was over and won the Nazi regime could be disposed of, or softened or civilized . . . but they didn't want them to get the tremendous power that would go with possession of such a weapon. Second, there were very bad relations on the whole between scientists and government and military in Germany. The scientists regarded the government simply as a source of revenue to be otherwise kept at arm's length.'

If we accept that most German scientists had misgivings about producing a bomb which may have unconsciously inhibited their work, does this mean that the full implication of their 'version' of events, developed at Farm Hall and later, should be accepted – that is, that if they had been fully committed they could have made a bomb on the same time scale as the Allies, and even that some of the scientists tried to sabotage the war effort? There is no evidence to substantiate any sabotage. There is, however, considerable evidence that the Germans did not understand the physics necessary to make a bomb, such as that revealed by Heisenberg's failure to describe the physics at Farm Hall, and in particular to identify the correct critical mass, and by Bothe's disregard of the potential of graphite as a moderator. As one of them said at Farm Hall, they also looked down on essential isotope separation work perhaps as more chemical engineering than physics.

The German project was, as admitted in the recorded conversations at Farm Hall, riven by personality differences and rivalries. Competition between the army, the Reich Research Council and the Kaiser Wilhelm Society fragmented rather than focused the German programme, and scientists competed against one another for scarce supplies of heavy water. Heisenberg was undoubtedly a great physicist, but he was not a great leader or project manager in the way that Groves was. Until the war he never oversaw any major project. He was not an enthusiastic experimentalist, as evinced by his near failure in his doctoral exams when questioned about how experiments and equipment worked. He was also not renowned for his skill at undertaking mathematical calculations to back up his brilliant insights. According to Rudolf Peierls, 'though a brilliant theoretician he was always very casual about numbers'.

The German project never employed more than about a hundred scientists. Even multiplying this figure ten or fifteen times to allow for technicians and other workers produces a total work force of only around 1 per cent of the Manhattan Project. Even had German scientists shown more commitment and pressed for greater resources from Speer and others, it is by no means clear they would have been forthcoming. The V-1 and V-2 rocket programmes and the jet aircraft programme had both begun before the war and had clear priority. In 1941, when the scientists seemed to have had greatest commitment, the very reason for that increase in commitment – the perceived proximity of a German victory – would have told against them because they could not claim a bomb would have been finished on a timescale to end the war.

One reason why Britain moved her work to the United States was a lack of resources. Another was fear that German bombing might destroy any facilities constructed. Similarly, the Germans at Farm Hall and elsewhere recognized that had they increased their resources, the British and Americans would have quickly become aware of this and destroyed their plants, just as they attacked the Vemork heavy water factory and the laboratories in Berlin. They were right. Not only did the Allies have a committed agent in Paul Rosbaud, they also had excellent photo-reconnaissance aircraft which would have been bound to identify any large-scale plant. They were also prepared to contemplate abduction or assassination of key personnel such as Heisenberg.

Therefore, leaving any moral queasiness on the

scientists' part aside, it is highly unlikely that Germany would have been able to make an atom bomb before she was defeated. She may well have been able to make a 'dirty bomb' which, attached to a V-1 or V-2 rocket, could have distributed radioactive material over London, as feared by several of the scientists at Los Alamos, or along the Normandy invasion beaches, as General Groves warned General Eisenhower. However, for whatever reason, in practice no-one in Germany seems to have considered this option.

A final question addresses the human element in the story. What if Niels Bohr had not mumbled? All an individual's attributes are important to whether they succeed in their chosen task. Leo Szilard said after his unsuccessful meeting with Secretary of State James Byrnes that it might have been better for the world if he, Szilard, had been Secretary of State and Byrnes a physicist. Hans Bethe said that the Manhattan Project 'changed everything; it took scientists into politics'. The requirement on scientists to understand the wider world applies in reverse, of course, to politicians and lay people with regard to science. Rudolf Peierls said in 1986, 'I'm afraid forty years ago we overestimated the capacity of those in power to understand the implications of what we had created.'

In fact, interaction across these boundaries had begun before Los Alamos at the time when Leo Szilard was first voicing his concerns about the nuclear bomb. Such interchanges required a whole new set of skills from the scientists. To be a good scientist had previously required people to be intelligent and knowledgeable in their own sphere. However, it did not require them to be a good communicator or good with people, to be politically aware, or to display good judgement of any kind in any area other than science, and even then only on a specialized subject. Atomic energy was perhaps the first subject among several which now exist, such as genetics and nanotechnology, to require scientists first to communicate their findings effectively and then to decide whether to express a view on their use, to present options or simply to say 'here they are, this is what they mean, it is for others to decide to what use they should be put'.

Robert Oppenheimer's view was simple. A scientist could not stop such a thing as the bomb: 'If you are a scientist you believe that it is good to find out how the world works; that it is good to find out what the realities are; that it is good to turn over to mankind at large the greatest possible power to control the world and to deal with it according to its lights and values.' However, individuals' personalities determine what attitude they take to such challenges. Their abilities outside science determine the success of their arguments. Niels Bohr was a great scientist and a warm, compassionate human being who found it easy to win the affection and respect, both scientific and general, of his colleagues. Yet when he tried to move out of his own sphere into the political to make the case for an international approach to the bomb he failed because he could not communicate properly. Within physics, the difficulty of understanding him in print and in speech was well known and affectionately tolerated. Robert Oppenheimer once said it was 'easy for even wise men not to know what Bohr was talking about'. A greater ability to communicate (and perhaps to listen) might have given Bohr more political influence. It would certainly have provided greater clarity about some of the major events such as the Copenhagen meeting in which he was involved. Similarly, if Szilard had been less eccentric or egocentric he might have won more allies and convinced more people. If Heisenberg had been more empathetic and less conceited he might have seen things differently.

Marie Curie insisted that 'in science we must be interested in things, not in persons'. William Penney, who retained no personal papers, went further: 'People are not important – history is too much about people.' Both were wrong. History, even the history of science, is inherently about people – how they thought, what they did with their thoughts and how they interacted with the individuals immediately around them, and then with society and the greater world order. All involved in this story, regardless of race, sex, creed, age or intellectual ability, had the potential to act individually. In thinking about history, but above all about the future, we should not depersonalize situations but remember our individual responsibility for them and the consequences for others. The plea of one Hiroshima survivor stands out: 'When I was younger they used to call us atomic bomb maidens – more recently they called us *hibakusha* [atom bomb survivors] . . . I don't like this special view of us. I'd like to stand up as an individual.'

GLOSSARY

Alpha particles (alpha rays) – one of the three types of radiation (see also **beta** and **gamma**) discovered around 1900. Alpha particles are helium nuclei made up of two protons and two neutrons. They are slow and heavy, compared to other forms of radiation, and cannot penetrate paper.

Artificial radioactivity – radioactivity induced by bombarding and thus destabilizing nuclei.

Atom – the basic unit of matter consisting of a single nucleus surrounded by orbiting electrons.

Atomic number – the number of protons in the nucleus of an atom. The number of protons is equal to the number of electrons orbiting the nucleus, and since an atom's chemical properties are determined by the number of such electrons, the atomic number establishes an atom's chemical identity.

Atomic weight – the mass of an atom effectively equivalent to the number of its protons and neutrons. Hydrogen's atomic weight is thus one.

Beta particles (beta rays) – radiation in the form of electrons travelling at high speed. They are lighter and faster than alpha particles and more penetrating, but can be stopped by a sheet of metal.

Chain reaction – a self-sustaining nuclear reaction triggered when a neutron induces a nucleus to fission thereby causing it to release energy and further neutrons which, in turn, cause further fissions.

Complementarity – the theory developed by Niels Bohr that seemingly conflicting or ambiguous findings may need to be placed side by side to create a full understanding of a phenomenon.

Critical mass – the amount of fissile material needed to sustain a chain reaction.

Deuterium – the isotope of hydrogen otherwise known as heavy hydrogen (see **heavy water**).

Electron – an elementary particle carrying one unit of negative electrical charge.

Enriched uranium – uranium with a higher content of the isotope U-235 than natural uranium.

Fissile material – any element containing an isotope with nuclei capable of undergoing fission. Uranium-235 and plutonium-239 are important fissile materials.

Fission – the splitting of the nucleus of a heavy atom (i.e. near the top end of the Periodic Table) accompanied by the release of energy and of atomic particles. It can occur spontaneously or be induced by external stimuli.

Fusion – the combination of two light nuclei to form a single heavier nucleus accompanied by a release of energy.

Gamma rays – the name originally given to the most penetrating of the three types of radiation discovered to be emitted from radioactive substances as they decay. Gamma rays consist of electromagnetic radiation, like light.

Gaseous diffusion – a method for enriching uranium (i.e. increasing the content of U-235) by pumping uranium hexafluoride gas through permeable membranes.

Graphite – an elemental form of carbon used as a moderator.

Half-life – the time taken for a radioactive material, as it decays, to reach half its previous radioactivity, i.e. in two half-lives it would be at a quarter of its original level. Half-lives range from fractions of a second to billions of years.

Heavy water – water containing significantly more than the natural proportion (1 in 6,500) of heavy hydrogen (deuterium) atoms to ordinary hydrogen atoms. Hydrogen atoms have one proton. Deuterium atoms contain one proton and one neutron. Heavy water slows neutrons down more effectively and absorbs them less than ordinary water and is therefore suitable for use as a moderator.

Ions and ionization – basically ions are electrically charged atoms and molecules. Ionization – the production of ions – which can occur in many ways, is associated with radioactivity because of the strong electrical disturbances caused when the products of radioactive decay pass through their surroundings.

Isotopes – different forms of the same element. Isotopes of the same element have the same number of protons in their nucleus but the number of neutrons varies.

Liquid drop model – the model of the atom developed by Niels Bohr who visualized it as resembling a droplet of liquid with nuclear forces playing the part of surface tension.

Moderator – a substance, e.g. heavy water or graphite, used in thermal reactors to slow neutrons down and thereby increase their chances of causing fission.

Natural uranium – uranium whose isotopic composition, as it occurs in nature, has not been altered.

Neutron – an uncharged, i.e. electrically neutral, particle found in the nucleus of every atom heavier than hydrogen. It has almost the same mass as a proton but is very slightly heavier.

Nuclear energy – the energy released by a nuclear reaction such as fission.

Nucleus – the positively charged central core of the atom. It carries over 99.9 per cent of the atom's mass but occupies only a tiny part of its volume. All nuclei consist of protons and neutrons except for the nucleus of hydrogen, which contains only one proton.

Periodic Table – the table classifying elements according to their **atomic number**. The number derives from the number of protons in their nucleus. Thus, hydrogen with its single proton is at number one in the table.

Plutonium – an element with atomic number 94 formed in nuclear fission reactors.

Positron – the antiparticle of the electron. It has the same mass but carries a positive charge.

Proton – a positively charged particle found in the nucleus of every atom.

Quantum mechanics – a mathematical system developed by Werner Heisenberg and others based on matrix algebra and used to describe the properties of matter at the atomic level.

Quantum theory – the theory first postulated by Max Planck in 1900 that energy is released in discrete bursts – 'quanta' – and not continuously. It was subsequently applied to other phenomena.

Radiation – emitted energy and particles, e.g. the energy and particles released as nuclei disintegrate or decay.

Radioactivity – the term used to describe the disintegration or decay of nuclei usually accompanied by the emission of particles and energy.

Transuranic element – an element of higher atomic number and larger mass than the heaviest of the naturally occurring elements, uranium – for example plutonium.

Uncertainty principle – the principle defined by Werner Heisenberg in 1927 that one cannot measure precisely and simultaneously atomic properties such as momentum and position.

Uranium – the heaviest naturally occurring element. It has the atomic number 92.

Uranium 235 – an isotope of uranium, the atomic nucleus of which contains 92 protons and 143 neutrons. Natural uranium contains approximately 0.7 per cent by weight of U-235, which is capable of fission with thermal neutrons.

Uranium 238 – an isotope of uranium, the atomic nucleus of which contains 92 protons and 146 neutrons. This isotope comprises approximately 99.3 per cent by weight of natural uranium. It is not capable of fission with thermal neutrons but can absorb them to form plutonium-239, a fissile isotope of plutonium.

Uranium hexafluoride – a gaseous compound of uranium and fluorine used in the gaseous diffusion enrichment process.

X-rays – the name given to the form of penetrating radiation discovered by Wilhelm Röntgen in 1896. They are, in fact, electromagnetic waves, similar to light but much shorter in wavelength.

NOTES AND SOURCES

To simplify the notes I have used the following abbreviations and designations to identify some of the main sources:

AIP (American Institute of Physics, College Park, Maryland, US)

BBC (interview and programme transcripts in the production files of the BBC Written Archives Centre, Caversham, Berkshire, UK)

CCC (Churchill College, Cambridge, UK; among the papers consulted, JC denotes James Chadwick, LM denotes Lise Meitner and NF denotes Norman Feather)

CUL/R (Cambridge University Library, Cambridge, UK, the Ernest Rutherford Papers)

CUL/S (Cambridge University Library, Cambridge, UK, Henry Stimson's diary on microfilm; the original is in Yale University)

LIV (Liverpool University Physics Department Archive)

LOC (Library of Congress, Washington DC, US)

NARA (US National Archives and Records Administration, Maryland, US)

PRO (UK National Archive, Kew, London, UK; individual file numbers are given in each case)

UCLA/BL (University of California, Bancroft Library)

Prologue

13: 'an airplane . . . blue sky . . . A parachute . . . falling . . . indescribable light': account of Futaba Kitayama, published in *Bombing Eye-Witness Accounts* and available on the Hiroshima Peace Memorial Museum website.

13: 'a giant purple . . . terribly alive . . . Fellows . . . history': P. W. Tibbets, *Mission Hiroshima*, p. 227.

13: 'When I . . . peeled off . . . Suddenly . . . inaction': account of Futaba Kitayama, in op. cit.

14: 'dried . . . to eat': account of Dr Hiroshi Sawachika from www.inicom.com/hibakusha/hiroshi.html.

14: 'cried . . . after all': P. Fussell, *Kansas City Star and Times*, 30 August 1981.

14: 'This is . . . history': Truman, *Year of Decisions*, p. 421.

14: 'This revelation . . . comprehension': *The Times*, 7 August 1945 (PRO/PREM/8/109 contains the text and earlier drafts).

14: 'torturing . . . moment': P. Monck, quoted in Boyer, *By the Bomb's Early Light*, p. 16.

16: 'witches' sabbaths': Pflaum, *Grand Obsession*, p. 160.

17: 'wholly new . . . physics . . . filled . . . away': P. Jordan, quoted in Jungk, *Brighter Than a Thousand Suns*, p. 9.

17: 'It was . . . creation': these quotes are from Oppenheimer's 1953 BBC Reith Lectures reproduced in his book *Science and the Common Understanding*, p. 37.

17: 'moonshine': Rutherford, quoted in *The Times*, 12 September 1933.

19: 'physicists . . . sin': *Los Alamos Science*, Winter/Spring, 1983, p. 25.

19: 'not completely . . . guilt': BBC interview with Oppenheimer for *The Building of the Bomb*.

19: 'killed . . . subject': M. Oliphant, quoted in Snow, *Variety of Men*, p. 10.

19: 'maid's work' (fn): Lanouette, *Genius in the Shadows,* p. 274.

22: 'bombardment . . . cities': *The Times,* 2 September 1939.

23: 'may . . . country': the Frisch–Peierls memorandum (see chapter nine notes for detailed reference).

Chapter One – 'Brilliant in the Darkness'

24: 'brilliant . . . darkness': *Lancet*, 21 November 1907.

26: 'one of . . . youth': M. Curie, *Pierre Curie*, p. 81.

27: 'during these years . . . future work . . . the habit . . . work': ibid. p. 82.

27: 'icy . . . criticism': letter of 18 March 1888, in E. Curie, *Madame Curie*, p. 80.

28: 'very cold . . . lacked coal . . . to . . . bedcovers': M. Curie, op. cit., p. 84.

28: 'It was . . . all liberty': ibid. p. 85.

29: 'the tender . . . hours': Pflaum, *Grand Obsession*, p. 40.

29: 'a kiss . . . need': Quinn, *Marie Curie*, p. 114.

30: 'women . . . rare . . . when . . . struggle': Pierre Curie's entry in his diary as a young man of twenty-two, M. Curie, op. cit., p. 36.

30: 'his simplicity . . . confidence . . . a surprising kinship': ibid. p. 34.

30: 'It . . . legitimate': ibid. p. 35.

31: 'little queen': letter from M. Curie to M. Sklodovski of 10 November 1897, in E. Curie, op. cit., p. 158.

33: 'X-ray proof underwear': Larsen, *Cavendish Laboratory*, p. 31.

35: 'I hear . . . rays': Nye, *Before Big Science*, p. 148.

35: 'We do not . . . *in toto*!': *Punch*, 25 January 1896.

36: 'the unconscious . . . genius . . . shining out . . . background': W. Crookes, *Proceedings of the Royal Society, Obituary Notices of Fellows Deceased*, A. 83, xxii, 1909–10.

37: 'much . . . study of it': M. Curie, op. cit., p. 89.

38: 'The element . . . find it': E. Curie, op. cit., p. 166.

38: 'with passionate . . . substance': ibid. p. 167.

40: 'must be enormous': paper of 26 December 1898 to the French Academy of Sciences.

40: 'one of us . . . radioactivity . . . property': ibid.

40: 'miserable . . . shed': M. Curie, op. cit., p. 92.

40: 'extremely . . . personnel': ibid. p. 47.

41: 'Sometimes . . . radium': ibid. p. 92.

42: 'Our precious . . . enchantment': ibid. p. 49.

44: 'Are . . . oil?': H. C. Bolton, quoted in Badash, *Scientists and the Development of Nuclear Weapons*, p. 17.

44: 'butterfly . . . radium': E. Curie, op. cit., p. 243.

45: 'Madame . . . burns': ibid. p. 207.

Chapter Two – 'A Rabbit from the Antipodes'

48: 'That's . . . dig': Eve, *Rutherford*, p. 11.

49: 'would have . . . mathematics': Brown, *The Neutron and the Bomb,* p. 45.

49: 'in a few years . . . poured': Maxwell's inaugural lecture of 25 October 1871, manuscript, CUL.

50: 'a most radiating smile': letter from Rutherford to Mary Newton of 2 December 1896, in Eve, op. cit., p. 41.

50: 'not fossilized at all': letter from Rutherford to Mary Newton of 3 October 1895, in ibid. p. 15.

50: 'we've got . . . mighty deep': A. Balfour, quoted in ibid. p. 14.

51: 'jolly . . . almost see them': ibid. p. 43.

51: 'my usual . . . self-consciousness': letter from Rutherford to Mary Newton of 25 January 1896, in ibid. p. 25.

52: 'he gave . . . I abhor': letter from Rutherford to Mary Newton of May 1896, in ibid. p. 35.

52: 'the alarming . . . our dips': letter from Rutherford to Mary Newton of 10 April 1896, in ibid. p. 32.

53: 'the assumption . . . startling one': J. J. Thomson's lecture to the Royal Institution on 30 April 1897.

53: 'pulling their legs': J. J. Thomson, *Recollections and Reflections*, p. 341.

54: 'to do ... Yankees!': letter from Rutherford to Mary Newton of 30 July 1898, in Eve, op. cit., p. 54.

54: 'to find ... matter ... before ... warpath ...': letter from Rutherford to Mary Newton of 25 January 1896, in Pais, *Inward Bound*, p. 39.

54: 'the best sprinters': letter from Rutherford to his mother of 5 January 1902, in Eve, op. cit., p. 80.

57: 'A good ... research work': letter from Rutherford to Mary Newton of August 1896, in ibid. p. 39.

58: 'in the ... suspicion': Hahn, *My Life*, p. 73.

58: 'in a geometrical ... with time': *Philosophical Magazine*, January 1900.

59: 'an indefatigable ... and important': Frederick Soddy Papers, Bodleian Library.

59: 'standing there ... thing ... Rutherford ... argon gas ... For ... they are': quoted in Howarth, *Pioneer Research on the Atom*, pp. 83–4.

60: 'incurable suicide mania': Nye, *Before Big Science*, p. 156.

60: 'playful suggestion ... a wave ... smoke': letter from Sir William Dampier to Rutherford of 26 July 1903, in Eve, op. cit., p. 102.

63: Marie rejected ... times ahead: quotes in this paragraph are from E. Curie, *Madame Curie*, pp. 231–3.

64: 'I wish ... work': Wilson, *Rutherford*, p. 241.

65: 'I see ... apparatus': Campbell, *Rutherford*, pp. 343–4.

66: 'almost as ... hit you': Chadwick's Rutherford Memorial Lecture, 1954, *Proceedings of the Royal Society*.

66: 'Go home ... my boy': Eve, op. cit., p. 152.

66: 'obviously ... spirits ... he now ... looked like': H. Geiger, in Chadwick (ed.), *The Collected Papers of Lord Rutherford*, p. 297.

67: 'most shattering': Chadwick's oral history, AIP.

Chapter Three – Forces of Nature

69: 'child . . . educated': letter from E. Ramstedt to M. Curie of 14 December 1911, in Rayner-Canham, *A Devotion to their Science*, p. 21.

69: 'very wan . . . figure': letter from Rutherford to his mother of 14 October 1910, in Eve, *Rutherford*, p. 191.

70: 'It was . . . under one': Pais, *Inward Bound*, p. 134.

71: 'The solution . . . suddenly': Brian, *Einstein*, p. 61.

72: 'quietly . . . unity': Snow, *Variety of Men*, p. 75.

72: 'witches' sabbath': Pflaum, *Grand Obsession*, p. 160.

72: 'much . . . just . . . people . . . passionateness . . . sparkling intelligence': Quinn, *Marie Curie*, p. 302.

73: 'Madame Curie . . . being a woman . . . in the . . . person': letter from Rutherford to B. Boltwood of 20 November 1911, in Badash (ed.), *Rutherford and Boltwood*, pp. 257–8.

73: 'pure fantasy': Pflaum, op. cit., p. 164.

74: 'plodder . . . working . . . man': Rayner-Canham, op. cit., p. 45.

74: 'something . . . winter': ibid. p. 168.

75: 'odious blackmail': Quinn, op. cit., p. 299.

75: 'the Vestal . . . radium . . . an . . . Langevin's': Pflaum, op. cit., p. 174.

75: 'It's . . . do it': Quinn, op. cit., p. 323.

76: 'The defence . . . brain': Pflaum, op. cit., p. 177.

76: 'Madame . . . like': Klein et al. (eds), *The Collected Papers of Albert Einstein*, p. 544 (author's translation).

77: 'made sense . . . pompous talk': Moore, *Niels Bohr*, p. 39.

78: 'Of course not . . . in it': Pais, op. cit., p. 210.

80: 'the most . . . met': Moore, op. cit., p. 44.

80: 'it is . . . possible': letter from Rutherford to Bohr of 25 March 1913, CUL/R.

80: 'big eyes . . . discoveries': Nye, *Before Big Science*, p. 170.

81: 'to give . . . true': Frisch, *What Little I Remember*, p. 93.

84: 'slightly . . . easy-going': Hahn, *My Life*, p. 37.

84: 'beery days': ibid. p. 73.

84: Hahn . . . floor: the quotes in this paragraph are from ibid. pp. 65–6.

85: 'the only . . . grasp': ibid. p. 77.

85: 'much-chewed specimens': ibid., p. 74.

85: 'the German Marie Curie' : Frisch, op. cit., p. 3.

86: 'in great . . . a man!'": L. Meitner, 'Looking Back', *Bulletin of the Atomic Scientists*, November 1964.

87: Chadwick had arrived . . . explain it to him: all quotes in these three paragraphs are from Chadwick's oral history, AIP.

88: 'his services . . . Turkish bullet': Wilson, *Rutherford*, p. 344.

89: 'To hate . . . blood': M. Curie, *Pierre Curie*, p. 106.

90: 'in a pitiable . . . profoundly . . . perturbed': Hahn, op. cit., p. 120.

90: 'so numbed . . . whole thing': ibid. p. 122.

90: 'it was . . . sooner': ibid. p. 118.

90: 'a higher form of killing': Harris and Paxman, *A Higher Form of Killing*, p. xiii.

91: 'to do . . . future wars': S. Lindqvist, *A History of Bombing*, item 96 (no page numbers).

91: 'to control . . . occupation': ibid. item 102.

Chapter Four – 'Make Physics Boom'

93: 'If . . . the war': Wilson, *Rutherford*, p. 405.

93: 'man . . . projectiles': Snow, *Variety of Men*, p. 6.

93: 'like . . . dark': Frisch, *What Little I Remember*, p. 63.

93: That same year . . . departed: all quotes in this paragraph are from Oliphant, *Rutherford*, pp. 22 and 128.

94: The 1920s . . . Germans: all quotes in this paragraph are from Rayner-Canham, *A Devotion to their Science*, p. 4.

95: 'treated . . . guests': Rees, *Horror in the East*, p. 13.

96: 'the simpleness . . . obtain': letter from H. Nagaoka to E. Rutherford, 22 February 1911, CUL/R.

97: 'the greatest . . . world': Quinn, *Marie Curie*, p. 384.

97: 'the sister . . . Prometheus': ibid. p. 391.

97: 'very uncouth': author's interview with Bertrand Goldschmidt, 3 April 2002.

99: 'a cow': ibid.

99: 'the young . . . Irène': ibid.

99: 'a violent . . . over': Elizabeth Rona, quoted in Rayner-Canham, op. cit., p. 212.

99: 'shaking . . . clothing': a student, quoted in ibid. p. 108.

99: 'promises miracles': Quinn, op. cit., p. 410.

100: 'Can . . . rays?': ibid. p. 412.

101: 'the stronghold of physics': W. Heisenberg, *Physics and Beyond*, p. 62.

102: 'it . . . in Germany': Rhodes, *The Making of the Atomic Bomb*, p. 174.

103: 'eggs . . . good': Sime, *Lise Meitner*, p. 77.

104: 'to do . . . prices': Hahn, *My Life*, p. 138.

104: 'while . . . food-coupons': ibid.

104: 'the secrets of nature . . . whole new . . . contradictions': Pascual Jordan, quoted in Jungk, *Brighter Than a Thousand Suns*, p. 9.

104: 'henceforth . . . war': Born, *My Life*, p. 193.

105: 'make . . . boom': Moore, *Niels Bohr*, p. 97.

105: 'the cocoon . . . periods': W. Heisenberg, op. cit., p. 1.

106: 'prophet of nature': ibid. p. 28.

106: 'Atomic Housing Officer': Frisch, op. cit., p. 20.

106: It was . . . afternoon': all quotes in this paragraph are from W. Heisenberg, op. cit., pp. 19 and 38.

107: Also . . . politics': all quotes in this paragraph are from ibid. pp. 44–5.

108: 'through . . . before me': ibid. p. 61.

109: 'a coherent . . . physics': ibid. p. 62.

109: 'The more . . . crap': quoted on AIP website (Quantum Mechanics).

110: 'The more . . . versa': ibid.

111: 'the impossibility . . . instruments': Moore, op. cit., p. 160.
111: 'It . . . single moment': Bryson, *A Short History*, p. 131.
111: That same year, 1927 . . . his uncertainty principle: all quotes in these five paragraphs are from Teller, *Memoirs*, pp. 2, 5, 33, 39 and 47.

Chapter Five – Days of Alchemy

115: 'a heroic . . . creation': Robert Oppenheimer, quoted in Jungk, *Brighter Than a Thousand Suns*, p. 9.
116: 'in Russia . . . Rutherford': *Biographical Memoirs of Fellows of the Royal Society* (D. Shoenberg's memoir of Kapitza).
117: 'fly . . . cathedral': Pharr Davis, *Lawrence and Oppenheimer*, p. 17.
117: 'proton merry-go-round': Herken, *Brotherhood of the Bomb*, p. 4.
117: 'I'm going . . . famous': Nye, *Before Big Science*, p. 213.
117: 'a four-inch . . . octopus': ibid. p. 30.
118: 'to break up atoms': *New York Times*, quoted in ibid. p. 31.
119: 'to modernise . . . an abstraction': E. Wigner's oral history, AIP.
119: 'It took . . . interested in': J. Chadwick's oral history, AIP.
121: 'just kept . . . pegging away . . . did . . . silly experiments': ibid.
122: 'I don't believe it': undated notes entitled 'Discovery of the Neutron', CCC/JC.
122: 'knew in his bones': *Biographical Memoirs of Fellows of the Royal Society* (Sir H. Massey's and N. Feather's memoir of Chadwick).
122: 'particles . . . charge 0': Wilson, op. cit., p. 550.
123: 'Kapitza . . . vision': M. Oliphant, 'The Beginning: Chadwick and the Neutron', *Bulletin of the Atomic Scientists*, December 1982.

123: 'to be . . . fortnight': Snow, *The Physicists*, p. 85.

123: 'immediately . . . physicist': Brown, *The Neutron and the Bomb*, p. 108.

123: 'The reason . . . nucleus': letter from Chadwick to N. Feather of 25 July 1968, CCC/NF.

124: The discovery . . . hidden riches': all quotes in this paragraph are from Quinn, *Marie Curie*, p. 427.

125: 'with . . . hold of': author's interview with Hans Bethe, 28 April 2002.

125: 'that would . . . area . . . The experts . . . New York': Nicolson, *Public Faces*, p. 17.

126: But in . . . million': all quotes in this paragraph are from Kurzman, *Blood and Water*, pp. 6–7.

126: 'it never . . . pours': letter from Rutherford to Bohr of 21 April 1932, CUL/R.

126: 'do what . . . ago': V. Bowden, quoted in Campbell, *Rutherford*, p. 438.

127: 'normally . . . atom!': Snow, *Variety of Men*, p. 1.

127: 'Dr. Livingston . . . Whoopee!" ': Herken, op. cit., p. 5.

128: 'literally . . . room': Pharr Davis, op. cit., p. 42.

128: It was . . . elements': all quotes in this paragraph are from ibid. pp. 45 and 48.

129: 'both . . . innocent': Herken, op. cit., p. 11.

129: 'an abnormally . . . boy': Rouzé, *Robert Oppenheimer*, p. 13.

129: 'As appears . . . application': letter from P. W. Bridgman to Rutherford of 24 June 1925, in Kimball-Smith and Weiner (eds), *Robert Oppenheimer – Letters and Recollections*, p. 77.

130: 'That's bad': Powers, *Heisenberg's War*, p. 170.

130: 'The point . . . is': Pharr Davis, op. cit., p. 21.

130: 'the awful . . . excellence': letter to F. Fergusson of 23 January 1923, in Kimball-Smith and Weiner (eds), op. cit., p. 92.

131: 'part . . . physics': Kimball-Smith and Weiner (eds), op. cit., p. 98.

131: 'they are . . . successful': ibid. p. 100.

131: 'a vile . . . seventy': ibid. p. 135.

131: 'I would . . . criticism': ibid. p. 130.

132: 'I have . . . together': Pharr Davis, op. cit., p. 25.

132: 'unbelievable . . . life': Herken, op. cit., p. 12.

132: 'nasty gory': Kimball-Smith and Weiner (eds), op. cit., p. 133.

132: 'knew . . . did': Pharr Davis, op. cit., p. 51.

133: 'We . . . experiments': letter from R. Oppenheimer to F. Oppenheimer of autumn 1932, in Kimball-Smith and Weiner (eds), op. cit., p. 159.

133: 'The experimenters . . . again': J. Cockcroft, quoted in Hendry (ed.), *Cambridge Physics*, p. 79.

134: 'catching . . . countries': Holloway, *Stalin and the Bomb*, p. 15.

134: 'Comrade . . . Rutherford': Peierls, *Bird of Passage,* p. 95.

135: 'the world . . . not of it': Powers, op. cit., p. 172.

137: In 1932 . . . superiority: all quotes in this paragraph are from Rees, *Horror in the East*, p. 23.

Chapter Six – Persecution and Purge

139: 'vision . . . bathed . . . woke up': W. Heisenberg, *Physics and Beyond*, p. 125.

139: Perhaps to forget . . . snow-line: all quotes in these two paragraphs are from ibid. pp. 125 and 127.

141: 'We . . . Fatherland': Sime, *Lise Meitner*, p. 139.

141: 'All . . . family': ibid. p. 140.

142: 'Jewish . . . pawn': E. Heisenberg, *Inner Exile*, p. 48.

142: 'spirit . . . chair': *Physics World*, December 2001.

142: 'White Jew . . . representatives . . . themselves': Jungk, *Brighter Than a Thousand Suns*, p. 74.

143: 'ugly . . . quietly': W. Heisenberg, op. cit., p. 173.

143: 'complete idiot': *Physics World*, December 2001.

143: 'civil liberty . . . law': Sime, op. cit., p. 137.

144: 'My . . . young ones': Williams, *Klaus Fuchs*, p. 36.

144: 'No . . . dress': Snow, *Variety of Men*, p. 87.

144: 'It's . . . sciences': Jungk, op. cit., p. 46.

144: 'no . . . rabble': Sime, op. cit., p. 138.

144: 'whipped . . . frenzy': Powers, *Heisenberg's War*, p. 39.

144: 'finely . . . futile': W. Heisenberg, op. cit., p. 150.

145: 'Since . . . hateful': Sime, op. cit., p. 143.

145: 'inner exile': E. Heisenberg, op. cit.

145: 'the old values': W. Heisenberg, op. cit., p. 151.

145: 'after the catastrophe': ibid. p. 154.

145: 'at least . . . stay on': ibid. p. 152.

146: 'If . . . former': Sime, op. cit., pp. 145–6.

146: 'Haber . . . him': ibid. p. 156.

147: 'going backwards . . . way': Pflaum, *Grand Obsession*, p. 298.

148: 'My . . . neutron': ibid. p. 302.

149: 'With . . . time': Quinn, *Marie Curie*, p. 429.

150: The discovery . . . Livingston': all quotes in this paragraph are from Pharr Davis, *Lawrence and Oppenheimer*, p. 60.

150: 'the expression . . . life': Quinn, op. cit., p. 430.

151: 'We're back . . . laboratory': E. Curie, *Madame Curie*, p. 384.

151: 'La Patronne . . . Langevin': all quotes in this paragraph are from the author's interview with Bertrand Goldschmidt, 3 April 2002.

151: 'of rapid . . . radiations . . . in . . . hand': E. Curie, op. cit., pp. 400–1.

152: 'scientists . . . character': Jungk, op. cit., p. 48.

152: 'talking moonshine': *The Times*, 12 September 1933.

152: 'Not only . . . to become': Moore, *Niels Bohr*, p. 213.

152: 'a blind . . . ducks': ibid.

153: 'great . . . forehead . . . enormous . . . eyes': Snow, op cit., p. 84.

153: 'A disaster ... surroundings': Born, *My Life*, p. 38.

153: 'declined violently ... he did not ... poison-gas': letter from M. Born to J. Chadwick of 11 August 1954, quoted in Oliphant, *Rutherford*, p. 60.

153: 'a sort ... exchange': Frisch, *What Little I Remember*, p. 53.

153: 'They ... world': Teller, *Memoirs*, pp. 120–1.

154: 'God ... me': Frisch, op. cit., p. 76.

154: 'a continuing ... Germany': Kimball-Smith and Weiner (eds), *Robert Oppenheimer – Letters and Recollections*, p. 196.

155: 'living ... volcano': Jungk, op. cit., p. 29.

155: 'could ... Hitler' and 'we ... us': ibid. pp. 57–8.

156: 'A reasonable ... number': Badash, *Kapitza, Rutherford*, p. 110.

156: 'I am ... alive': Wilson, *Rutherford*, p. 528.

157: 'It was ... equipment': J. Chadwick's oral history, AIP.

157: 'He was ... very well': ibid.

158: 'only ... thread': letter from Lady Rutherford to Chadwick of 19 October 1937, CCC/JC.

158: 'I feel ... water!': Oliphant, op. cit., p. 110.

158: 'stupefied ... did ... things': Snow, op. cit., p. 14.

158: 'one of ... time': ibid. p. 79.

159: 'in quite ... found it': Bohr's obituary of Rutherford in *Nature*, 1937.

159: 'it was ... nature': Jungk, op. cit., p. 59.

159: 'keep ... matter': Powers, op. cit., p. 53.

159: 'a steamroller ... obstacles': Segrè, *Enrico Fermi*, p. 56.

160: 'ancient ... fields': Sime, op. cit., p. 164.

160: 'the counter ... madly': Fermi, *Atoms in the Family*, p. 98.

161: 'so fascinating ... outstanding chemist': Sime, op. cit., pp. 164–5.

162: 'conceivable ... elements': Jungk, op. cit., p. 61.

162: 'It said ... to me': Goldsmith, *Frédéric Joliot-Curie*, p. 71.

163: 'plain dishonest': E. Segrè's oral history, AIP.

163: 'assumption ... really absurd': I. Noddack, quoting

O. Hahn, in Rayner-Canham, *A Devotion to their Science*, p. 223.

Chapter Seven – 'Wonderful Findings'

164: 'a sapphire . . . street': Kurlansky, *The Basque History of the World*, p. 202.

165: Strict . . . appearing': all quotes in this paragraph are from Gilbert, *A History of the Twentieth Century, Vol. II*, p. 163.

166: 'The American . . . humanity': US government statement in the *New York Times*, 23 September 1937.

167: 'still . . . today': Jungk, *Brighter Than a Thousand Suns*, p. 65.

167: 'This damned . . . wrong': Pflaum, *Grand Obsession*, p. 335.

167: 'dear friend': J. Chadwick's oral evidence, AIP.

167: 'The Jewess . . . institute': Sime, *Lise Meitner*, p. 184.

168: 'Hahn doesn't . . . thrown me out': Lise Meitner's diary, 21 March 1938, CCC/LM.

168: 'I too . . . nerve': Sime, op. cit., p. 185.

169: 'you . . . your stay': ibid. p. 187.

170: 'a magic . . . enhanced': Sime, op. cit., p. 97.

170: 'the . . . Lise Meitner': Hahn, *My Life*, p. 148.

171: 'It is considered . . . the Interior': letter from W. Frick to C. Bosch of 16 June 1938, in Sime, op. cit., pp. 195–6.

172: 'We agreed . . . brought back': Hahn, op. cit., p. 149.

172: 'left Germany . . . purse': letter from L. Meitner to G. von Ubisch of 1 July 1947, CCC/LM.

172: 'a beautiful . . . emergency': Hahn, op. cit., p. 149.

173: 'uprooted . . . at all': Rayner-Canham, *A Devotion to their Science*, pp. 183–4.

173: 'You have . . . hafnium': Sime, op. cit., p. 205.

173: 'The shot . . . missed you': letter from M. von Laue to L. Meitner of November 1958, quoted in Sime, op. cit., p. 205.

174: 'One dare . . . forward': Sime, op. cit., p. 209.

174: Despite her eminence . . . embarrassment': all quotes in this paragraph are from Rayner-Canham, op. cit., p. 184.

176: 'she urgently . . . undertaken': Sime, op. cit., p. 229.

176: 'a mechanical doll . . . inside': letter from L. Meitner to O. Hahn of 5 December 1938, CCC/LM.

176: 'forces . . . easy': ibid.

177: 'perhaps . . . explanation': Hahn, op. cit., p. 151.

177: 'physically . . . out of this': letter from O. Hahn to L. Meitner of 21 December 1938, CCC/LM.

177: 'we poor . . . physics people': O. Hahn interview, AIP.

177: 'very odd . . . impossible': Hahn, op. cit., p. 151.

177: 'short . . . bossy': Sime, op. cit., p. ix.

177: 'knew . . . unexpected results': Rayner-Canham, op. cit., p. 186.

177: 'sort of . . . around': O. Frisch, oral history, AIP.

178: 'In the course . . . great force': Rayner-Canham, op. cit., p. 186.

178: 'The charge . . . all fitted!': Frisch, *What Little I Remember*, p. 116.

179: 'caught . . . tail': O. Frisch, oral history, AIP.

179: *'perhaps* . . . break up': Sime, op. cit., p. 456.

179: 'I am . . . result . . . wonderful findings': letter from L. Meitner to O. Hahn of 3 January 1939, CCC/LM.

180: 'the secret . . . years': Teller, *Memoirs*, p. 141.

180: 'none . . . it': Hahn, op. cit., p. 164.

180: 'smote . . . hand': Frisch, op. cit., p. 116.

180: 'We were . . . fools': Snow, *Variety of Men*, p. 89.

180: 'Fission': Frisch, op. cit., p. 117.

180: 'was knocked . . . camp': O. Frisch, oral history, AIP.

181: 'Otto . . . preceding note': Krafft, *Im Schatten der Sensation*, p. 319 (author's translation).

181: 'I have . . . at it . . . got hold . . . solution . . . it . . . simple': L. Rosenfeld interview, AIP.

182: 'the state . . . description': Moore, *Niels Bohr*, p. 236.

182: 'that's impossible ... make bombs': Nye, *Before Big Science*, p. 218.

183: 'peace in our time': Gilbert, op. cit., p. 203.

183: 'sparkling ... originality': Goldschmidt, *Atomic Rivals*, p. 46.

184: 'That's impossible ... I did it': Weart and Szilard (eds), *Leo Szilard*, pp. 9–10.

185: 'miserably ... potatoes ... chain reaction': E. Wigner's oral history, AIP.

185: 'sort of ... reaction': Weart and Szilard (eds), op. cit., p. 17.

186: 'all we ... beryllium ... everything ... flashes ... very little ... grief': ibid. p. 55.

186: That night ... chillingly real': all quotes in this paragraph are from Teller, op. cit., p. 142.

Chapter Eight – 'We May Sleep Fairly Comfortably in Our Beds'

188: 'One ... nature ... I ... new discoveries': E. Curie, *Madame Curie*, pp. 238–9.

189: By this ... Dante: all quotes in this paragraph are from Fermi, *Atoms in the Family*, pp. 139, 143 and 151–2.

190: 'Fermi ... precautions': Lanouette, *Genius in the Shadows*, p. 181.

190: 'had ... other ... extremely ... else ... could ... experiment ... Szilard's ... assistant': E. Segrè, *Enrico Fermi*, pp. 107–8.

191: 'I went to see ... excited about it"': Weart and Szilard (eds), *Leo Szilard*, p. 54.

192: 'a gruff ... elephant': all quotes in this paragraph are from author's interview with B. Goldschmidt, 3 April 2002.

192: 'Obviously ... governments': Goldschmidt, *Atomic Rivals*, p. 47.

192: 'sufficient . . . ideas': Goldsmith, *Frédéric Joliot-Curie*, p. 73.

193: 'we were . . . realised': Nye, *Before Big Science*, p. 219.

193: 'slave . . . the Szilard . . . France': author's interview with B. Goldschmidt, 3 April 2002.

193: 'Hitler . . . the neutron': Jungk, *Brighter Than a Thousand Suns*, pp. 50–1.

194: 'unsurpassable advantage': Walker, *German National Socialism*, p. 18.

194: 'started . . . off': author's interview with B. Goldschmidt, 3 April 2002.

195: 'Why . . . butter': Pflaum, *Grand Obsession*, p. 341.

195: 'A little . . . disappear': interview with G. Uhlenbeck, who was then sharing an office with Fermi, Kevles, *The Physicists*, p. 324.

195: 'uranium . . . explosive . . . the probabilities . . . this . . . the bare . . . disregarded': the quotations in this paragraph are from G. B. Pegram's letter to Admiral S. C. Hooper of 16 March 1939, in Segrè, op. cit., p. 111.

195: 'There's . . . outside': Bernstein (ed.), *Hitler's Uranium Club*, p. 4. The Farm Hall transcripts are also reproduced in Frank, *Operation Epsilon*. Their reference at the PRO is WO208/5019, and at NARA, record group 77, MED, entry 22, box 163. The latter is more legible and fuller and is the basis of the transcript reproduced in both books.

196: 'although . . . goodwill': Segrè, op. cit., p. 111.

196: 'openness . . . tampered with': Teller, *Memoirs*, p. 143.

196: 'Europe . . . occupation . . . the doom . . . idea': Fermi, op. cit., pp. 144–5.

197: 'you . . . factory': Teller, op. cit., p. 143.

198: 'philosopher's . . . blow . . . Patchogue': *New York Times*, 5 May 1939.

198: 'Physicists . . . Landscape': *Washington Post*, 29 April 1939.

198: 'I . . . practices': Jungk, op. cit., p. 75.

199: Heisenberg's . . . traitor': all quotes in this paragraph are from W. Heisenberg, *Physics and Beyond*, pp. 169–70.

200: 'could . . . channels . . . no principles . . . grant you . . . Germany . . . him': letter from Max Dresden to *Physics Today*, May 1991.

200: 'Every . . . environment . . . people . . . them': W. Heisenberg, op. cit., p. 171.

200: 'there . . . weapons . . . the war . . . built': ibid. p. 170.

201: 'lively and busy . . . a calm . . . stores': filmed survivor interview in *The Restoration of the Hypocentre Salugakucho*.

202: 'our . . . device"': Hahn, *My Life*, p. 184.

203: 'This was . . . this occasion': Weart and Szilard (eds), op. cit., p. 83.

204: 'great ability . . . anonymity': Jungk, op. cit., p. 85.

205: 'I served . . . pillar box': Snow, *Variety of Men*, p. 89.

205: 'We really . . . limited to that': Jungk, op. cit., p. 86.

205: Sachs . . . with him: all quotes in this paragraph are from the letter from Einstein to Roosevelt of 2 August 1939, in Segrè, op. cit., p. 113.

207: Roosevelt . . . different course': all quotes in these two paragraphs are from Jungk, op. cit., pp. 110–11.

208: The cautionary . . . action!': all quotes in these paragraphs are from Lanouette, op. cit., p. 210.

209: 'the fear . . . by them': letter from Churchill to the Secretary of State for War of 5 August 1939, in Churchill, *The Gathering Storm*, p. 301.

210: 'I . . . our beds': letter from Lord Hankey to E. Appleton of 12 December 1939, CAB/104/221, PRO.

Chapter Nine – A Cold Room in Birmingham

211: 'I am . . . help you': letter from J. Chadwick to L. Meitner of 11 September 1939, CCC/LM.

212: 'porridge': Powers, *Heisenberg's War*, p. 51.

212: 'stinking . . . steamer': Rowlands, *120 Years of Excellence*, p. 18.

212: 'witnessed . . . suffering . . . as a way . . . people': author's interview with J. Rotblat, 9 January 2002.

213: 'first . . . practice . . . I . . . domination': J. Rotblat, *Bulletin of the Atomic Scientists*, August 1985.

213: 'great . . . one': author's interview with J. Rotblat, 9 January 2002.

214: 'the worst . . . English': ibid.

214: 'they . . . building': Rowlands, op. cit., p. 17.

214: 'and discovered . . . horse': transcript of J. Rotblat's interview of 7 January 1994 with C. D. King and A. Brown, LIV.

214: 'very . . . himself': ibid.

215: 'Oh good . . . wife': ibid.

215: 'worked . . . bomb . . . the only . . . retaliate . . . never . . . Germans': J. Rotblat, *Bulletin of the Atomic Scientists*, August 1985.

216: 'the only . . . ourselves . . . he . . . conscience': author's interview with J. Rotblat, 9 January 2002.

216: 'the last minute': ibid.

216: 'there was . . . them': transcript of J. Rotblat's interview of 28 October 1992 with Dr Edwards and C. D. King, LIV.

217: 'very, very shaky . . . to go . . . time . . . that we . . . bomb': transcript of J. Rotblat's interview of 7 January 1994 with C. D. King and A. Brown, LIV.

217: Chadwick's . . . it again': all quotes in these two paragraphs are from J. Chadwick's oral history, AIP.

218: 'he just grunted': transcript of J. Rotblat's interview of 7 January 1994 with C. D. King and A. Brown, LIV.

218: 'It was . . . apart': J. Chadwick, *Liverpool Daily Post*, 4 March 1946.

218: Chadwick . . . each other': all quotes in this paragraph are from the transcript of J. Rotblat's interview of 7 January 1994 with C. D. King and A. Brown, LIV.

219: 'it seems . . . conditions': letter from J. Chadwick to
E. Appleton of 5 December 1939, CAB/21/1262, PRO.

220: Rudolf . . . house: all quotes in these two paragraphs are
from Peierls, *Bird of Passage*, pp. 91 and 99.

221: In 1939 . . . combine: all quotes in this paragraph come from
O. Frisch's oral history, AIP, except 'to ponder . . . real physicist',
which comes from *Biographical Memoirs of Fellows of the Royal
Society* (R. Peierls' Biographical Memoir of O. Frisch).

222: Most . . . fission: all quotes in this paragraph are from
Frisch, *What Little I Remember*, p. 123.

222: 'the water . . . bedside': ibid. p. 125.

222: 'there are . . . at first': Gowing, UKAEA official historian,
lecture 'How Nuclear Power Began', p. 12.

223: 'Suppose . . . about a pound': Peierls, op. cit., p. 154.

223: 'such a lot': O. Frisch's oral history, AIP.

224: 'thousands . . . explosive . . . We . . . having"': Peierls, op.
cit., p. 154.

224: 'Look . . . about that?': O. Frisch's oral history, AIP.

The Frisch–Peierls memorandum was in two parts – one
technical and one non-technical. The technical part was reproduced first as Appendix I to Gowing, *Britain and Atomic
Energy*. The non-technical part was for a time lost, but then
found among Sir Henry Tizard's papers and first published in
Clark, *Tizard*. Both parts are reproduced as Appendix II of
Szasz, *British Scientists*. All quotes in the last three paragraphs
of this chapter come from the Frisch–Peierls memorandum with
the exception of 'the order of magnitude was right', which
comes from Peierls, op. cit., p. 154.

Chapter Ten – Maud Ray Kent

226: On the morning . . . France: all quotes in this paragraph
are from the author's interview with B. Goldschmidt, 3 April
2002.

228: 'At any . . . Freiss': Goldsmith, *Frédéric Joliot-Curie*, p. 86.

231: 'my mother . . . laboratory': Pflaum, *Grand Obsession*, p. 362.

231: The collier . . . Castle: all quotes in this paragraph are from L. Kowarski's interview with the *Radio Times*, 20 September 1973.

233: 'Met . . . Kent': Sime, *Lise Meitner*, p. 284.

234: The Maud . . . himself: all quotes in this paragraph are from Frisch, *What Little I Remember*, pp. 128 and 130.

235: 'the answers . . . questions': Peierls, *Bird of Passage*, p. 155.

235: 'to decipher . . . were': *Biographical Memoirs of Fellows of the Royal Society* (Biographical Memoir of F. Lindemann – Lord Cherwell).

237: 'if things . . . World': letter from M. Oliphant to J. Chadwick, 22 June 1940, CCC/JC.

237: Nuclear . . . shrapnel': all quotes in this paragraph are from Frisch, op. cit., p. 133.

237: 'Koventrieren . . . principle . . . so . . . time': Gilbert, *A History of the Twentieth Century, Vol. II*, p. 352.

237: 'burned . . . flames': *The Times*, in Lindqvist, *A History of Bombing*, item 165.

238: 'ridiculous': transcript of J. Rotblat's interview of 7 January 1994 with C. D. King and A. Brown, LIV.

238: 'putting . . . know': Frisch, op. cit., p. 142.

239: 'turning . . . bird . . . How . . . want? . . . Frisch and Chips': ibid. pp. 132 and 137.

239: 'the effect . . . zero': Peierls, op. cit., p. 157.

239: Chadwick . . . life: all quotes in this paragraph are from J. Chadwick's oral history, AIP.

241: In July . . . plutonium: all quotes in this paragraph are from the Maud Report, CAB/104/227, PRO, and also AB/16/266, PRO.

242: 'Although . . . improvement': Churchill's note to General Ismay, 30 August 1941, PREM/3/139/8A, PRO.

Chapter Eleven – 'Hitler's Success Could Depend on It'

243: 'unsurpassable advantage': Walker, *German National Socialism*, p. 18.

243: While . . . nothing': all quotes in this paragraph are from Ermenc (ed.), *Atomic Bomb Scientists*, p. 97.

244: 'a theoretical . . . Heisenberg': Powers, *Heisenberg's War*, p. 14.

244: 'decent . . . Party': Ermenc (ed.), op. cit., p. 67.

245: 'the slightest chance . . . be done': Bernstein (ed.), *Hitler's Uranium Club*, p. 132.

245: Otto Hahn . . . commit suicide': all quotes in this paragraph are from ibid. p. 31.

246: 'all . . . time': W. Heisenberg, *Physics and Beyond*, p. 172.

246: 'strangely . . . me': Powers, op. cit., p. 13.

246: 'to work . . . energy': W. Heisenberg, op. cit., p. 172.

246: 'Well . . . for that': Bernstein (ed.), op. cit., p. 32.

246: Heisenberg's . . . clear conscience': all quotes in this paragraph are from W. Heisenberg, op. cit., p. 174.

247: Heisenberg's role . . . explosives': all quotes in this paragraph are from Bernstein (ed.), op. cit., pp. xiii–xxiv, except for 'the most . . . power': Powers, op. cit., p. vii, and 'the explosive . . . magnitude', which is the author's translation of Heisenberg's secret report *Geheimdokumente zum deutschen Atomprogramm* in the Deutsches Museum.

248: 'I . . . water': W. Heisenberg, op. cit., p. 180.

250: 'abandoned . . . idea . . . prematurely': ibid.

253: Strassmann . . . Valerie: information and the quote in this paragraph are from A. Wolffenstein's memoirs in the Jüdisches Museum, Berlin, her testimony to Yad Vashem in Jerusalem and her letters reprinted in Krafft, *Im Schatten der Sensation*, p. 46 (author's translation).

255: 'impressive . . . man': Frisch, *What Little I Remember*, p. 71.

255: 'not quite . . . discipline': letter from O. Frisch to K. Fuchs of 21 March 1942, AB/1/574, PRO.

255: 'former human being . . . Fizzl is in Berlin': Powers, op. cit., pp. 92–3.

256: 'We . . . the thing': F. Reiche's oral history, AIP.

257: 'They . . . grateful': ibid.

257: 'Hitler's . . . it': Powers, op. cit., p. 58.

257: Most . . . reactions': all quotes in this paragraph are from Holloway, *Stalin and the Bomb*, pp. 53–4.

259: On 13 April . . . West: all quotes in this paragraph are from Bix, *Hirohito*, p. 397.

Chapter Twelve – 'He Said "Bomb" in No Uncertain Terms'

261: 'amazed and distressed': Oliphant, 'The Beginning: Chadwick and the Neutron', *Bulletin of the Atomic Scientists*, December 1982.

262: 'he said . . . up to us': Pharr Davis, *Lawrence and Oppenheimer*, p. 112.

262: 'distinguished . . . foreigners': Weart and Szilard (eds), *Leo Szilard*, p. 146.

263: 'put in wraps': Gowing, *Britain and Atomic Energy*, p. 116.

264: 'extreme secrecy': letter from Dr C. Darwin to Lord Hankey of 2 August 1941, CAB/104/227, PRO.

264: 'Are . . . single blow?': ibid.

264: 'may . . . this country': Frisch–Peierls memorandum (see chapter nine).

264: 'bombardment . . . cities': *The Times*, 2 September 1939.

265: 'complete . . . matters': memo from V. Bush to J. Conant of 9 October 1941, Bush–Conant files, OSRD, S1 Record Group 227, M 1392, NARA.

266: 'It appears . . . conducted': letter from Roosevelt to

Churchill of 11 October 1941, PREM/3/139/8A, PRO.

266: 'It was . . . bomb': Irving, *The Virus House*, p. 93.

266: 'panic reaction': interview with *Der Spiegel*, 3 July 1967, in Powers, *Heisenberg's War*, p. 112.

266: 'perhaps . . . related to it': Cassidy, *Uncertainty*, p. 436.

267: Houtermans' findings . . . with uranium"': all quotes in these three paragraphs are from W. Heisenberg, *Physics and Beyond*, pp. 180–2.

269: 'A feeling . . . unobserved': Cassidy, op. cit., p. 440.

270: Despite . . . walked out: all quotes in this paragraph are from Moore, *Niels Bohr*, p. 218.

270: Bohr . . . themselves': all quotes in this paragraph are from Pais, *Niels Bohr's Times*, p. 483.

270: 'a background . . . Denmark': draft letter from N. Bohr to W. Heisenberg (never sent), 1957 or 1958, Niels Bohr Archive.

271: Heisenberg provided . . . but could not: all quotes in these two paragraphs are from W. Heisenberg's letter to R. Jungk, in Jungk, *Brighter Than a Thousand Suns*, pp. 103–4.

273: 'a tremendous . . . conflict': W. Heisenberg, op. cit., p. 182.

273: 'hostile': Margrethe Bohr, quoted in letter from S. Goudsmit to A. Hermann, 18 October 1946, Goudsmit Papers, Folder 100, Box 11, AIP.

273: 'hope and belief': letter from R. Ladenburg to S. Goudsmit, 23 October 1946, Goudsmit Papers, Folder 138, Box 14, AIP.

273: Nearly . . . atomic weapons': all quotes in these two paragraphs are from the draft letter from N. Bohr to W. Heisenberg (never sent), 1957 or 1958, Niels Bohr Archive.

274: 'fondest . . . happy years': draft letter from N. Bohr to W. Heisenberg (never sent), 30 November 1961, Niels Bohr Archive.

274: 'I have . . . great dangers': handwritten note (undated) from N. Bohr to W. Heisenberg (never sent), Niels Bohr Archive.

275: 'what purpose . . . behind': draft letter from N. Bohr to W. Heisenberg (undated), Niels Bohr Archive.

275: 'quite incomprehensible . . . that German . . . they could': draft letter from N. Bohr to W. Heisenberg (undated), Niels Bohr Archive.

275: 'you . . . direction': further draft letter from N. Bohr to W. Heisenberg (undated), Niels Bohr Archive.

275: 'less . . . did not': *New York Review of Books*, 17 December 1964.

275: 'concerning . . . than we': Cassidy, op. cit., p. 442.

275: 'we hoped . . . on a bomb': letter from C-F. von Weizsäcker to the author of 10 December 2002.

275: 'saw . . . this idea': E. Heisenberg, *Inner Exile*, p. 79.

276: 'as far . . . on London?"': Hans Bethe's contribution to French and Kennedy (eds), *Niels Bohr – A Centenary Volume*, p. 233.

276: 'was clearly . . . reactor': Bernstein, *Hans Bethe*, p. 77.

276: 'It was . . . misunderstanding': author's interview with H. Bethe, 28 April 2002.

277: 'Bohr . . . mumbled': ibid.

277: 'They seem . . . Bohr': Bernstein (ed.), *Hitler's Uranium Club*, p. 155.

277: 'Undoubtedly . . . towards them': A. Bohr's interview with BBC on 16 January 1965, for the programme *The Building of the Bomb*.

278: 'He said . . . Nazis . . . unbelievable . . . naive': Bernstein, *Hans Bethe*, pp. 74–5.

279: 'entirely filled . . . victory': letter from L. Meitner to P. Scherrer of 26 June 1945, CCC/LM.

279: 'it was a mistake': letter from L. Meitner to M. von Laue of 20 April 1942, CCC/LM.

279: 'I have often . . . nothing about': letter from M. von Laue to L. Meitner of 26 April 1942, CCC/LM.

Chapter Thirteen – 'We'll Wipe the Japs Out of the Maps'

280: 'a specious ... about it': Gowing, *Britain and Atomic Energy*, p. 109.

280: 'I need ... this matter': Churchill's note to Roosevelt of December 1941, PREM/3/139/8A, PRO.

281: 'God's butler': B. Goldschmidt, *Atomic Rivals*, p. 129.

281: 'disturbed ... enemy': note of a meeting of 27 November 1941, PREM/3/139/8A, PRO.

281: 'in this ballet ... superiority': Goldschmidt, op. cit, p. 133.

281: By late ... Roosevelt: all quotes in these two paragraphs are from the third report of the National Academy of Sciences Committee contained in the Bush–Conant files, OSRD, record group 227, M 1392, NARA.

284: 'We'll ... maps': Fermi, *Atoms in the Family*, p. 173.

285: 'the fruits ... quickly': Kurzman, *Day of the Bomb*, p. 93.

285: 'saving ... foreigners ... the iron ... Japan': material in possession of Hiroshima Peace Memorial Museum.

286: 'Isawa ... Singapore': Hersey, *Hiroshima*, p. 12.

287: 'missed ... bells': author's interview with Takeko Kagawa, Hiroshima, April 2004.

288: 'only ... future': Bernstein (ed.), *Hitler's Uranium Club*, p. xxvii.

289: 'Pure ... force ... through ... force': Irving, *The Virus House*, p. 99.

289: 'I received ... everybody': Goebbels, *Diaries*, entry for 21 March 1942, p. 96.

290: 'by no means ... support': Speer, *Inside the Third Reich*, p. 226.

290: 'ridiculously tiny': interview given by A. Speer to *Der Spiegel*, 3 July 1967.

291: 'no orders ... decision ... As ... period': W. Heisenberg, *Physics and Beyond*, p. 183.

Chapter Fourteen – 'V.B. OK'

292: 'V.B. OK . . . I think . . . safe': Bush–Conant files, OSRD, record group 227, M 1392, NARA.

293: 'the whole . . . department': letter from V. Bush to Roosevelt, 9 March 1942, in ibid.

294: Groves . . . soldiers': all quotes in these two paragraphs are from Groves, *Now It Can Be Told*, pp. 3–5.

295: 'I fear . . . soup': ibid. p. 20.

296: 'very keen mind': Norris, *Racing for the Bomb*, p. 89.

296: 'This is . . . budge': ibid. p. 53.

296: 'that it . . . doing': Groves, op. cit., p. 21.

297: 'It was . . . wagon': ibid. p. 28.

298: 'simple . . . decisions': ibid.

298: 'an even greater . . . Columbus': ibid. p. 38.

298: 'Do you . . . making bombs"': article by P. Morrison in the *Scientific American*, August 1985.

299: Groves . . . all routes: all quotes in these two paragraphs are from Groves, op. cit., pp. 39–41.

299: 'that scientists . . . me . . . Who cares?': Ermenc (ed.), *Atomic Bomb Scientists*, p. 47.

300: 'atomic . . . science': Groves, op. cit., p. xiii.

300: 'There is . . . wouldn't it?': Groueff, *Manhattan Project*, p. 35.

300: 'the unique . . . assistance': Groves, op. cit., p. 46.

301: 'My reputation . . . project': Teller, *Memoirs*, p. 204.

301: 'they were . . . capability': Groves, op. cit., p. 47.

302: 'the entirely . . . involved': ibid. p. 57.

302: 'The Italian . . . friendly': Fermi, *Atoms in the Family*, p. 198.

303: It was . . . straw covering: E. Fermi's account in the *Chicago Sun-Times* on 23 November 1952 for the first of a series entitled 'The Atom and You'.

304: 'about . . . catastrophic': Groves, op. cit., p. 69.

305: 'dogs . . . bark': Holloway, *Stalin and the Bomb*, p. 78.

306: 'to realise . . . bomb': ibid. p. 88.

Chapter Fifteen – 'The Best Coup'

308: 'The Best Coup': British War Cabinet report on SOE activities, May 1945, PREM/3/139/4, PRO.
308: 'we have just . . . be ready': letter from Compton to Bush of 22 June 1942, Industrial and Social Branch, OSRD, S-1, NARA.
311: 'a hard blow': Haukelid, *Skis Against the Atom*, p. 6.
313: 'a . . . way': Kurzman, *Blood and Water*, p. 139.
316: 'astonishingly small': Haukelid, op. cit., p. 112.
316: 'the English bandits . . . this war': ibid. p. 162.

Chapter Sixteen – Beautiful and Savage Country

319: 'This . . . place': E. McMillan, quoted in Kimball-Smith and Weiner (eds), *Robert Oppenheimer – Letters and Reflections*, p. 236.
319: 'a lovely spot': letter from R. Oppenheimer to J. H. Manley of 6 November 1942, ibid.
319: 'Oppenheimer . . . Groves': author's interview with Hans Bethe, 28 April 2002.
320: 'playing . . . United States': memo from de Silva to B. Pash of 2 September 1943, in Norris, *Racing for the Bomb*, p. 268.
320: 'his potential . . . project': Groves, *Now It Can Be Told*, p. 63.
321: 'Oppenheimer . . . lieutenant-colonel': author's interview with Hans Bethe, 28 April 2002.
321: 'came . . . pointless"': ibid.
322: 'as raw . . . scar': Eleanor Jette, quoted in Hales, *Atomic Spaces*, p. 72.

323: 'two . . . achieved': Groves, op. cit., p. 157.

325: 'the fission . . . doing it': Bernstein, *Hans Bethe*, p. 73.

325: Much . . . electronic': all quotes in this paragraph come from the author's interview with Hans Bethe, 28 April 2002.

326: 'I was . . . Bethe': Teller, *Memoirs*, p. 177.

326: ' "little . . . detailed': ibid. p. 176.

326: 'jutted . . . intensity': Fermi, *Atoms in the Family*, p. 218.

326: 'marked . . . friendship': Teller, op. cit., p. 178.

327: 'Sergeant . . . captain': Groves, op. cit., p. 161.

328: The trip . . . disturbing echoes: quotes in these two paragraphs are from Ruth Marshak and Eleanor Jette, in Hales, op. cit., pp. 71–2.

329: 'self-reliance . . . stubbornness': Fermi, op. cit., p. 229.

330: 'No . . . course': Peierls, *Bird of Passage*, p. 193.

330: 'Apparently . . . records': Groves, op. cit., p. 166.

330: 'when . . . go free': Badash et al. (eds), *Reminiscences of Los Alamos*, p. 93.

331: 'it . . . living in it': Fermi, op. cit., p. 230.

331: 'The isolation . . . dancing': Segrè, *Enrico Fermi*, p. 139.

332: 'the utterly . . . landscape': author's interview with P. Morrison, 30 September 2002.

332: 'a wonderful . . . New Yorkers': author's interview with R. Christy, 17 July 2002.

333: 'Nobody . . . crazy': Lanouette, *Genius in the Shadows*, p. 255.

333: 'research . . . knowledge': Groves, op. cit., p. 96.

334: 'on . . . checked': Teller, op. cit., p. 195.

334: 'There . . . tough town': Leona Marshall Libby, quoted in Hales, op. cit., p. 105.

335: 'A physicist . . . Los Alamos': author's interview with Hans Bethe, 28 April 2002.

335: 'unless . . . men': letter from R. Oppenheimer to E. Fermi, 25 May 1943, Oppenheimer papers, LOC.

335: 'each . . . else': Groves, op. cit., p. 140.

336: 'fought . . . won': Teller, op. cit., p. 172.

336: 'Enrico ... escort ... as ... saboteur': Fermi, op. cit., pp. 212–13.

Chapter Seventeen – 'Mr Baker'

338: 'for the present ... help': Kurzman, *Blood and Water*, p. 51.

339: Passengers ... silk (fn): M. Oliphant's description of his flight comes from Cockburn and Ellyard, *Oliphant*, pp. 114–15.

339: 'like death': Brown, *The Neutron and the Bomb*, p. 246.

340: 'the dominant ... a dictator': J. Chadwick in continuation of a memo by W. Akers of 24 September 1943, AB/1/129, PRO.

340: 'could ... problem': Peierls, *Bird of Passage*, p. 179.

340: 'I ... bomb': Sime, *Lise Meitner*, p. 305.

340: 'any ... justified': ibid. p. 58.

340: 'I hoped ... will': Rayner-Canham, *A Devotion to their Science*, p. 188.

341: 'the attitude ... Englishman': Groves, op. cit., p. 144.

341: 'Dear Fucks' (fn): letter from R. Peierls to K. Fuchs of 20 August 1943, AB/1/574, PRO.

342: 'I was ... doubts': K. Fuchs' confession to W. Skardon, 27 January 1950, reproduced as Appendix A in Williams, *Klaus Fuchs*, pp. 180–6.

342: 'I cannot ... importance': letter from R. Peierls to K. Fuchs of 10 May 1941, AB/1/574, PRO.

343: 'In the Los Alamos ... reverence': Fermi, *Atoms in the Family*, p. 222.

344: 'a very ... cause': Brown, op. cit., p. 242.

344: 'We are ... for you': Moore, *Niels Bohr*, p. 301.

344: A few days ... conscious once more: Margrethe Bohr gave an account of their escape to the BBC in an interview for the 1965 programme *The Building of the Bomb*.

348: 'I've ... Bohr': Moore, op. cit., p. 324.

348: 'For many . . . possible': Segrè, *Enrico Fermi*, p. 319.

349: 'Those poor bastards . . . It was . . . to be': Lindqvist, *A History of Bombing*, item 202.

349: 'Are . . . things?' (fn): Simpson, *A Mad World, My Masters*, p. 222.

349: '*I want . . . women*' (fn): PREM/3/89, PRO.

Chapter Eighteen – Heavy Water

351: 'Snow . . . we went': K. Haukelid, *Skis Against the Atom*, p. 195.

353: They considered . . . complex: a report of the visit of Professor Harteck to the Dolomites in November 1943, describing the German interest in producing heavy water in Italy, is among the *Geheimdokumente zum deutschen Atomprogramm* in the Deutsches Museum.

353: 'We . . . aborted?': Kurzman, *Blood and Water*, p. 210.

354: 'we must . . . reprisals': ibid. p. 217.

354: 'Matter . . . Greetings': Haukelid, op. cit., p. 186.

354: 'There were . . . carried': ibid. p. 188.

355: 'The bitterly . . . the march': ibid. p. 192.

355: 'illicit things . . . It was . . . the bilge': ibid. p. 193.

356: 'shaking . . . puzzled him': ibid. p. 194.

Chapter Nineteen – Boon or Disaster?

360: 'looking . . . way . . . to certify . . . solved': author's interview with P. Morrison, 30 September 2002.

361: 'an utter failure': author's interview with Hans Bethe, 28 April 2002.

361: 'a most important key': ibid.

361: 'His presentation . . . killer"': Peierls, *Bird of Passage*, p. 201.

362: 'over-organised': E. Teller, quoted in Rhodes, *The Making of the Atomic Bomb*, p. 539.

362: 'the best . . . proceed': author's interview with Hans Bethe, 28 April 2002.

362: 'The influx . . . construction': Fermi, *Atoms in the Family*, p. 206.

362: 'We . . . stay there': Peierls, op. cit., p. 187.

363: 'two . . . negroes . . . transport . . . segregated': ibid. p. 183.

363: 'incessant action': Fermi, op. cit., p. 223.

363: 'an attractive . . . eyeglasses . . . sparingly . . . words': ibid. p. 209.

363: ' "Penny-in-the-slot . . . question': author's interview with Hans Bethe, 28 April 2002.

363: 'controlled schizophrenia . . . establish . . . society': K. Fuchs' confession to W. Skardon of 27 January 1950, reproduced in full as Appendix A in Williams, *Klaus Fuchs*, pp. 180–6.

364: 'didn't . . . thing': Feynman, *Surely You're Joking*, p. 120.

367: 'great pressure . . . them': Peierls, op. cit., p. 200.

367: 'to interpose . . . the bomb': Pharr Davis, *Lawrence and Oppenheimer*, p. 227.

367: 'The Tech Area . . . at all': Hales, *Atomic Spaces*, p. 212.

368: As the Manhattan . . . his body: the catalogue of experiments on humans is detailed in ibid. pp. 297–8.

369: 'we had . . . monitored': author's interview with P. Morrison, 30 September 2002. Hacker's *The Dragon's Tail* gives fuller details of the radiation safety arrangements at Los Alamos.

370: 'glowing continuously . . . reflected . . . critical . . . the reaction . . . second . . . if I . . . been fatal': Frisch, *What Little I Remember*, pp. 161–2.

370: 'we had . . . reaction': P. Morrison's article in the *Scientific American*, August 1995.

370: 'much . . . technique': Peierls, op. cit., pp. 200–1.

371: 'the complex ... work': Groves, *Now It Can Be Told*, p. 288.

371: 'When I saw ... conditions': author's interview with J. Rotblat, 9 January 2002.

372: 'the real ... Soviets': J. Rotblat's article in the *Bulletin of the Atomic Scientists*, August 1985.

372: 'a terrible ... ally': J. Rotblat's interview with C. D. King and A. Brown, 7 January 1994, LIV.

372: Rotblat ... of all: all quotes are from author's interview with J. Rotblat, 9 January 2002.

373: 'What ... or physics?': quoted in Jungk, p. 174.

373: 'ought ... mortal crimes ... I did not ... head': Churchill's telegram of 20 September 1944 to Lord Cherwell, PREM/3/139/8A, PRO.

373: 'It was ... schoolboys': *Biographical Memoirs of Fellows of the Royal Society* (R. V. Jones, Obituary of W. Churchill).

374: 'the suggestion ... accepted ... Enquiries ... Russians': aide-mémoire of discussion between Churchill and Roosevelt (Hyde Park Agreement), 18 September 1944, PREM/3/139/8A, PRO.

374: While Bohr ... immediately: all quotes in these two paragraphs are from author's interview with J. Rotblat, 9 January 2002, except for 'within ... truth', which comes from J. Rotblat's article in the *Bulletin of the Atomic Scientists*, August 1985.

376: 'it ... ignore': letter from J. Chadwick to T. Allibone of 29 January 1945, CCC/JC.

377: 'the salad ... suspicion': letter from R. Campbell to C. Barnes of 29 January 1945, CAB/126/259, PRO.

377: 'to learn ... Joliot-Curie ... a transcript ... everything': author's interview with B. Goldschmidt, 3 April 2002.

377: 'However ... Joliot-Curie': all quotes in this paragraph are from ibid.

378: 'Thank you ... very well': ibid.

378: 'unusually captivating': Herken, *Brotherhood of the Bomb*, p. 90.

379: 'extremely . . . valuable': Williams, *Klaus Fuchs*, p. 82.

380: 'the more . . . atmosphere': Kurzman, *Day of the Bomb*, p. 175.

381: In July 1944 . . . preparations': all quotes in this paragraph are from ibid. p. 100.

382: 'The beasts . . . strike back': material in the possession of the Hiroshima Peace Memorial Museum.

Chapter Twenty – 'This Thing is Going to be Very Big'

All quotes from Paul Tibbets come from his account *Mission Hiroshima* unless otherwise stated.

385: 'the United States . . . high explosive': Thomas and Witts, *Ruin from the Air*, p. 6.

387: 'the most . . . my life': ibid. p. 37.

387: 'too young': Groves, quoted in Norris, *Racing for the Bomb*, p. 319.

390: 'even . . . extra turn': Thomas and Witts, op. cit., p. 147.

391: 'the semi-liquid . . . dust . . . Dead . . . asleep': Lindqvist, *A History of Bombing*, items 214 and 215.

391: 'How many . . . the number?': Gilbert, *A History of the Twentieth Century, Vol. II*, p. 642.

393: 'comfort . . . pre-war life': author's interview with Takeko Kagawa, Hiroshima, March 2004.

393: 'higher beings . . . their . . . Empress': author's interview with Yoko Kono, Hiroshima, March 2004.

Chapter Twenty-One – 'Germany Had No Atomic Bomb'

Quotes attributed to Sam Goudsmit are from his book *Alsos: The Failure in German Science* unless otherwise stated.

Similarly, quotes from Boris Pash come from his account *The Alsos Mission* unless otherwise noted.

396: 'Germany . . . Bomb': Goudsmit, *Alsos*, p. 71.

396: 'our . . . uranium': Hahn, *My Life*, p. 157.

398: 'the Germans . . . troops': Groves, *Now It Can Be Told*, p. 207.

400: 'would be . . . to me . . . any difficulties': letter from Heisenberg to D. Coster of 16 February 1943, in Powers, *Heisenberg's War*, p. 326.

404: 'By far . . . a kidnapping': letter from V. Weisskopf to R. Oppenheimer of 28 October 1942, Oppenheimer Papers, Weisskopf Folder, Box 77, LOC.

404: 'interesting . . . matter': letter from R. Oppenheimer to V. Weisskopf of 29 October 1942, ibid.

407: 'like . . . firm land': W. Heisenberg, *Physics and Beyond*, p. 191.

408: 'actively . . . nationalistic': report on interrogation of Heisenberg, 11 May 1945, Alsos papers among the Manhattan Papers, NARA.

Chapter Twenty-Two – 'A Profound Psychological Impression'

The records of the first, second and third meetings of the Target Committee are contained in top-secret correspondence of the MED, file 5D, Roll 1, M 1109, NARA. The record of the Interim Committee is on file 100, Roll 8, Harrison Bundy files, M 1108, NARA: V – Problems of Control and Inspection; VI – Russia; VIII – Effect of the Bombing on the Japanese and their Will to Fight; and IX – Handling of Undesirable Scientists.

409: 'A Profound . . . Impression': minutes of Interim Committee of 31 May 1945.

409: 'More ... governments': Gilbert, *A History of the Twentieth Century, Vol. II*, p. 711.

409: 'the development ... destructive power ... so powerful ... scale': Truman, *Year of Decisions*, p. 10.

410: 'to make ... effective': Groves, *Now It Can Be Told*, p. 266.

410: 'I had set ... air raids': ibid. p. 267.

413: 'a personal ... of himself': H. Stimson's diary, entry 2 May 1945, CUL/S.

413: 'ought to ... many people': letter from K. Darrow to E. Lawrence of 9 August 1945, UCLA/BL.

413: 'perhaps ten minutes': letter from E. Lawrence to K. Darrow of 17 August 1945, UCLA/BL.

413: 'sufficiently ... useless': ibid.

414: 'the reason ... unfortunate ... I was ... over-ruled': Groves, op. cit., pp. 273–5.

415: 'he could ... in accord': J. Byrnes, quoted in Giovannitti and Freed, *The Decision to Drop the Bomb*, p. 109.

416: 'The subject ... eccentric': Lanouette, *Genius in the Shadows*, p. 249.

417: 'I didn't ... I was': Weart and Szilard (eds), *Leo Szilard*, p. 181.

417: 'he was ... a coward': Norris, *Racing for the Bomb*, p. 526.

417: 'flabbergasted ... assumption ... manageable': Weart and Szilard (eds), op. cit., p. 184.

418: Nevertheless ... off the hook: all quotes in this paragraph are from file 76, Roll 6, Harrison Bundy files, M1108, NARA.

418: Meanwhile ... to the President: all quotes in these two paragraphs are from Teller, *Memoirs*, pp. 204–8.

421: 'to make ... of it ... the guys ... pilot': Rees, *Horror in the East*, pp. 122–3.

422: 'hoped ... the other': Keegan, *The Second World War*, p. 575.

422: 'a royal ... card': H. Stimson's diary, entries 14 and 15

May 1945, CUL/S.

422: 'didn't . . . test': the proofs of his book *The Days of Their Power* (Davies Papers, Box 100, p. 100), quoted in Alperovitz, *The Decision to Use the Atomic Bomb*, p. 153. These files are closed to the public.

423: 'His Majesty's . . . the war': *Foreign Relations of the US*, document 582, p. 876.

423: 'under incredible . . . meeting': *US Atomic Energy Commission – In the Matter of J. Robert Oppenheimer*, p. 31.

423: 'just standing there . . . dumbfounded . . . Look . . . blowed up': transcript of BBC TV programme *The Day the Sun Blowed Up*, 1975.

425: 'be . . . accidentally': Rhodes, *The Making of the Atomic Bomb*, p. 666.

425: 'unbelievable noise . . . hundreds . . . water': Segrè, *Enrico Fermi*, p. 146.

426: 'the first . . . cracked': J. Chadwick's note of 23 July 1945 to Sir John Anderson, CAB/126/250, PRO.

427: 'suddenly . . . hills': Otto Frisch's eyewitness account of the Trinity test, 16 July 1945, CAB/126/250, PRO. J. Chadwick praised it as 'the best' description.

428: 'To us . . . the consequences': BBC radio broadcast by R. Peierls on 'Atomic Energy and its Present Potentialities', 1948.

428: 'a little . . . made': Szasz, *The Day the Sun Rose Twice*, p. 89.

428: 'I am . . . worlds': Pharr Davis, *Lawrence and Oppenheimer*, p. 240.

428: 'walk . . . strut': Rhodes, op. cit., p. 676.

428: 'only . . . nothing happened . . . dribbling . . . from his hand . . . away': Groves, op. cit., p. 296.

429: 'Operated . . . expectations': MED, Roll 1, M 1109, NARA.

429: 'I hope . . . of it': Truman's diary, quoted in Ferrell (ed.), *Off the Record*, pp. 52–3.

429: 'babies ... born ... Now I know ... same way': H. Stimson's diary, entry 22 July 1945, CUL/S.

429: 'completely ... the Russians!': Danchev and Todman (eds), *War Diaries of Field Marshal Lord Alanbrooke*, p. 709.

430: 'It is ... Japan': Ehrman, *Grand Strategy, Vol. 6*, p. 292.

431: 'casually ... Japanese"': Truman, op. cit., p. 416.

431: 'Stalin ... anything"': Holloway, *Stalin and the Bomb*, p. 125.

432: '1. The 509 ... project staff': the directive to bomb to General C. Spaatz, 25 July 1945 – Roll 1 M 1109 5B, NARA.

432: 'ignore it ... the war': there is considerable debate about the precise translation of Prime Minister Suzuki's statement of 28 July 1945 – cf. Alperovitz, op. cit., pp. 408–9; R. Rhodes, op. cit., p. 403; and Bix, op. cit., p. 501. However, the academic consensus is that this translation fairly reflects Suzuki's intentions. Japan would not pursue the peace offer.

432: 'a laughable matter': Kurzman, *Day of the Bomb*, p. 403.

Chapter Twenty-Three – 'An Elongated Trash Can with Fins'

Quotes by Paul Tibbets are from his book *Mission Hiroshima* unless otherwise attributed. Similarly, quotes from Bob Caron are from his article in *Veterans of Foreign Wars Magazine*, November 1959, unless stated otherwise.

434: 'the bathroom ... the island': L. Cheshire, quoted in Morris, *Cheshire*, p. 196.

436: 'an elongated ... with fins': Jacob Beser, quoted in Thomas and Witts, *Ruin from the Air*, p. 266.

436: 'straightened up ... very special ... I must ... man's will': L. Cheshire, quoted in Morris, op. cit., pp. 209–10.

438: The main briefing ... Truman himself': the quotes in these two paragraphs come from Miller and Spitzer, *We*

Dropped the A-Bomb, pp. 10–18.

439: 'they just . . . over there': Norris, *Racing for the Bomb*, p. 418.

440: 'a military . . . cortege': Morris, op. cit., p. 212.

440: 'very angry': Thomas and Witts, op. cit., p. 288.

441: 'Guard . . . end': quoted by *New York Times* journalist W. Laurence (*Dawn Over Zero*, p. 209), who was assigned to the project and present on Tinian.

441: 'to ask . . . beforehand': Thomas and Witts, op. cit., p. 294.

441: 'It's . . . accept it': Miller and Spitzer, op. cit., p. 27.

442: 'there was . . . tail': Thomas and Witts, op. cit., p. 300.

Chapter Twenty-Four – 'It's Hiroshima'

Quotes from Paul Tibbets are from his book *Mission Hiroshima* unless otherwise attributed. Similarly, quotes from Bob Caron are from his article in *Veterans of Foreign Wars Magazine*, November 1959, unless stated otherwise.

444: 'It's Hiroshima': Tibbets, op. cit., p. 220.

446: 'The bomb . . . Knock wood': Laurence, *Dawn Over Zero*, p. 221. Laurence had asked Lewis to keep the log.

448: 'there will . . . our target': ibid.

448: 'the ring . . . towards us': Thomas and Witts, *Ruin from the Air*, p. 325.

449: 'beautifully horrible': *Veterans of Foreign Wars Magazine*, November 1959.

449: 'A column . . . smoke': Thomas and Witts, op. cit., p. 326.

450: 'what . . . it worked . . . My God . . . done?': ibid.

450: 'Clear . . . base': Thomas and Witts, op. cit., p. 328.

450: 'withdrawn and meditative': Christman, *Target Hiroshima*, p. 194.

450: 'Even . . . warriors': Laurence, op. cit., p. 219.

450: 'I had ... surrender': Tibbets, op. cit., p. 229.

Chapter Twenty-Five – 'Mother Will Not Die'

452: 'Mother ... die' and other quotes: evidence of Futaba Kitayama referenced in the prologue.

453: 'Why ... fall down?': Hersey, *Hiroshima*, p. 27. This contains Mrs Nakamura's story.

457: Down ... become silent': Dr Hiroshi Sawachika's story is taken from his account at
www.inicom.com/hibakusha/hiroshi.html.

460: 'the product ... physics': Kurzman, *Day of the Bomb*, p. 418.

460: 'not that ... weapon': Rees, *Horror in the East*, p. 141.

461: 'to continue ... planned ... delivered on': Giovannitti and Freed, *The Decision to Drop the Bomb*, p. 264.

461: 'We tried ... war effort': ibid. p. 271.

462: 'the colour ... evil ... quality': Morris, *Cheshire*, p. 222.

462: 'not comprise ... sovereign ruler': MAGIC file/1233, 10 August 1945, RG 457, NARA.

462: 'from total extinction': Bix, *Hirohito*, p. 526.

463: 'Having found ... young Americans': Clemens (ed.), *Truman Speaks*, p. 69.

Chapter Twenty-Six – 'A New Fact in the World's Power Politics'

The academic study of American editorials is quoted in Alperovitz, *The Decision to Use the Atomic Bomb*, p. 427. PRO file CAB/126/191 provides an extensive, considered analysis of world press reaction.

465: 'We were packing ... six months later': Gilbert, *The Day the War Ended*, p. 401.

465: 'when the bombs . . . after all': P. Fussell, *Kansas City Star and Times*, 30 August 1981.

466: 'acutely depressing . . . wasted': Werth, *Russia at War*, p. 1,037.

466: 'Hiroshima . . . destroyed': Holloway, *Stalin and the Bomb*, p. 132.

466: 'I was . . . revered': Sakharov, *Memoirs*, p. 92.

466: 'if a child . . . refused': Holloway, op. cit., p. 132.

467: 'I wonder . . . here? . . . Microphones . . . respect': Bernstein (ed.), *Hitler's Uranium Club*, p. 78.

467: 'Hahn . . . incredulity': ibid. p. 115.

468: 'told me . . . about it': transcript of Hahn's interview for the BBC programme *Too Near the Sun*, 1965.

468: Major Rittner . . . consoles me': all quotes in this paragraph are from Bernstein (ed.), op. cit., pp. 116–17.

469: 'Each one . . . unimportant': ibid. p. 120.

469: 'and only . . . margins': ibid. p. 321.

469: 'I believe . . . succeeded': ibid. p. 122.

469: 'I don't . . . succeed': ibid.

469: 'At the bottom . . . bomb': ibid. p. 123.

469: 'I thank God . . . uranium bomb . . . an inhuman weapon': ibid. p. 125.

469: 'sabotaged . . . to do so': ibid. p. 127.

469: 'At 2 a.m. . . . to bed': ibid. p. 322.

470: 'a ton': ibid. p. 129.

470: 'version/Lesart . . . mostly silent': ibid. p. 333.

471: 'not much . . . moment': Pharr Davis, *Lawrence and Oppenheimer*, p. 251.

471: 'sweet technology': E. Teller, quoted in his obituary, *The Economist*, 20 September 2003.

471: 'gazing . . . hell': Herken, *Brotherhood of the Bomb*, p. 257.

472: 'If it is . . . grant clearance': *US Atomic Energy Commission – In the Matter of J. Robert Oppenheimer*, p. 726.

473: 'would be astounded': ibid. p. 165.

474: 'Had I known . . . finger!': D. Bodanis, $E = mc^2$, p. 218.

475: 'had plumbed . . . peer into': Rowlands, *120 Years of Excellence*, p. 23.

476: 'a belligerent pacifist': Biographical Memoirs of Fellows of the Royal Society.

477: Colleagues . . . statement': all the quotes in this paragraph are from Peierls, *Bird of Passage*, pp. 223–4.

478: 'I have . . . reality': Williams, *Klaus Fuchs*, p. 177.

479: 'very nasty': author's interview with Hans Bethe, April 2002.

479: 'suppressing the past . . . suppressed past': Sime, *Lise Meitner*, p. x.

480: 'Let it be . . . between nations . . . Let . . . war': Tibbets, *Mission Hiroshima*, p. 6.

Epilogue

The debate about what difference it would have made had Rutherford not lived is described in Snow, *Variety of Men,* pp. 6–7, from which the quotes are taken. Snow took part in the discussion.

483: 'quietly . . . unity': ibid. p. 75.

484: 'The main . . . throughout the war . . . essential': Szasz, *British Scientists*, p. 97.

484: 'I think . . . taken their place': quoted in transcript of interview for BBC TV programme *The Building of the Bomb*, 1965.

484: The extent of Fuchs' help to the Russians is discussed inter alia in Holloway, *Stalin and the Bomb*, pp. 222–3, and in Szasz, *British Scientists*, p. 94.

486: 'catch up and overtake': Holloway, op. cit., p. 133.

487: 'just another military weapon': Szasz, *British Scientists*, p. 77.

488: 'between . . . voted': E. Teller's introduction to Groves, *Now It Can Be Told*, p. v.

488: 'without . . . anything': Teller, *Memoirs*, p. 277.

488: 'Mr. Roosevelt . . . to do it': Ermenc, *Atomic Bomb Scientists*, p. 252.

489: J. W. Dower's *War Without Mercy* discusses racial issues in the Pacific War in great detail.

489: 'assigned . . . annihilated': *Leatherneck*, March 1945.

489: The issue of *Life* is 22 May 1944.

489: 'yellow . . . rarin' . . . meat': Dower, op. cit., p. 85.

489: 'It seemed to him . . . bad . . . as far . . . original islands': note from an official in the British Embassy in Washington to the Foreign Office of 6 August 1942, in PREM 4/42/9, PRO.

490: 'sensible . . . than ours': Cronin, *Paris on the Eve*, p. 316.

491: The sidelining . . . at the bottom: all quotes in this paragraph are from Bryson, *A Short History*, p. 123.

492: 'really absurd': Rayner-Canham, *A Devotion to their Science*, p. 223.

492: The decision to use the atomic bomb is the subject of a major study of that name by G. Alperovitz, a key reference work for all research in this area.

493: 'the experience . . . Japan': Truman, *Year of Decisions*, p. 412.

493: 'in the days . . . came in': J. F. Byrnes, *U. S. News and World Report*, 15 August 1960.

493: 'after . . . kill': Alperovitz, *The Decision to Use the Atomic Bomb*, p. 274.

493: 'Japan . . . air blockade': Leahy, *I Was There*, p. 304.

494: 'to shock . . . action . . . out of . . . defence': Bix, *Hirohito*, p. 525.

494: 'there . . . [making peace]': ibid. p. 751.

494: 'in a sense . . . circumstances': Yongi Mitsumasa, quoted in ibid. p. 509.

494: 'When one . . . saved Japan': Szasz, *The Day the Sun Rose Twice*, p. 150.

495: 'the greatest ... of the Throne': Grew, *Turbulent Era*, vol. II, p. 1,429.

495: 'I personally ... of acceptance': memo included in Stimson's diary for 2 July 1945, CUL/S.

495: 'might not ... national existence': Ehrman, *Grand Strategy*, pp. 302–3.

496: 'Subject ... government': *Foreign Relations of the US Vol. II*, pp. 1,268–9.

496: 'I spoke ... put it in': Stimson's diary, entry 24 July 1945, CUL/S.

497: The available sources ... to remain: the Gallup Poll is cited in Bix, op. cit., p. 544.

497: 'like ... toboggan': *Look Magazine*, 13 August 1963.

497: 'At no time ... be justified': Stimson's article in the February 1947 edition of *Harper's Magazine* was extensively cleared in advance with former colleagues, cf. Alperovitz, pp. 450–79.

499: 'conflict ... Nazis': Teller, op. cit., p. 233.

499: 'I could not ... enthusiastically': ibid. p. 231.

499: 'the right ... chose that': Bernstein (ed.), *Hitler's Uranium Club*, p. 155.

499: 'this great ... we are': letter from S. Goudsmit to R. Peierls of 21 January 1977, Goudsmit Papers, Series III, Box 10, Folder 97, AIP.

500: 'An invention ... make': transcript of F. Houtermans' interview for the BBC TV programme *The Building of the Bomb*, 1965.

500: 'in the West ... possibilities ... the possibility ... do with it': transcript of R. Peierls' interview for the BBC TV programme *The Building of the Bomb*, 1965.

500: 'In the first place ... arm's length': transcript of O. Frisch's interview for the BBC TV programme *The Building of the Bomb*, 1965.

502: 'though ... numbers': *New York Review of Books*, 22 April 1993.

503: 'changed ... politics': Szasz, *British Scientists*, p. 103.

503: 'I'm afraid ... created': *Guardian*, 2 August 1986.

504: 'If you are ... values': Oppenheimer made these remarks in his farewell speech to the Association of Los Alamos Scientists on 2 November 1945, quoted in full in Kimball-Smith and Weiner (eds), op. cit., pp. 315–75.

504: 'easy ... about': Szasz, op. cit., p. 78.

505: 'in science ... persons': E. Curie, *Madame Curie*, p. 233.

505: 'People ... people': author's interview with Lorna Arnold, 11 December 2001.

505: 'When I was younger ... individual': transcript of interview for BBC TV programme *To Die, To Live*, 1975.

PICTURE CREDITS

AIP = American Institute of Physics
S&S = Science & Society Picture Library
SPL = Science Picture Library
WL = Wellcome Library, London

First section
Pierre and Marie Curie: WL; advertisement: Musée Curie, Paris

Ernest Rutherford: SPL; J. J. Thomson: S&S; Lise Meitner and Otto Hahn: AIP; Rutherford's Cambridge laboratory: Hulton Deutsch Collection/ CORBIS; Peter Kapitza: SPL; crocodile on the Mond Laboratory: SPL; Albert Einstein and Marie Curie: SPL; Werner Heisenberg and Niels Bohr: AIP; Robert Oppenheimer and Ernest Lawrence: AIP

Ujina: Hiroshima Municipal Archives; Shintenchi: Australian War Memorial; all other images from *The Spirit of Hiroshima* published by the Hiroshima Municipal Archives

Solvay Conference: S&S; nuclear pile: SPL; Yoshio Nishina, Seishi Kikuchi and Niels Bohr: AIP; Rudolf Peierls and his wife Genia: AIP; Fritz Strassmann: AIP; Leo Szilard: SPL

Joachim Ronneberg, the Rjukan heavy water plant and the damaged cells: Norway's Resistance Museum

Second section
Japanese propaganda cartoon: Hiroshima Municipal Archives

Los Alamos main gate: SPL; Robert Oppenheimer and General Leslie Groves: © Bettmann/CORBIS; Otto Frisch: SPL; housing at Los Alamos: © Bettmann/CORBIS; Trinity test: SPL

Atomic bomb explodes over Hiroshima: Australian War Memorial; Col P. W. Tibbets in front of *Enola Gay*: © Bettmann/CORBIS

BIBLIOGRAPHY

Archives Consulted

American Institute of Physics (Niels Bohr Library), College Park, Maryland, US. Goudsmit Correspondence and transcripts of taped oral histories – Hans Bethe (interviewed by Charles Weiner and Jagdish Mehra, 1966, and by Charles Weiner, 1967); James Chadwick (interviewed by Charles Weiner, 1969); Otto Frisch (interviewed by Charles Weiner, 1967); Michael Polanyi, 1962; Fritz Reiche, 1962; Emilio Segrè (interviewed by Charles Weiner, 1967); Eugene Wigner (interviewed by Charles Weiner and Jagdish Mehra, 1966).

Bancroft Library, University of California, Berkeley US-Ernest Lawrence Papers.

BBC Written Archives Centre, Caversham, Berkshire, UK (relevant programmes are: *To Die, To Live* (Horizon series, Files TX/0608 and T68/1011); *The Man and the Atom – Recollections of Niels Bohr* (Files TXOS/02/1968); *The Building of the Bomb* and *Too Near the Sun* (File T14/1916/1); *The Day the Sun Blowed Up* (File T6A/508/1).

Bodleian Library, Oxford University, UK – Frederick Soddy Papers.

Bohr Archive, Copenhagen, Denmark – Niels Bohr/Werner Heisenberg Correspondence.

Cambridge University Library, Cambridge, UK – Ernest Rutherford Papers.

Churchill College Archives, Cambridge University, UK – James Chadwick, Norman Feather and Lise Meitner Papers.

Deutsches Museum, Munich, Germany – *Geheimdokumente zum*

deutschen Atomprogramm [Secret Documents of the German Atomic Programme], 1938–45.

Hiroshima City Museum of History and Traditional Crafts, Hiroshima, Japan.

Hiroshima Municipal Archives, Hiroshima, Japan.

Hiroshima Peace Memorial Museum, Hiroshima, Japan.

Hiroshima Prefecture Archive, Hiroshima, Japan.

Jüdisches Museum, Berlin, Germany – Andrea Wolffenstein testimony.

Library of Congress, Washington DC, US – Robert Oppenheimer Papers.

Liverpool University Physics Department – interviews with Joseph Rotblat, 1992 and 1994.

Radiation Effects Research Facility, Hiroshima, Japan.

Royal Society, London, UK (including Biographical Memoirs of Fellows).

UK National Archives, Public Record Office, Kew, London, UK.

US Government Records, National Archives and Records Administration (NARA), Maryland, US.

Yad Vashem, Jerusalem, Israel – Fritz Strassmann citation.

Books and Government Publications

Alperovitz, G., *The Decision to Use the Atomic Bomb*, New York, Knopf, 1995.

Alvarez, L. W., *Adventures of a Physicist*, New York, Basic Books, 1987.

Badash, L., *Kapitza, Rutherford and the Kremlin*, London, Yale University Press, 1985.

——(ed.), *Rutherford and Boltwood – Letters on Radioactivity*, New Haven, Yale University Press, 1969.

——*Scientists and the Development of Nuclear Weapons*, New Jersey, Humanities Press, 1995.

——Hirschfelder, J. O., and Broida, H. P. (eds), *Reminiscences of Los Alamos – 1943–1945*, Dordrecht (Holland), D. Reidel, 1980.

Bernstein, J., *Hans Bethe, Prophet of Energy*, New York, Basic Books, 1980.

——(ed.), *Hitler's Uranium Club – The Secret Recordings at Farm Hall*, New York, Springer-Verlag, 2001.

——*Oppenheimer: Portrait of an Enigma*, Chicago, Ivan R. Dee, 2004.

Bethe, H. A., *The Road from Los Alamos*, New York, Simon and Schuster, 1991.

Bix, H. P., *Hirohito and the Making of Modern Japan*, New York, HarperCollins, 2000.

Bodanis, D., $E = mc^2$, London, Macmillan, 2000.

Born, M., *My Life and My Views*, New York, Charles Scribner's Sons, 1968.

Boyer, P., *By the Bomb's Early Light*, Chapel Hill, University of North Carolina Press, 1994.

Bragg, M., *On Giants' Shoulders*, London, Hodder and Stoughton, 1998.

Brian, D., *Einstein – A Life*, New York, John Wiley and Sons, 1996.

Brooks, G., *Hitler's Nuclear Weapons*, London, Leo Cooper, 1992.

Brown, A., *The Neutron and the Bomb – A Biography of Sir James Chadwick*, Oxford, Oxford University Press, 1997.

Bryant, A., *The Triumph in the West – 1943–46* (based on the diary and autobiographical notes of Field Marshal Viscount Alanbrooke), London, Collins, 1959.

Bryson, B., *A Short History of Nearly Everything*, London, Doubleday, 2003.

Byrnes, J., *All in One Lifetime*, London, Museum Press, 1960.

——*Speaking Frankly*, Westport, Connecticut, Greenwood Press, 1974 (originally published in 1947).

Campbell, J., *Rutherford: Scientist Supreme*, Christchurch, New Zealand, AAS Publications, 1999.

Cassidy, D., *Uncertainty – The Life and Science of Werner Heisenberg*, New York, W. H. Freeman, 1992.

Chadwick, Sir J. (ed.), *The Collected Papers of Lord Rutherford*, vol. II, London, George Allen and Unwin, 1963.

Christman, A., *Target Hiroshima*, Annapolis, Maryland, Naval Institute Press, 1998.

Churchill, W. S., *The Gathering Storm*, London, Cassell, 1948.

Clark, R., *Tizard*, London, Methuen, 1965.

Clemens, C. (ed.), *Truman Speaks*, Missouri, International Mark Twain Society, 1946.

Cockburn, S., and Ellyard, D., *Oliphant*, Kent Town, South Australia, Axiom Books, 1982.

Compton, A. H., *Atomic Quest – A Personal Narrative*, New York, Oxford University Press, 1956.

Cornwell, J., *Hitler's Scientists*, London, Viking, 2003.

Cronin, V., *Paris on the Eve – 1900–1914*, London, Collins, 1989.

Curie, E., *Madame Curie*, New York, Pocket Books, 1959.

Curie, M., *Pierre Curie* (including M. Curie's Autobiographical Notes), New York, Dover Publications, 1963 (republication of work first published by Macmillan in 1923.)

——and Curie, I., *Curie Correspondance, Choix de Lettres, 1905–1934*, Paris, Les Éditeurs Français Réunis, 1974.

Danchev, A., and Todman, D. (eds), *War Diaries of Field Marshal Lord Alanbrooke*, London, Weidenfeld and Nicolson, 2001.

Dawidoff, N., *The Catcher Was a Spy – The Mysterious Life of Moe Berg*, New York, Pantheon Books, 1994.

Dower, J. W., *War Without Mercy*, New York, Pantheon Books, 1986.

Dyson, F., *Disturbing the Universe*, New York, Harper and Row, 1979.

Ehrman, J., *Grand Strategy, Vol. 6, October 1944 to August 1945*, London, HMSO, 1956.

Ermenc, J. J. (ed.), *Atomic Bomb Scientists – Memoirs, 1939–1945*, Westport, Connecticut, Meckler, 1989.

Eve, A. S., *Rutherford – The Life and Letters*, Cambridge, Cambridge University Press, 1939.

Fermi, L., *Atoms in the Family*, Chicago, University of Chicago Press, 1954.

Ferrell, R. H. (ed.), *Off the Record – The Private Papers of H. S. Truman*, New York, Harper and Row, 1980.

Feynman, R., *Surely You're Joking, Mr Feynman*, New York, Bantam Books, 1986.

Foreign Relations of the US – The Conference of Berlin (Potsdam), 1945, Vols I and II, Washington DC, USGPO, 1960.

Frank, Sir C. (intro), *Operation Epsilon, The Farm Hall Transcripts*, Bristol, Institute of Physics, 1993.

French, A. P., and Kennedy, J. P. (eds), *Niels Bohr – A Centenary Volume*, Cambridge, Mass., Harvard University Press, 1985.

Frisch, O., *What Little I Remember*, Cambridge, Cambridge University Press, 1979.

Gilbert, Sir M., *The Day the War Ended*, London, HarperCollins, 1995.

——*A History of the Twentieth Century, Vols I and II*, London, HarperCollins, 1997 and 1998.

Giovannitti, L., and Freed, F., *The Decision to Drop the Bomb*, London, Methuen, 1967.

Gleich, J., *Richard Feynman and Modern Physics*, London, Little Brown, 1992.

Goebbels, J., *Diaries, 1941–43*, London, Hamish Hamilton, 1948.

Goldschmidt, B., *Atomic Rivals*, London, Rutgers University Press, 1990.

Goldsmith, M., *Frédéric Joliot-Curie*, London, Lawrence and Wishart, 1976.

Goudsmit, S. A., *Alsos, The Failure in German Science*, London, Sigma Books, 1947.

Gowing, M., *Britain and Atomic Energy 1939–1945*, London, Macmillan, 1964.

Grew, J. C., *Turbulent Era – A Diplomatic Record of Forty Years – 1904–45, Vols I and II*, London, Hammond, 1953.

Groueff, S., *Manhattan Project*, New York, Bantam, 1968.

Groves, L. R., *Now It Can Be Told*, New York, Da Capo, 1983.

Hachiya, M., *Hiroshima Diary – The Journal of a Japanese Physician, August 6 – September 30, 1945*, Chapel Hill, University of North Carolina Press, 1995.

Hacker, B. C., *The Dragon's Tail – Radiation Safety in the Manhattan Project, 1942–46*, Berkeley, University of California Press, 1987.

Hahn, O., *My Life*, New York, Herder and Herder, 1970.

Hales, P. B., *Atomic Spaces – Living on the Manhattan Project*, Urbana, Illinois, University of Illinois Press, 1997.

Harman, P. M. (ed.), *The Scientific Letters and Papers of James Clerk Maxwell, Vol. II (1862–1873)*, Cambridge, Cambridge University Press, 1995.

Harris, R., and Paxman, J., *A Higher Form of Killing*, London, Chatto and Windus, 1982.

Haukelid, K., *Skis Against the Atom*, Minot, North Dakota, North American Heritage Press, 1989.

Heilbron, J. L., *H. G. J. Moseley*, Berkeley, California University Press, 1974.

Heisenberg, E., *Inner Exile*, Boston, Mass., Birkhauser, 1984.

Heisenberg, W., *Physics and Beyond*, London, George Allen and Unwin, 1971.

Hendry, J. (ed.), *Cambridge Physics in the Thirties*, Bristol, Adam Hilger, 1984.

Herken, G., *Brotherhood of the Bomb*, New York, Henry Holt, 2002.

Hersey, J., *Hiroshima*, London, Penguin, 1946.

Holloway, D., *Stalin and the Bomb*, London, Yale University Press, 1994.

Howes, R. H., and Herzenberg, C. L., *Their Day in the Sun – Women of the Manhattan Project*, Philadelphia, Temple University Press, 1999.

Howorth, M., *Pioneer Research on the Atom – The Life Story of Frederick Soddy*, London, New World Publications, 1958.

Hughes, J., *The Manhattan Project*, Cambridge, Icon Books, 2002.

Irving, D., *The Virus House*, London, William Kimber, 1967.

Jungk, R., *Brighter Than a Thousand Suns*, New York, Harcourt, 1958.

Keegan, J., *The Second World War*, London, Hutchinson, 1989.

Kennedy, D. M., *Freedom from Fear*, New York, Oxford University Press, 1999.

Kevles, D. J., *The Physicists*, New York, Vintage Books, 1979.

Kimball-Smith, A., and Weiner, C. (eds), *Robert Oppenheimer – Letters and Recollections*, Stanford, Stanford University Press, 1980.

Klein, M. J., Kox, A. J., and Schulman, R. (eds), *The Collected Papers of Albert Einstein, Vol. V – The Swiss Years: Correspondence, 1902–1914*, Princeton, NJ, Princeton University Press, 1993.

Krafft, F., *Im Schatten der Sensation – Leben und Wirken von Fritz Strassmann*, Weinheim (Germany), Verlag Chemie, 1981.

Kramisch, A., *The Griffin*, London, Macmillan, 1987.

Kurlansky, M., *The Basque History of the World*, New York, Walker Books, 1999.

Kurzman, D., *Blood and Water – Sabotaging Hitler's Bomb*, New York, Henry Holt, 1997.

——*Day of the Bomb*, New York, McGraw-Hill, 1986.

Lanouette, W. (with Szilard, B.), *Genius in the Shadows*, Chicago, University of Chicago Press, 1994.

Larsen, E., *The Cavendish Laboratory*, New York, Franklin Watts, 1962.

Laurence, W., *Dawn Over Zero*, New York, Knopf, 1946.

Leahy, W. D., *I Was There*, London, Victor Gollancz, 1950.

Lifton, R. J., *Death in Life*, Chapel Hill, University of North Carolina Press, 1991.

Lindqvist, S., *A History of Bombing*, London, Granta, 2001.

Marx, J. L., *Seven Hours to Zero*, New York, Macfadden-Bartell, 1969.

Medawar, J., and Pyke, D., *Hitler's Gift*, London, Piatkus, 2000.

Miller, M., and Spitzer, A., *We Dropped the A-Bomb*, New York, Thomas Y. Crowell, 1946.

Moore, R., *Niels Bohr*, London, Hodder and Stoughton, 1967.

Morris, R., *Cheshire*, London, Viking, 2000.

Moss, N., *Klaus Fuchs*, London, Grafton, 1987.

Nicolson, H., *Public Faces*, London, Constable, 1932.

Norris, R. S., *Racing for the Bomb – General Leslie R. Groves, the Manhattan Project's Indispensable Man*, South Royalton, Vermont, Steerforth, 2002.

Nye, M. J., *Before Big Science*, Cambridge, Mass., Harvard University Press, 1996.

Oliphant, M., *Rutherford – Recollections of the Cambridge Days*, London, Elsevier, 1972.

Oppenheimer, J. R., *Science and the Common Understanding*, Oxford, Oxford University Press, 1954.

Pais, A., *Einstein Lived Here*, New York, Oxford University Press, 1994.

——*Inward Bound*, New York, Oxford University Press, 1986.

——*Niels Bohr's Times – In Physics, Philosophy and Polity*, Oxford, Clarendon Press, 1991.

Pash, B., *The Alsos Mission*, New York, Charter, 1980.

Peierls, R., *Bird of Passage*, Princeton, NJ, Princeton University Press, 1985.

Pflaum, R., *Grand Obsession – Marie Curie and her World*, New York, Doubleday, 1989.

Pharr Davis, N., *Lawrence and Oppenheimer*, New York, Simon and Schuster, 1968.

Planck, M., *A Scientific Autobiography*, London, Williams and Norgate, 1950.

Powers, T., *Heisenberg's War*, New York, Da Capo Press, 2000.

Quinn, S., *Marie Curie*, London, Heinemann, 1995.

Rayner-Canham, M. F. and G. W., *A Devotion to their Science*, Montreal, McGill-Queen's University Press, 1997.

Rees, L., *Horror in the East*, London, BBC Books, 2001.

Rhodes, R., *The Making of the Atomic Bomb*, New York, Simon and Schuster, 1986.

Rose, P. L., *Heisenberg*, Berkeley, University of California Press, 1998.

Rouzé, M., *Robert Oppenheimer – The Man and his Theories*, London, Souvenir Press, 1964.

Rowlands, P., *120 Years of Excellence – The University of Liverpool Physics Department*, Liverpool, U-P L Communications/PD Publications, 2001.

Sakharov, A., *Memoirs*, London, Hutchison, 1990.

Sanger, S. L., *Working on the Bomb – An Oral History of Hanford*, Portland, Oregon, Portland State University Press, 1995.

Segrè, E., *Enrico Fermi, Physicist*, Chicago, University of Chicago Press, 1970.

Sime, R. L., *Lise Meitner – A Life in Physics*, Berkeley, University of California Press, 1996.

Simpson, J., *A Mad World, My Masters*, London, Pan, 2001.

Snow, C. P., *The Physicists*, London, Macmillan, 1981.

——*Variety of Men*, London, Macmillan, 1968.

Speer, A., *Inside the Third Reich*, London, Weidenfeld and Nicolson, 1990.

Szasz, F. M., *British Scientists and the Manhattan Project*, London, Macmillan, 1992.

——*The Day the Sun Rose Twice*, Albuquerque, University of New Mexico Press, 1984.

Takayama, H. (ed.), *Hiroshima in Memoriam and Today*, Hiroshima, The Society for the Publication of *Hiroshima in Memoriam and Today*, 1973.

Teller, E., *Memoirs*, Cambridge, Mass., Perseus, 2001.

Thomas, G., and Morgan Witts, M., *Ruin from the Air*, London, Book Club Associates, 1977.

Thomson, J. J., *Recollections and Reflections*, London, Bell and Sons, 1936.

Tibbets, P. W., *Mission Hiroshima*, New York, Stein and Day, 1985.

Truman, H. S., *Memoirs – Vol. I, Year of Decisions*, New York, Doubleday, 1955.

US Atomic Energy Commission – In the Matter of J. Robert Oppenheimer, Transcript of Hearing Before Personnel Security Board, Washington D.C., April 12, 1954, through May 6, 1954, Washington, DC, USGPO, 1954.

Various, *Bombing Eye-Witness Accounts*, Hiroshima, Hiroshima City, 1950.

Various (survivors of Hiroshima and Nagasaki), *Hibakusha*, Tokyo, Kosei, 1986.

Walker, M., *German National Socialism and the Quest for Nuclear Power, 1939–49*, Cambridge, Cambridge University Press, 1989.

Weart, S. R., and Szilard, G. W. (eds), *Leo Szilard, His Version of the Facts – Selected Recollections and Correspondence – Vol. II*, Cambridge, Mass., MIT Press, 1978.

Weizsäcker, C.-F., *The Politics of Peril*, New York, Seabury Press, 1978.

Werth, A., *Russia at War – 1941–45*, London, Barrie and Rockliff, 1964.

Williams, R. C., *Klaus Fuchs, Atom Spy*, Cambridge, Mass., Harvard University Press, 1987.

Wilson, D., *Rutherford*, London, Hodder and Stoughton, 1983.

Other publications (for details of specific articles quoted, see Notes and Sources).

Journals and Magazines
Angewandte Chemie
Biographical Memoirs of Fellows of the Royal Society
Bulletin of the Atomic Scientists
The Economist
Harper's
Illustrated London News
Lancet
Leatherneck (US Marine Corps Magazine)
Life
Look Magazine
Los Alamos Science
Nature
New Republic
New Scientist
New York Review of Books
Observer
Philosophical Magazine
Physics Today
Physics World
Punch
Radio Times
Scientific American
Der Spiegel
Time
U.S. News and World Report
Vanity Fair
Veterans of Foreign Wars Magazine

Newspapers (main sources)

United Kingdom
Daily Express
Daily Mail
Daily Mirror
Daily Telegraph

Guardian
Independent
Liverpool Daily Post
The Times

United States
Baltimore Sun
Chicago Sun-Times
Kansas City Star and Times
Los Angeles Times
New York Times
Washington Post

Others
Frankfurter Allgemeine Zeitung
Le Monde

Miscellaneous

Text of Michael Frayn's play *Copenhagen*.
Reprint of lecture by M. Gowing, UKAEA official historian, 'How Nuclear Power Began', University of Southampton, 1987.
Film, *The Restoration of the Hypocentre Salugakucho*, Knack Images Production Centre, 2004.

INDEX

NOTE: page numbers in italics refer to illustrations; those followed by italic g refer to glossary items.

WILFUL MURDER
The Sinking of the Lusitania
by Diana Preston

On May 7th, 1915, the *Lusitania*, a passenger ship, was torpedoed by a German U-boat in the Atlantic. 1,200 people died. *Wilful Murder*, the first book to look at this tragedy in its full historical context, is also the first to place the human dimension at its heart. Through first-hand accounts, we relive the splendour of the liner setting sail and the horror of its final moments.

Using British, American and German research material, Diana Preston answers many of the unanswered and controversial questions surrounding the *Lusitania*: why didn't Cunard heed warnings that the ship was a German target? Had Cunard's offices been infiltrated by German agents? What was really in the *Lusitania*'s hold, and was she armed? Did international outrage change the outcome of the First World War?

And perhaps most importantly, was the *Lusitania* sacrificed to bring America into the war? Engrossing and brilliantly researched, *Wilful Murder* casts dramatic new light on one of the world's most famous maritime disasters.

'A COMPLEX STORY OF HEROISM AND GREAT COURAGE . . . COMPULSIVELY READABLE'
Independent on Sunday

'IT IS NOT EASY, NOWADAYS, TO WRITE AN ORIGINAL BOOK ON THE FIRST WORLD WAR . . . BUT PRESTON HAS SUCCEEDED'
Norman Stone, *Sunday Times*

'VERY GOOD . . . PRESTON HAS DONE AN EXTRAORDINARY AMOUNT OF WORK, PARTICULARLY IN TRACING THE MEMORIES OF SURVIVORS'
Sunday Times

'SETS A STANDARD WHICH OTHER BOOKS HAVE NOT ACHIEVED'
Irish Independent

'CLEAR AND EFFECTIVE . . . BENEFITS FROM EXHAUSTIVE RESEARCH'
TLS

0 552 99886 9

CORGI BOOKS

A PIRATE OF EXQUISITE MIND
The Life of William Dampier
Explorer, Naturalist and Buccaneer
by Diana and Michael Preston

'GRIPPING AND WELL-RESEARCHED . . . AN IMPRESSIVE
ACHIEVEMENT'
Guardian

William Dampier is one of England's forgotten heroes. In 1676,
he started his career as a poor buccaneer, preying on ships on the
Spanish Main. He could easily have ended up on the gallows for
piracy. Instead, his sense of adventure and curiosity about the world
around him led him to become the first person to circumnavigate the
world three times, and to map the winds and the currents of the
world's oceans. He landed in Australia eighty years before Cook and
visited the Galapagos Islands one hundred and fifty years before
Darwin. He wrote the first bestselling travel books, which inspired
Defoe's *Robinson Crusoe* and Swift's *Gulliver's Travels*, and enriched
the English language with many new words, from 'barbeque' and
'avocado' to 'sub-species'.

A curious man in a curious age, now all but forgotten in his native
country, William Dampier combined a swashbuckling life of adventure
with remarkable scientific achievements. In *A Pirate of Exquisite Mind*,
Diana and Michael Preston reveal, in a compelling narrative, the story
of a uniquely English hero.

'LIVELY . . . EXTRAORDINARY LIFE – AN UNLIKELY
COMBINATION OF PLUNDERING AND PIONEERING
ACHIEVEMENTS IN NATURAL HISTORY AND EXPLORATION'
Sunday Times

'THIS LONG OVERDUE BIOGRAPHY WONDERFULLY BRINGS
TO LIFE ONE OF THE MOST IMPORTANT EXPLORERS OF
THE SEVENTEENTH CENTURY'
Nathaniel Philbrick, author of *In the Heart of the Sea*

'THIS ELOQUENTLY ENTHUSIASTIC BIOGRAPHY, BESIDES
CHARTING DAMPIER'S ASTONISHING ACHIEVEMENTS,
OFFERS FASCINATING INFORMATION ABOUT HIS TIMES'
The Age, Melbourne

0 552 77210 0

CORGI BOOKS

A SELECTED LIST OF NON-FICTION TITLES AVAILABLE FROM TRANSWORLD PUBLISHERS

15108 4	WITNESS TO WAR: DIARIES OF THE SECOND WORLD WAR IN EUROPE	Richard Aldrich	£9.99
14718 4	PICKING UP THE PIECES	Paul Britton	£7.99
99704 8	A SHORT HISTORY OF NEARLY EVERYTHING	Bill Bryson	£8.99
14869 5	OUT OF THE DARK	Linda Caine and Robin Royston	£6.99
99923 7	THE MYSTERY OF CAPITAL	Hernando de Soto	£8.99
99981 4	A PROFOUND SECRET	Josceline Dimbleby	£8.99
15085 1	NELSON'S PURSE	Martyn Downer	£8.99
15094 0	IN THE COMPANY OF HEROES	Michael J. Durant	£6.99
99545 2	ISRAEL: A HISTORY	Martin Gilbert	£14.99
99850 8	THE RIGHTEOUS	Martin Gilbert	£8.99
50692 7	THE ARCANUM	Janet Gleeson	£6.99
81247 5	THE MONEYMAKER	Janet Gleeson	£6.99
12555 5	IN SEARCH OF SCHRÖDINGER'S CAT	John Gribbin	£8.99
15022 3	INSIDE DELTA FORCE	Eric Haney	£6.99
14789 3	THE DREADFUL JUDGEMENT	Neil Hanson	£7.99
17521 1	A BRIEF HISTORY OF TIME	Stephen Hawking	£8.99
15408 3	THEIR KINGDOM COME	Robert Hutchison	£7.99
99958 X	ALMOST LIKE A WHALE	Steve Jones	£9.99
15049 5	OVERWORLD	Larry J. Kolb	£7.99
81522 9	1421: THE YEAR CHINA DISCOVERED THE WORLD	Gavin Menzies	£9.99
99886 9	WILFUL MURDER: THE SINKING OF THE LUSITANIA	Diana Preston	£7.99
77210 0	A PIRATE OF EXQUISITE MIND: THE LIFE OF WILLIAM DAMPIER	Diana and Michael Preston	£8.99
81469 9	ALPHA AND OMEGA	Charles Seife	£9.99
77100 7	TO THE HEART OF THE NILE	Pat Shipman	£8.99
99982 2	THE ISLAND AT THE CENTRE OF THE WORLD	Russell Shorto	£7.99
99750 1	SPEAKING FOR THEMSELVES: THE PERSONAL LETTERS OF WINSTON AND CLEMENTINE CHURCHILL	Mary Soames ed.	£15.00
99941 5	THE COMMON THREAD	John Sulston & Georgina Ferry	£7.99
14989 6	ADAM'S CURSE	Bryan Sykes	£7.99
81492 3	HUMAN INSTINCT	Robert Winston	£8.99